Microbial Community Studies in Industrial Wastewater Treatment

Focusing on microbial community structure in the field of wastewater treatment, this book highlights structural analyses in relation to changes in physico-chemical parameters. It further covers physiological analyses of microbial communities, enrichment of pure cultures of key species in relation to changes in physico-chemical parameters, and analyses and modelling of consequences of changes in the microbial community structure. Based on 16S rRNA gene sequencing, groups of bacteria that perform nitrogen fixation, nitrification, ammonification and other biochemical processes are covered for an entire wastewater treatment plant bioreactor along with temporal dynamics of bacterial communities.

Features:

- Describes the state-of-the-art techniques and the application of omics tools in wastewater treatment reactors (WWTRs).
- Includes both theoretical and practical knowledge on the fundamental roles of microorganisms in WWTRs.
- Discusses environmental microbial community proteomics.
- Covers relating function and community structure of complex microbial systems using neural networks.
- Reviews the economics of wastewater treatment and the development of suitable alternatives in terms of performance and cost-effectiveness.

This book is aimed at graduates and researchers in biological engineering, biochemical engineering, chemistry, environmental engineering, environmental microbiology, systems ecology and environmental biotechnology.

Microbial Community Studies in Industrial Wastewater Treatment

Edited by
Maulin P. Shah

CRC Press
Taylor & Francis Group
Boca Raton London New York

CRC Press is an imprint of the
Taylor & Francis Group, an **informa** business

Cover image: ©Shutterstock

First edition published 2023
by CRC Press
6000 Broken Sound Parkway NW, Suite 300, Boca Raton, FL 33487-2742

and by CRC Press
4 Park Square, Milton Park, Abingdon, Oxon, OX14 4RN

CRC Press is an imprint of Taylor & Francis Group, LLC

ISBN: 9781032341880 (hbk)
ISBN: 9781032406565 (pbk)
ISBN: 9781003354147 (ebk)

DOI: 10.1201/9781003354147

Typeset in Times
by codeMantra

Contents

Preface

Treatment of wastewater discharged from various sources such as municipal, industrial and agricultural activities and mining sites is a global challenge. Furthermore, the disposal of raw wastewater into natural water streams put the ecology at the risk of detrimental contamination. Generally, organic (hydrocarbons) and inorganic contaminants (heavy metals, inorganic salts and mineral acids) are the main sources of pollution in water, affecting environment and health. In order to remove these harmful contaminants, treatment of wastewater by biological means is the most sustainable mode over chemical wastewater treatment. Microbial reactors or treatment systems employing either bacterial, microalgae or fungal species are widely accepted in biological wastewater treatment systems. The selection of suitable species or consortia is an important prerequisite for the performance of any microbial reactors. Moreover, factors such as native microbial community, presence of organic and inorganic components in the wastewater, and physico-chemical parameters such as pH, ambient or regulated temperature, and light (for photosynthetic microorganisms) play key roles in the performance of wastewater treatment process. These factors, directly or indirectly, affect the initial microbial community present in the reactor, by aiding the wastewater treatment process in a positive or negative way. Furthermore, advanced molecular techniques play an important role in the identification of variance in the microbial community during the whole wastewater treatment process. Molecular tools such as denaturing gradient gel electrophoresis (DGGE), confocal laser scanning microscopy (CLSM), ribosomal spacer analysis (RISA), amplified ribosomal DNA restriction analysis (ARDRA), fluorescent in situ hybridisation (FISH) and terminal restriction fragment length polymorphism (t-RFLP) have been developed for understanding microbial community during wastewater treatment.

Editor

Dr. Maulin P. Shah is Chief Scientist and Head of the Industrial Wastewater Research Lab, Division of Applied and Environmental Microbiology Lab at Enviro Technology Ltd., Ankleshwar, Gujarat, India. His work focuses on the impact of industrial pollution on the microbial diversity of wastewater following cultivation-dependent and cultivation-independent analysis. His major work involves isolation, screening, identification, and genetically engineering high-impact microbes for the degradation of hazardous materials. His research interests include biological wastewater treatment, environmental microbiology, biodegradation, bioremediation, and phytoremediation of environmental pollutants from industrial wastewaters. He has published more than 250 research papers in national and international journals of repute on various aspects of microbial biodegradation and bioremediation of environmental pollutants. He is the editor of more than 150 books of international repute (Elsevier, Springer, RSC, and CRC Press, De Gruyter, Nova Sciences). He is an active editorial board member in top-rated International journals.

1 Metagenomics
A Powerful Lens Viewing the Microbial World

Perumalla Srikanth
Amity University

Sivakumar Durairaj
SIMATS Deemed University

CONTENTS

DOI: 10.1201/9781003354147-1

1.1 INTRODUCTION

Soil microbial networks do scratch environmental benefits, which are essential for all time on the planet, including carbon cycling, different supplements, and supporting plant development. Astoundingly, numerous advantageous capacities of the soil microbiome are as of now undermined owing to changes in the environment, designs, soil debasement, and ineffective land the board rehearses (Amundson et al., 2015). As of late, there may be expanded awareness on top of things of soil microbial community to re-establish environment work (Calderon et al., 2017). The prospect of overseeing environment administrations and bioprospecting digestion of soil microbes would be conceivable through a more noteworthy understanding of the collaboration of soil microbiomes in various conditions. Investigation on soil microbiomes stays an awesome errand, be that as it may, because the majority of soil microorganisms have not yet been confined and sub-atomic subtleties fundamental to their capacities are generally mysterious (Jansson and Hofmockel, 2018).

Microorganisms are delicate to their natural pH, and contrasts in soil pH have been utilised to clarify variety in local area organisation, environmental variety, and local area gathering. The acidic soil conditions select individuals from the Acidobacteria and cause a general reduction in bacterial variety. The genotype of different microbes is varied at various pH levels in different soils (Tripathi et al., 2018). Nonetheless, the general ecological prompts that impact the structure and variety of the dirt bacterial local area stay a matter for guess and speculation (Yavitt et al., 2020).

Studies zeroing in on fungal networks as a rule focus on the atomic ribosomal inside interpreted spacer (ITS), the prescribed contagious scanner tag, and allot scientific categorisation utilising the UNITE electronic data set (Nilsson et al., 2019). The ordered data within UNITE information base depends on the arrangement of the NCBI taxonomy, and it is enhanced through adjustments from Index Fungorum and MycoBank. However, there is no agreement concerning that locale to focus for depicting the arbuscular mycorrhizal fungi (AMF) people group, established researchers for the most part arrangements and the V4 districts of 18S ribosomal DNA due to accessibility of the MaarjAM data set and a few preliminaries, which enhances the most identified AMF families. The MaarjAM data set supplies openly accessible Glomeromycota grouping information and related metadata, yet they have no curated ordered data like UNITE for ITS arrangements. Accordingly, the inquiry is raised of which information base to question to precisely relegate scientific classification to 18S AMF successions. Just two data sets other than MaarjAM are accessible to relegate ordered data to 18S contagious groupings. PHYMYCO-DB is another data set of 12,670 parasitic 18S successions, amongst which 1,400 are 18S arrangements of Glomeromycota; however, ordered data was refreshed since 201. Also, 920 of these 1,400 AMF arrangements have been distinguished at the family level. The Silva data covers 56,554 successions of Glomeromycota, yet ordered data is not up-to-date: the Glomeraceae family is addressed distinctly in about 5 genera, and at present, Glomeraceae family has 12 genera (Stefani et al., 2020).

The high-throughput grouping-based AMF investigation has an issue that an exorbitant number of arrangements were delivered by employing Illumina® MiSeq innovation. MiSeq reagent units v2 and v3 are some of the examples that create up to 30 and 50 million matched-end successions, separately. Contrasting the other important soil microbiota specifically, AMF have a restricted animal-type extravagance (333 species truly depicted; amfphylogeny.com) and a complete number of animal varieties dependent on sub-atomic information assessed between 1,700 and 2,700. The outcome of the investigation is that the number of Glomeromycota species in either the soil or a root framework is next to any rate when compared to either Ascomycota or Basidiomycota species. The millions of arrangements through Illumina MiSeq innovation are the most part pointless excess in tests zeroing in on AMF in the danger of yielding a huge extent of non-AMF groupings if groundworks are not completely explicit to Glomeromycota. Furthermore, AM parasitic DNA is under-represented within the natural DNA extricates (Stefani et al., 2020).

Assessment of specific epidemiologic group of microbes on earth could be done by the group of researchers under the specific research conditions. Infections are the foremost bountiful organic element on Earth, contaminate both soil and water environment, and are extensively perceived as vital microbial controllers' networks and cycles (Koonin, 2020). Therefore, researchers have gone to metagenomic sequencing to recuperate and consider the genomes of crude infections (Emerson, 2018). Ordinarily, DNA or RNA is far away from an ecological example, divided, and afterwards sequenced, producing an enormous number of short peruses, which are collected into contigs. Metagenomic viral-based contigs utilise computational instruments, and therefore, the calculations utilise an assortment of viral-explicit arrangement highlights and marks (Kief et al., 2020). Instead of microorganisms and Archaea, numerous viral genomes are also adequately studied with the help of metagenomics and recently using the nanopore (Beaulaurier, 2020) or PacBio advances.

However, metagenome binning could be needed for infections with outstandingly huge genomes, for example, monster infections (Schulz, 2020).

The gathering of infections from metagenomes is testing (Smits, 2014), and therefore, the fulfilment of collected contigs can shift generally, going from short parts to finish or approaching total genomes (Roux, 2019). Little genome sections may antagonistically influence downstream investigations including assessment of viral variety and have expectation or ID of centre qualities inside viral ancestries. Viral contigs could likewise be gotten from incorporated proviruses, where viral succession could be fringed on one among the two sides by areas ranging in the host genome. Host defilement additionally antagonistically influences downstream investigations, particularly the assessment of viral genome size, portrayal of viral quality substance, and recognisable proof of viral-encoded metabolic qualities (Nayfach et al., 2020).

The investigation of the genomes of microbes would provide the basic information of particular microbes genera's and helps for doing classification of different microbes. Most approaches of metagenomics incorporate cutting-edge DNA sequencing, polymerase chain reaction (PCR), RT-PCR, and microarrays. Metagenomic research might be helpful for differentiating the varied microorganisms, infections, permitted DNA within the indigenous habitat, and DNA successions within the creatures.

Metagenomics utilises the knowledge of nucleic acids possessed by living organisms, which encode different protein items; along these lines, creatures do not need to be refined yet might be distinguished by a selected quality grouping based on either protein or metabolite. Metagenomics is a methodology utilised in genomics. Genomics focuses on one living being, though metagenomics manages a combination of DNA from various organic entities, "quality animals" (i.e., infections, viroids, plasmids, and then forth), and additionally free DNA (Clark and Pazdernik, 2016).

Metagenomics features a wide selection of utilisations, and various investigations use metagenomic examinations. The investigations range from distinguishing and classifying huge-scope indigenous habitats like seas or soil to limited scope however similarly complex studies of human gut microbes are also having a wide range of scope to explore new microbes. The investigations of metagenomes examine the microbial populaces, recognise new valuable qualities from the climate, test the microorganisms' structure harmonious associations with their hosts, and analyse similar quality across an assortment of animal types to characterise how related the standard family is in nature (Clark and Pazdernik, 2016; Shah, 2021).

Metagenomics has been utilised to discover novel anti-infectious agents, compounds that biodegrade contaminations and proteins that make novel items (Table 1.1). Verifiably, examining organisms in the different climates has recognised numerous valuable items. A novel methodology was recognised newly for identifying antimicrobial drugs during the year 2002. Furthermore, the team of Gillespie created a library of various DNA pieces disconnected from the refined soil samples. Each part was replicated into an articulation route and afterwards inserted into *Escherichia coli*. Every microbe communicated the potentials initiate on the arbitrary DNA piece (Clark and Pazdernik, 2016). Identification of genes and antibiotics through gene mining is presented in Table 1.1.

TABLE 1.1
Identification of Genes and Antibiotics through Gene Mining

Gene	Environmental sample
Esterases	Urania remote ocean hypersaline anoxic bowls in the Mediterranean Sea
Lipases	Soil from the Madison, soil outside the University of Göttingen, Germany; Agricultural Research Station, USA
Amylases	Soil from the Madison, soil outside the University of Göttingen, Germany; Agricultural Research Station, USA
Chitinases	Estuarine and waterfront seawater collected from Delaware Bay, USA
Turbomycin A and B	Soil from the Madison, Wisconsin, USA
Polyketide synthases	Soil from an arable field in La Côte Saint André, France
Nutrient biosynthesis	Sandy soil from a seashore in Kavalla, Greece; volcanic soil in Mt. Hood, Oregon, USA
4-Hydroxybutyrate dehydrogenase	Soils from sugar beet field in Göttingen, River Nieme valley, Germany

1.2 BACKGROUND OF METAGENOMICS

Advances in innovation have essentially determined the current transformation in general microbiology, especially gut microbiota investigation. The improvement of microbial culture-free techniques furnished the opportunity to manage new theories and standards. At a time when biotechnology was at early stage, the molecular strategy came out, aided by plans for the general deployment of ribosomal RNA grade 16S PCR for biological research. In this way, the utilisation of measurable PCR and advancement of the clusters with groundworks commenced the microbiota examinations. The high-throughput sequencing or NGS arrived at microbiology in the 2000s, and researchers (Kozinska et al., 2019) participated in using bioinformatics approaches for next-level research activities. The semiquantitative strategy of denaturing slope gel electrophoresis was demonstrated, which is a very powerful technique for the contrasting microbiota syntheses, and it is used as a standard for screening the bacterial networks before NGS strategies evolved (Moreno-Indias et al., 2020; Shah, 2020).

The advancements in the metagenomics and genomics were pyrosequencing, driven via the 454-sequencing stage from Roche. An individual nucleotide expansion cycle, where pyrophosphate is delivered after the DNA polymerisation response that is changed within a radiant sign recognised by the machine and converted into nucleotide successions, is the basic innovation of metagenomics. Nonetheless, regardless of the progressive idea of this innovation for metagenomics, it is presently out of date. A closely resembling innovation was the particle downpour stage (Moreno-Indias et al., 2020).

The Ion S5 and Ion Torrent Personal Genome Machine can recognise deviations in hydrogen potential. Nevertheless, read length decline is an additional theory in a specific stage for diminishing sequencing costs. These are the conditions of Illumina development that have gotten conceivably the utmost standard headways inferable from its straightforwardness and exceptional yield. The reason for Illumina science is reversible-end sequencing through the association of fluorescently named nucleotides. DNA parts of a cell, wherein the sequencing reaction happens by nucleotide addition. Exactly when a stamped nucleotide is united into the models, fluorescent molecules luminesce after photon-induced excitation (Moreno-Indias et al., 2020).

Although these advances are most ordinarily utilised for understanding the metagenome, the improvement of sequencing has kept on settling the identified inclinations of these methodologies and giving a superior compromise among yield, cost, and read length. PacBio (created by Pacific Biosciences) depends on single-molecule real-time (SMRT) sequencing. This method of SMRT sequencing is based on (i) zero-mode waveguides – that permit light to enlighten just the lower part of a well, wherein a DNA polymerase template complex is immobilised, and (ii) phosphor-linked nucleotides that empower a valuation of the powerless perplexing as the DNA polymerase creates normal DNA strand. PacBio SMRT sequencing offers different strategies, via lower-throughput, higher blunder rate, and greater expense per base. Nanopore sequencing is the most recent innovation with the compact gadget MinION that gives long peruses and fits in a pocket. Nanopore innovation depends on nanoscale openings. Sequencing of DNA depends on the strand DNA section via the nanopore (Moreno-Indias et al., 2020).

1.3 CLASSIFICATION OF METAGENOMICS

The classification of metagenomics is an amplicon-based strategy that incorporates 16S ribosomal RNA for microscopic organisms, by using metagenomic shotgun sequencing method.

1.3.1 Shotgun Metagenomic Sequencing

The investigation of shotgun metagenomic can recognise most of the life forms (culturable and unculturable microorganisms) in an ecological example. A group of scientists was made the biodiversity profile, which is additionally connected with utilitarian structure examination of organic entity ancestries. Researches predominantly worked on soil microbes initially, later on advanced sequencing profundity likewise permits the location of uncommon taxa. The shotgun metagenomic investigation is done by two methods, viz. (i) sequence-based screens that portray the microbial variety and genomes of a specific ecological example and (ii) functional screens that distinguish the quality of useful items from the different hereditary (Madhavan et al., 2017). At the beginning of the metagenomic examination, it is important to know the bacteriological variety and the number of microbial species in the natural environment. The metagenome of refined soil microbial tests revealed the additional information about bacterial source is one of the classical examples in the shotgun metagenomic investigation. Thus, for legitimate inclusion, further information should be created.

1.3.2 16S Sequencing

Even in the 16S sequencing, amplicon-based 16S sequencing is mostly used for taxonomic assessments. The investigation cost of shotgun metagenomic sequencing is more when compared with 16S sequencing for identifying the microbial species. Furthermore, amplicon-based 16S sequencing is also used for identifying the microbial diversity in soil and the human gut. During the shotgun metagenomic sequence analysis, some degree of divergence is accepted and the same is used in OTU. One of the drawbacks of shotgun metagenomic sequencing is that the organisms have a 16S rRNA gene. These are classified into the same species, though they come from different species. The 16S analysis used the OTUs, which are not able to differentiate the strains and related species. Weinstock (2012) studied the OUT-based taxonomic level of both *Escherichia coli* O157: H7 and *E. coli* K-12 by 16S analyses (Weinstock, 2012). Moreover, *Shigella flexneri* can also be distinguished from *E. coli* using 16S analysis. The identification at the species level is less precise using 16S sequence analysis when compared with 16S rRNA sequencing (Ranjan et al., 2016).

The most appropriate markers for parasitic phylogenetic investigations are ITS1 and ITS2, which are based on variable groupings, multicopy nature, and rationed preliminaries (Cuadros-Orellana et al., 2013). The amplicon-based 18S rRNA is the most important segment of parasitic cells and contains rationed and hypervariable districts. ITS DNA is sandwiched between the 18S and 5.8S rRNA qualities with a various level of grouping variety. The investigations of parasites utilise the 18S rRNA, and the ITS area is broadly utilised for dissecting contagious variety in ecological examples (Bromberg et al., 2015). The investigations of microbial organisms frequently originated in the atomic ribosomal quality bunch, which incorporates the 18S or little subunit, 5.8S subunit, and 28S or huge subunit rRNA qualities.

Mothur is utilised to perform ordered and useful investigations on the identification of genes and antibiotics. The different data's such as MG-RAST, QIIME and in QIIME, UNITE utilised through parasitic rDNA ITS arrangements. Regardless of rRNA quality, amplicon-based studies are performed with a clear competence. B. Various Studies on Nitrogen Obsession and Nitrogenase-Reductase Properties. Arbuscular mycorrhizal parasites are identified within the harmonious connection and depend on examinations of the contagious SSU rRNA quality. The list of important tools in metagenomics is presented in Table 1.2.

TABLE 1.2
Important Tools in Metagenomics

Category	Tools	References
Assembly	IDBA-UD	Peng et al. (2012)
	MEGAHIT	Li et al. (2015)
	MetaVelvet	Namiki et al. (2012)
	MetaVelvet-SL	Sato and Sakakibara (2014)
	Ray Meta	Boisvert et al. (2012)
	SOAPdenovo2	Luo et al. (2012)
	Omega	Haider et al. (2014)
	metaSPAdes	Nurk et al. (2017)
	MetAMOS	Treangen et al. (2013)
Binning	SCIMM and PHYSCIMM	Kelley and Salzberg (2010)
	Phymm and PhymmBL	Brady and Salzberg (2009)
	CONCOCT	Alneberg et al. (2014)
	IMG/MER 4	Markowitz et al. (2013)
	MEGAN	Huson and Weber (2013)
	MetaCluster	Wang et al. (2014)
Gene prediction	MetaGeneMark	Zhu et al. (2010)
	Prodigal	Hyatt et al. (2010)
	FragGeneScan	Rho et al. (2010)
	Glimmer-MG	Kelley and Salzberg (2011)
	Prokka	Seemann (2014)
Functional 16S analysis	PICRUSt	Langille et al. (2013)
Databases	Ribosomal database (RDP)	Cole et al. (2006)
	SEED	Overbeek et al. (2005)
	eggnog	Powell et al. (2014)
	PFAM	Bateman et al. (2004)
	TIGRFAM	Haft et al. (2003)
	UNITE	Kõljalg et al. (2013)

1.4 INVESTIGATION OF METAGENOME

1.4.1 Shotgun Metagenomic Sequencing Analysis

Shotgun metagenomics analysis is used to examine both the known and unknown species. A new get-together of more limited peruses might be done to get genomic contigs, trailed via a platform that is regularly performed to become a more minimised and brief perspective on the sequenced natural examples. Realising full gatherings of the total genomes present locally is troublesome, and is once in a while conceivable in straightforward networks, or with exceptionally profound sequencing (Venter et al., 2004). The knowledge of total arrangements of protein-coding qualities in the sequenced genomes must be known in shotgun metagenomic analysis before implementation.

1.4.2 Construction of Metagenomic Athenaeum

From the dwelling of diverse microbial environments, DNA was isolated. Due to more availability of microbial species, more metagenomic studies were carried out in the early days in the soil and seawater. In addition to the isolation of microorganisms from the soil and seawater, more microorganisms are also isolated from aquatic sediments, biofilms, and wastewater. Several biocatalysts like hydrolases, laccases, and xylanases are extracted from microbial DNAs that are used for constructing metagenomic libraries (Sabree et al., 2009).

The microbial environment influences the size, quality, and measure of isolated microbial DNA. The planktonic networks are equipped for dealing with huge capacities of water for focusing adequate microbial biomass that acquires sufficient DNA to fabricate libraries. Contaminating synthetic compounds and chemicals routinely remain in the water, making DNA solubilization reasonably easy without large amounts of contaminants. Interestingly, inorganic soil, humic acids, sediments, and biological pollutants are used for DNA extraction.

Interaction of eliminating impurities decides on the size of the DNA and clonability because a significant number of cycles that viably eliminate toxins that restrain cloning likewise shear the DNA. Actual dissociation of microorganisms from the semisolid framework, commonly named "cell division", may yield a cell pellet, particularly with a high-atomic-weight (>20 kb) DNA. Agarose grid immobilisation cells further lessen DNA shear powers, and electrophoresis encourages partition of high-atomic-weight DNA in humic acids and DNases. Different business units could be utilised to extricate DNA in soil and other semisolid grids. Smearing different abstraction strategies to a DNA test may yield negligibly tainted DNA. A fast DNA spin arrangement trailed by a hexadecyl trimethyl ammonium bromide extraction produces boundless DNA (Sabree et al., 2009). The steps in the metagenomics process are shown in Figure 1.1.

1.4.3 Rendering Vectors

The 20 kb little DNA parts should be cloned into fosmids, cosmids, or bacterial fake chromosomes (BACs) to produce the maximum quality. pCC1FOS and pWE15 vectors could be utilised to clone huge DNA sections in different networks of microbes.

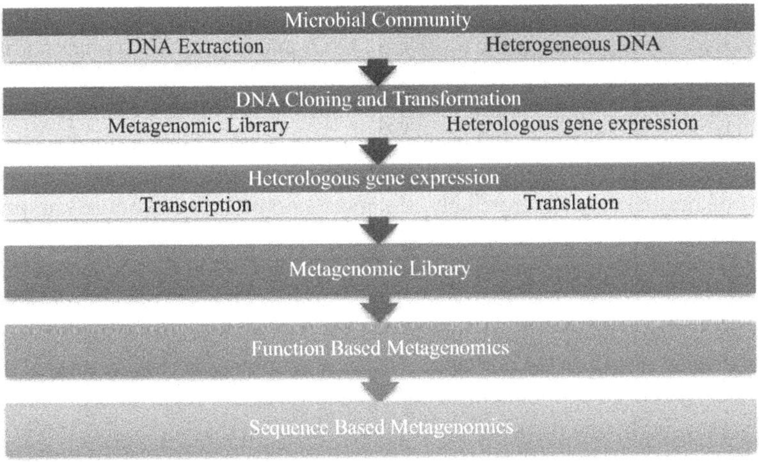

FIGURE 1.1 Steps in the metagenomics process.

The pCC1FOS vector produces more benefits when it acts as a proper host. The pCC1FOS duplicate number may be constrained by the growth of arabinose to expand the yield of DNA. Bacteriological detecting signals, anti-infection obstruction causes, antibiosis, colour creation, and eukaryotic development balancing features were distinguished in metagenomic libraries built with pCC1FOS (Sabree et al., 2009).

1.4.4 Investigation of Metagenomics

Investigations on metagenomics helps to gathering genomic data in organisms without refining them. Rather than useful screening, this methodology depends on succession investigation to give the premise to expectations about work. Enormous datasets are currently indexed in the "Natural Genomic Sequence" data set, and every sequencing is more valuable than an account of the gathered information from the assorted conditions. In the ecological succession environment, the grouping-based techniques could help to study the microbial metagenomics. A few investigations utilise the quality of premium or "anchor" for distinguishing metagenomic clones of revenue for additional examination. A library is developed and separated using the PCR to intensify the anchor through metagenomics. The ribosomal RNA quality possessed by the anchors may have metabolic quality. The anchor replicas are either sequenced or further broken down to give data about the genomic setting of the anchor. The clones of 16S rRNA qualities were sequenced to give an image of local area individual varieties (Sabree et al., 2009).

1.4.5 Construction of Clones

The assurance of target embed size, least number of library clones, and cloning vector are managed through a sort of qualities. The sequencing is normally controlled by little addition of clones, though fruitful practical examinations can be performed on little and huge supplement clones. The metagenomic collections established in plasmid DNA are steadily kept up to 10 kb of DNA and require 3–20 times additional clones contrasted with reference library built-in fosmids (30–40 kb supplements) to get equivalent inclusion of a similar microbial local area. Mine waste microbes having very limited information and a group of researchers was examined profoundly with an enormous supplement metagenomic library, creating it conceivable to cover one part of the genome. Interestingly, the metagenome of an extremely unpredictable local area, for example, that in the soil must be examined, not comprehensively sequenced. With the present innovation, a local area could be sequenced rapidly more and economically little than enormous addition clones, yet future innovation advancement may change this. Conversely, for activity-based investigations, giant complements are ideal, as the probability that the action of interest is encoded by any clone is clearly related to complement size, and if the action is encoded by a bunch of qualities, they are bound to be caught in a huge addition (Sabree et al., 2009).

1.4.6 Investigation of G+C Content

Absolute people group variety of an abnormal climate can be resolved based on their guanine and cytosine content (G+C) as the vast majority of phylogenetically associated gatherings change just inside 3%–5% of their G+C content. The centrifugation process depends up on G+C content of DNA of a whole local area. Along these lines, the general wealth of the microbial local area as a factor of G+C substance can be resolved. Extra strategies, for example, denaturing inclination gel electrophoresis or enhanced ribosomal DNA limitation examination, could be utilised for a predominant evaluation of the microbial local area in abnormal climate. Nonetheless, a significant G+C content investigation based on local area variety evaluation is its coarse degree of phylogenetic goal, a prerequisite of huge measure of DNA (50 μg), and season of fruition (4 days). Given the examination of gas chromatography (GC) content, an investigation anticipated the strength of archaeal space under abnormal conditions with an altogether higher G+C content Brocchieri, 2014).

1.4.7 Examination of Genomes

The examination of genomes between various creatures could be anticipated at the most precise level by the efficiency of the entire genome DDH. The microbial species depiction is of 70% DDH content. It is to be speculated that the microbial elements with 70% of DDH esteem for the most part have over 97% of arrangement similitudes in their 16S rRNA quality. The DDH strategy may apply to an unadulterated bacterial culture examination. Utilisation of the DDH procedure for the bacterial local area investigation includes the extraction of complete DNA from abnormal ecological examples, trailed by denaturation, and hatching underneath appropriate situations to develop their reassociation or hybridisation. Local area variety records can be determined by deciding on the half affiliation esteem, determined as C0t, where 'C0' represents the underlying grouping of single-abandoned DNA and 't' represents the necessary time. The level of reassociation between the specific DNA may be related to comparing genomic intricacy. A summed-up idea test with high

DNA grouping variety gives an adverse effect on DNA reassociation (Agrawal et al., 2015). Several studies utilised the DDH procedure for investigating microbes in local abnormal conditions. It is broadly utilised in novel species acknowledgement. Thus (Li et al., 2017), genomic DDHs were used to demonstrate new microbial animal types in climates with extreme cold, high UV radiation, and low complement (Dash and Das, 2018).

1.4.8 16S rRNA Quality Analysis

Early activities utilised rRNA to recognise and enlighten the clones in metagenomic libraries. This procedure was utilised to distinguish clones that contained DNA in planktonic marine Archaea. Already, 16S rRNA quality arrangements were recuperated in a few ecological examples, proposing the occurrence of Archaea within non-extreme conditions like soil and untamed water. A group of researchers worked on immediate cloning strategy to disconnect genomic DNA from microbes was instead of endeavouring to culture them. They separated seawater and arranged a fosmid library in the collected organic entities. The tested library was used to discover little subunit rRNA qualities from archaeal species. A few putative protein-encoding qualities had been distinguished, and others were not recognised in Archaea (Sabree et al., 2009).

1.4.9 PCR-Based Examination Method

This is a fractional local area examination method that uses the strategies of PCR in which the absolute DNA and/or RNA is utilised for interpretation of organisms (Figure 1.2). The PCR along with the combination of microbial quality marks in all life forms with VBNC partition. PCR intensification of rationed qualities like 16S rRNA is found on the whole prokaryotes. They are fundamentally and practically moderated (Hugenholtz, 2002) and are regularly indicated as the "best quality levels" for the phylogenetic microbial biology examinations. In light of phylogenetic likenesses with the identified microorganisms, as apparent in the thermology of the 16S rRNA arrangement, a sub-atomic grouping is relegated to novel separates.

Intermediate grade segregating subpopulations of microbial species using RNA polymerase beta subunit (rpoB), gyrase beta subunit (gyrB), recombinase A (recA), and heat stun protein (hsp60). The collection of clone techniques, hereditary fingerprinting, and microarrays of DNA, as summed up beneath, is generally used for breaking down PCR-enhanced ecological DNA. The library of clone strategy is used for phylogenetic variety investigation until late days. The PCR-enhanced phylogenetic marker successions in the DNA of an ecological environment are very much utilised for making a clone. The cloning procedure is implemented to PCR items by Taq polymerase, and it permits productive ligation into plasmid vectors with an overhanging 30-T (Chakraborty and Das, 2014).

The acquired groupings are contrasted, and the recognised successions are accessible from the universally perceived data set, for instance, GenBank, Ribosomal Database Project (RDP) and Green-qualities, second genome, and so forth. RDP alone has a vault of data on 3.4 million successions with the relevant information

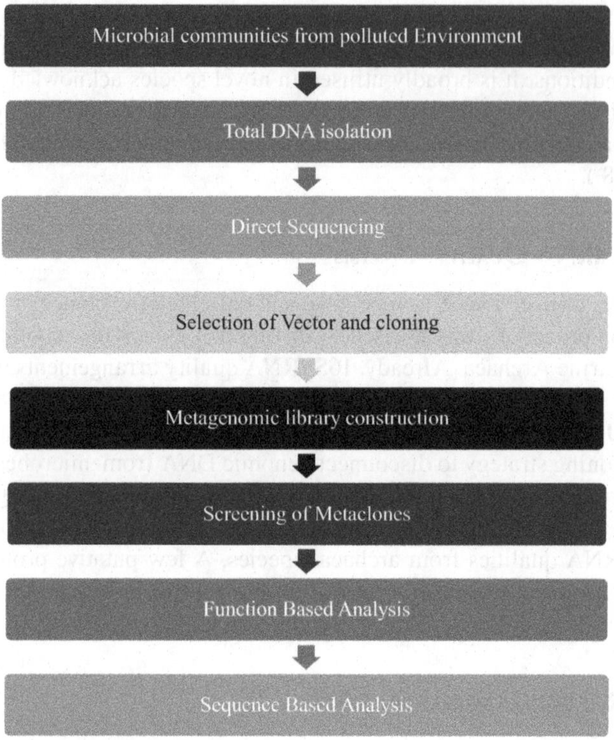

FIGURE 1.2 PCR-based examination method.

and is as a rule intermittently refreshed. Typically, the cloned arrangements are distinguished systematically from phylum to species level through cut-off estimations of 80% and 97%, respectively. The 16S rRNA clone library markers grant an underlying review of variety, and an enormous clone should be sequenced to archive half of the species wealth of a dirt example from a tainted site (Chakraborty and Das, 2014). 16S rRNA clone library qualities contain less than 1,000 groupings that uncover a little segment of the bacterial variety in an ecological environment. However, cloning strategies are dreary, tedious, and costly. The clone libraries are used for fundamental appraisal of microbial variety in defiled destinations. Fresher and modest techniques for narrow sequencing and expanding patterns in robotisation may acquire more noteworthy advancement in this strategy for microbial variety investigation. The PCR-based examination method is shown in Figure 1.2.

1.5 HEREDITARY FINGERPRINTING PROCEDURE

The hereditary fingerprinting procedure gives a particular profile of microbial networks dependent on the direct investigation of PCR-enhanced items in the ecological DNA. They depend on either detachment of amplicons after PCR intensification of phylogenetic (16S rRNA) or useful qualities utilising widespread or explicit groundworks. These methods are fast and permit concurrent examinations of numerous examples and don't give directly ordered personalities. Closeness and contrasts between the microbial networks are looked at utilising PC-helped bunch investigation (e.g., GelCompar) in light of the various microbial fingerprints. The genetic fingerprinting methods have a high-throughput configuration either denaturing or temperature gradient gel electrophoresis, single-strand adaptation polymorphism, irregular enhanced polymorphic DNA, intensified ribosomal DNA limitation examination, the terminal limitation section length polymorphism, length heterogeneity investigation by length heterogeneity PCR (LH-PCR), and ribosomal RNA (rRNA) intergenic spacer examination (Panigrahi et al., 2019).

1.5.1 16S rRNA Quality Check

The limitation profiles could be utilised to diagnose the stimulating clones that further assess the operational ordered units in the abnormal ecological examples. The ARDRA procedure prevents the grouping variety of 16S rRNA quality that creates a unique finger impression. The entire DNA is disengaged, and a PCR item is created by increasing the 16S rRNA from ecological DNA. The cultivable bacterial varieties in the dregs of Great Rann of Kachchh, India, were examined using the ARDRA procedure in abnormal climatic conditions. It is lagged through the control absorption of amplicon PCR through tetra shaper limitation endonuclease proteins for most of AluI and HaeIII. The created parts of unique finger impressions are further examined with either agarose or polyacrylamide gel electrophoresis, which produces the limitation profiles. During the investigations, 42 cultivable microbial confines were dispersed into 14 clades with a closeness coefficient of 1.0. The circulation size of microorganisms was observed as Firmicutes > Actinobacteria > Proteobacter ia (Dash and Das, 2018).

1.5.2 LH-PCR and TRFLP Investigations

The LH-PCR and TRFLP investigations are having the same approach. The diversity of nucleotide length arrangement is identified in LH-PCR investigation after the PCR assimilation. The presence of normal length polymorphisms inside the microorganisms produces length non-uniformity due to conversion process. In this approach, hypervariable locale of the 16S rRNA quality is concentrated for creating a novel profile that may be used to recognise the microbial networks. A colossal Archaea variety was noticed through LH-PCR investigation in the North Arm of Great Salt Lake (Almeida-Dalmet et al., 2015) in abnormal natural conditions (Dash and Das, 2018).

1.5.3 RAPD Method

The RAPD method utilises PCR techniques that enhance the DNA piece with short groundworks under a low-temperature environment. Accordingly, the quantity of DNA parts of shifted length is produced and the resultant partition on either agarose or polyacrylamide gels may misuse hereditary intricacies of a microbial local area with an abnormal climatic condition. Likewise, *Vibrio vulnificus* disengaged from several geographic districts, which created distinctive RAPD profiles with abnormal conditions. RAPD preliminaries focus on DNA successions in novel and tedious groupings through the subsequent amplicons that are to be utilised as a hereditary unique finger impression. There are numerous utilisations of the RAPD method, to refer to a multiple investigations of populace structure, mutagenesis, hereditary planning, and phylogenetics. RAPD surveys of *Bacillus cereus* and *Pseudomonas aeruginosa* were isolated in estuarine climates, predicting the presence of DNA cultivars under unusual climatic conditions (Nawas et al., 2016).

1.6 DDH-BASED MICROARRAY

DDH-based microarray strategy is generally used for evaluating the microbial local area within a climate. The PCR amplicons from the ecological DNA are hybridised using a DDH-based microarray strategy with comprehended sub-atomic tests present within the microarray. A positive hybridisation identified through confocal laser filtering magnifying lens connotes the presence of the objective DNA succession in the addressed example. The significant constraint of the DNA microarray procedure includes the conceivable cross-hybridisation and its non-usefulness in recognising new microbial strains because of the non-availability of comparing tests. The prevalent convenience of the strategy is resolved because a solitary cluster harbours a large number of focused DNA successions with high particularity. In any case, this may be utilised in two different ways, viz. (i) focusing on the nucleotide length varieties of 16S rRNA quality and (ii) focusing on the mark practical qualities explicit to a microorganism (Dash and Das, 2018).

1.6.1 Construction of rDNA Microarray

Development of relevant datasets containing microbial small subunit (SSU) and ribosomal DNA (rDNA) sequences. The high-thickness microarray investigation has been created with a more up-to-date, dependable, and moderate approach to dissect many identified microbes present in a climate. The PhyloChip was created to determine up to 8,471 individual microbial taxa at subfamily degrees or higher. PhyloChips have proved to have the most significant level of reproducibility and empower reliable investigation of a microbe's area by changing the ecological conditions as prevailing microbial species do not veil the discovery of uncommon species. PhyloChip has found a widespread application in abnormal conditions as its information shows a critical positive connection between diminished local area phylogenetic distance and expanded the microbial flora in the Great Salt Lake, Utah, USA (Parnell et al., 2011).

1.6.2 Steps in Gene Array Construction

The microarray strategy not only identifies the local area of microbial structure in the specific climate but also distinguishes the in situ metabolic potential local area. The FGA focuses on realised local metabolic qualities with changed capacities like smelling salts oxidation, methane oxidation, and nitrogen obsession in this sense, unlike the PhyloChip, the FGA assay detects the metabolic accumulation of microscopic

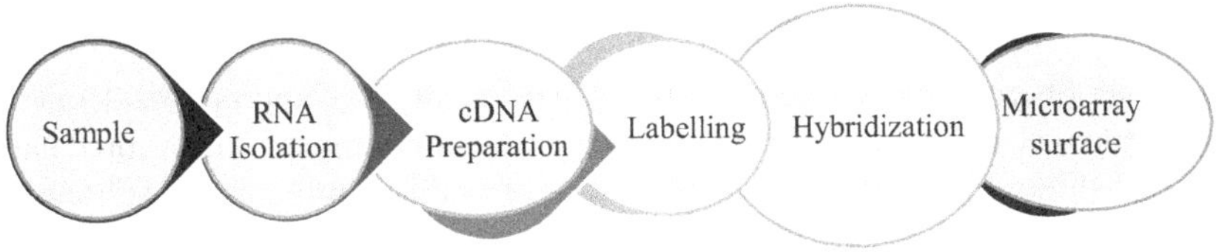

FIGURE 1.3 Steps in gene array construction.

organisms, enabling connections between local tissues of microbes and biological systems.The microarray comprises various tests of recognised metabolic qualities that are distinguished by hybridisation. A functional gene array of the hybridisation interaction is presented in Figure 1.3. The practical quality microarray is utilising GeoChip 2.0, which demonstrated helpfulness in abnormal conditions. The microbial networks from corrosive mine seepage test outcomes are represented by the ecological elements, useful key qualities, and utilitarian cycles, i.e., methane age, ammonification, denitrification, sulphite decrease, and natural toxin corruption (Xie et al., 2011).

1.7 MASSIVE PARALLEL SEQUENCING TECHNOLOGY

The most well-known economically accessible NGS stages incorporate 454 pyrosequencing, Illumina, and SOLiD. These stages vary as far as their operational methodology; nonetheless, the essential science of their basics stays as before which incorporates discontinuity of the objective DNA, the ligation of short connector particles followed by tying, and the location of the groupings generally by the pyrosequencing. In the examination, NGS manages in vitro library development and enhancement measures. Hence, this method gets simpler without dealing with the issues of state picking (Behjati and Tarpey, 2013).

1.8 METAPROTEOMICS

The term "metaproteomics" was first used by Wilmes and Bond (2004). Metaproteomics is characterised as an enormous scope portrayal of whole protein supplements of natural microbiota in a state of time (Wilmes and Bond, 2004). Biotransformation interaction of any type is happening in the environment, and the compounds of protein have trademark jobs. Consequently, metaproteome investigation gives a legitimate understanding of the usefulness of elements of a biological system. Protein examinations gave a considerably more dependable and immaculate image of useful microbial variety in contrast to the investigation of some other biomolecules. In this manner, proteomics investigation gives a lot of nitty-gritty data as long as the major and practical parts of the environment are subjective and quantitative.Furthermore, the proteome profiling changes of an environment give data concerning the physiological reactions of the occupant organisms to the changing biological system conditions (Chen et al., 2017).

1.8.1 Proteogenomics Approach

The limit of natural metaproteomics is the inaccessibility of appropriate data sets to look at and recognise the proteins in the ecological environment. The limited natural metaproteomics leads to the non-identification of available proteins and subsequently influences the task of their phylogenetic noticeability. Under the specific circumstance, "proteogenomics" becomes possibly the most important factor as a mixture of metaproteomic approaches. Once both proteins and DNA are extricated in a positive abnormal natural example and investigated all the while, the phylogenetic goal turns out to be more critical. Numerous Archaea and microscopic organisms have been broken down utilising the proteogenomics to yield helpful outcomes in characterising the microbial local area construction and capacity of the biological system. Also, developmental marks, strain-settled articulation examples, and compound biosynthesis have been settled by utilising the approach of proteomics in normal acidophilic microbial networks (Armengaud et al., 2013)

1.9 CONCLUSION

The molecular methodology now used for examination of microbial organisations can in like manner be used for unusual conditions. The best techniques used for revenue-focused environments are section from the fittingness of system and the hypothesis being attempted. Most investigations propose the usage of multidisciplinary approaches. The framework may be varied with the level of study. Molecular strategies of microbial phylogeny assessment not simply give a concise glance at the formative past but also propose new encounters to ponder microbial nature like their utilitarian credits, metabolic properties, flexible approaches, and work in biological enhancement cycling similar to their biotechnological application. The metagenomic assessments and NGS changed the normal examination of phylogeny because we were able to refine our investigations without creating cells under unusual conditions. Thus, sub-molecular procedures do not give a piece of authentic information to the microbial assortment, but are valuable in understanding their common cycles.

REFERENCES

Agrawal, P.K., Agrawal, S. & Shrivastava, R. (2015) Modern molecular approaches for analyzing microbial diversity from mushroom compost ecosystem. *3Biotech.* (5), 853–866.

Almeida-Dalmet, S., Sikaroodi, M., Gillevet, P.M., Litchfield, C.D. & Baxter, B.K. (2015) Temporal study of the microbial diversity of the north arm of Great Salt Lake, Utah, U.S. *Microorganisms.* (3), 310–326.

Alneberg, J., Bjarnason, B.S. & Bruijn, I.De. (2014) Binning metagenomic contigs by coverage and composition. *Nature Methods.* 11 (11), 1144–1146.

Amundson, R., Berhe, A.A., Hopmans, J.W., Olson, C., Sztein, A.E. & Sparks, D.L. (2015) Soil and human security in the 21st century. *Science.* 348, 1261071-1–1261071-6.

Armengaud, J., Hartmann, E.M. & Bland, C. (2013) Proteogenomics for environmental microbiology. *Proteomics.* 13, 2731–2742.

Bateman, A., Coin, L. & Durbin, R. (2004) The Pfam protein families database. *Nucleic Acids Research.* 32 (suppl_1), D138–D141.

Beaulaurier, J. (2020) Assembly-free single-molecule sequencing recovers complete virus genomes from natural microbial communities. *Genome Res.* 30, 437–446.

Behjati, S. & Tarpey, P.S. (2013) What is next generation sequencing? Archives of disease in childhood. *Education and Practice Edition.* 98, 236–238.

Boisvert, S., Raymond, F., Godzaridis, É., Laviolette, F. & Corbeil, J. (2012) Ray Meta: Scalable de novo metagenome assembly and profiling. *Genome Biology.* 13 (12), R122.

Brady, A. & Salzberg, S.L. (2009) Phymm and PhymmBL: Metagenomic phylogenetic classification with interpolated Markov models. *Nature Methods.* 6 (9), 673–676.

Brocchieri, L. (2014) The GC content of bacterial genomes. *Journal of Phylogenetics and Evolutionary Biology.* 2, e108.

Bromberg, J.S., Fricke, W.F., Brinkman, C.C., Simon, T. & Mongodin, E.F. (2015) Microbiota [mdash] implications for immunity and transplantation. *Nature Reviews Nephrology.* 11 (6), 342–353.

Calderon, K., Spor, A., Breuil, M-C., Bru, D., Bizouard, F., Violle, C. & Barnard, R.L. (2017) Philippot L: Effectiveness of ecological rescue for altered soil microbial communities and functions. *ISMEJ.* 11, 1–12.

Chakraborty, J. & Das, S. (2014) Characterization and cadmium resistant gene expression of biofilm-forming marine bacterium Pseudomonas aeruginosa JP-11. *Environmental Science and Pollution Research.* 21 (24), 14188–14201.

Chen, B., Zhang, D., Wang, X., Ma, W., Deng, S. & Zhang, P. (2017) Proteomics progresses in microbial physiology and clinical antimicrobial therapy. *European Journal of Clinical Microbiology and Infectious Diseases.* 36, 403–413.

Clark, D.P. & Pazdernik, N.J. (2016) *Environmental Biotechnology* (2nd edition). http://dx.doi.org/10.1016/B978-0-12-385015-7.00012-0.

Cole, J.R., Chai, B. & Farris, R.J. (2006) The ribosomal database project (RDP-II): Introducing myRDP space and quality controlled public data. *Nucleic Acids Research.* 35 (suppl_1), D169–D172.

Cuadros-Orellana, S., Leite, L.R. & Smith, A. (2013) Assessment of fungal diversity in the environment using metagenomics: A decade in review. *Fungal Genomics & Biology.* 3 (2), 1.

Dash, H.R. & Das, S. (2018) Molecular methods for studying microorganisms from atypical environments. *Methods in Microbiology.* https://doi.org/10.1016/bs.mim.2018.07.005.

Emerson, J.B. (2018) Host-linked soil viral ecology along a permafrost thaw gradient. *Nature Microbiology.* 3, 870–880.

Haft, D.H., Selengut, J.D. & White, O. (2003) The TIGRFAMs database of protein families. *Nucleic Acids Research.* 31 (1), 371–373.

Haider, B., Ahn, T.H. & Bushnell, B. (2014) Omega: An overlap-graph de novo assembler for metagenomics. *Bioinformatics.* 30 (19), 2717–2722.

Hugenholtz, P. (2002) Exploring prokaryotic diversity in the genomic era. *Genome Biology.* 3, reviews0003.1–reviews0003.8.

Huson, D.H. & Weber, N. (2013) Microbial community analysis using MEGAN. *Methods in Enzymology.* 531, 465–485.

Hyatt, D., Chen, G.L. & LoCascio, P.F. (2010) Prodigal: Prokaryotic gene recognition and translation initiation site identification. *BMC Bioinformatics.* 11 (1), 119.

Jansson, J.K. & Hofmockel, K.S. (2018) The soil microbiome—From metagenomics to metaphenomics. *Current Opinion in Microbiology.* 43, 162–168.

Kelley, D.R. & Salzberg, S.L. (2010) Clustering metagenomic sequences with interpolated Markov models. *BMC Bioinformatics.* 11 (1), 544.

Kief, K., Zhou, Z. & Anantharaman, K. (2020) VIBRANT: Automated recovery, annotation and curation of microbial viruses, and evaluation of viral community function from genomic sequences. *Microbiome.* 8, 90.

Kõljalg, U., Nilsson, R.H. & Abarenkov, K. (2013) Towards a unified paradigm for sequence-based identification of fungi. *Molecular Ecology*. 22 (21), 5271–5277.

Koonin, E.V. (2020) Global organization and proposed megataxonomy of the virus world. *Microbiology and Molecular Biology Reviews*. 84, e00061-19.

Kozinska, A., Seweryn, P. & Sitkiewicz, I. (2019) A crash course in sequencing for a microbiologist. *Journal of Applied Genetics*. 60 (1), 103.e111. https://doi.org/10.1007/s13353-019-00482-2.

Langille, M.G., Zaneveld, J. & Caporaso, J.G. (2013) Predictive functional profiling of microbial communities using 16S rRNA marker gene sequences. *Nature Biotechnology*. 31 (9), 814–821.

Li, D., Liu, C.M., Luo, R., Sadakane, K. & Lam, T.W. (2015) MEGAHIT: An ultra-fast single-node solution for large and complex metagenomics assembly via succinct de Bruijn graph. *Bioinformatics*. 31 (10), 1674–1676.

Li, Y., Ren, Y. & Jiang, N. (2017) Analysis of draft genome sequence of Pseudomonas sp. QTF5 reveals its benzoic acid degradation ability and heavy metal tolerance. *BioMed Research International*. 7. https://doi.org/10.1155/2017/4565960.

Luo, R., Liu, B. & Xie, Y. (2012) SOAPdenovo2: An empirically improved memory-efficient short-read de novo assembler. *Gigascience*. 1 (1), 18.

Madhavan, A., Sindhu, R., Parameswaran, B., Sukumaran, R.K. & Pandey, A. (2017) Metagenome analysis: A powerful tool for enzyme bioprospecting. *Applied Biochemistry and Biotechnology*, 1–16.

Markowitz, V.M., Chen, I.M.A. & Chu, K. (2013) IMG/M 4 version of the integrated metagenome comparative analysis system. *Nucleic Acids Research*. 42 (D1), D568–D573.

Moreno-Indias, I. & Tinahones, F.J. (2020) Metagenomics. Principles of nutrigenetics and nutrigenomics. https://doi.org/10.1016/B978-0-12-804572-5.00011-2.

Namiki, T., Hachiya, T., Tanaka, H. & Sakakibara, Y. (2012) MetaVelvet: An extension of Velvet assembler to de novo metagenome assembly from short sequence reads. *Nucleic Acids Research*. 40 (20), e155–e155.

Nawas, P.M.A., Ramasubburayan, R., Palavesam, A. & Immanuel, G. (2016) RAPD analysis of dominant denitrifying bacterial species in the estuarine environment of south west coast of India. *Indian Journal of Geo-Marine Science*. 45, 1696–1703.

Nayfach, S., Camargo, A.P., Schulz, F., Eloe-Fadrosh, E., Roux, S. & Kyrpides, N.C. (2020) Check V assesses the quality and completeness of metagenome-assembled viral genomes. *Nature Biotechnology*. https://doi.org/10.1038/s41587-020-00774-7.

Nilsson, R.H., Larsson, K-H.H. & Taylor, A.F. (2019) The UNITE database for molecular identification of fungi: Handling dark taxa and parallel taxonomic classifications. *Nucleic Acids Res*. 47, D259–D264. https://doi.org/10.1093/nar/gky1022.

Nurk, S., Meleshko, D., Korobeynikov, A. & Pevzner, P.A. (2017) metaSPAdes: A new versatile metagenomic assembler. *Genome Research*. 27 (5), 824–834.

Overbeek, R., Begley, T. & Butler, R.M. (2005) The subsystems approach to genome annotation and its use in the project to annotate 1000 genomes. *Nucleic Acids Research*. 33 (17), 5691–5702.

Panigrahi, S., Velraj, P. & Rao, T.S. (2019) Functional microbial diversity in contaminated environment and application in bioremediation. *Microbial Diversity in the Genomic Era*. https://doi.org/10.1016/B978-0-12-814849-5.00021-6.

Parnell, J.J., Rompato, G., Crowl, T.A., Weimer, B.C. & Pfrender, M.E. (2011) Phylogenetic distance in Great Salt Lake microbial communities. *Aquatic Microbial Ecology*. 64, 267–273.

Peng, Y., Leung, H.C., Yiu, S.M. & Chin, F.Y. (2012) IDBA-UD: A de novo assembler for single-cell and metagenomic sequencing data with highly uneven depth. *Bioinformatics*. 28 (11), 1420–1428.

Powell, S., Forslund, K. & Szklarczyk, D. (2014) eggNOG v4. 0: Nested orthology inference across 3686 organisms. *Nucleic Acids Research*. 42 (D1), D231–D239.

Ranjan, R., Rani, A., Metwally, A., McGee, H.S. & Perkins, D.L. (2016) Analysis of the microbiome: Advantages of whole genome shotgun versus 16S amplicon sequencing. *Biochemical and Biophysical Research Communications*. 469 (4), 967–977.

Rho, M., Tang, H. & Ye, Y. (2010) FragGeneScan: Predicting genes in short and error-prone reads. *Nucleic Acids Research*. 38 (20), e191–e191.

Roux, S. (2019) Minimum information about an uncultivated virus genome (MIUViG). *Nature Biotechnology*. 37, 29–37.

Sabree, Z.L., Rondon, M.R. & Handelsman, J. (2009) Metagenomics. *Genetics*, 622–632.

Sato, K. & Sakakibara, Y. (2014) MetaVelvet-SL: An extension of the Velvet assembler to a de novo metagenomic assembler utilizing supervised learning. *DNA Research*. 22 (1), 69–77

Schulz, F. (2020) Giant virus diversity and host interactions through global metagenomics. *Nature*. 578, 432–436.

Seemann, T. (2014) Prokka: Rapid prokaryotic genome annotation. *Bioinformatics*. 30 (14), 2068–2069.

Shah, M.P. (2020) *Advanced Oxidation Processes for Effluent Treatment Plants*. Elsevier.

Shah, M.P. (2021) *Removal of Emerging Contaminants through Microbial Processes*. Springer.

Smits, S.L. (2014) Assembly of viral genomes from metagenomes. *Frontiers in Microbiology*. 5, 714.

Stefani, F., Bencherif, K., Sabourin, S., Hadj-Sahraoui, A.L., Banchini, C., Séguin, S. & Dalpé, Y. (2020) Taxonomic assignment of arbuscular mycorrhizal fungi in an 18S metagenomic dataset: A case study with saltcedar (Tamarix aphylla). *Mycorrhiza*. https://doi.org/10.1007/s00572-020-00946-y.

Treangen, T.J., Koren, S. & Sommer, D.D. (2013) MetAMOS: A modular and open source metagenomic assembly and analysis pipeline. *Genome Biology*. 14 (1), R2.

Tripathi, B.M., Stegen, J.C., Kim, M., Dong, K., Adams, J.M. & Lee, Y.K. (2018) Soil pH mediates the balance between stochastic and deterministic assembly of bacteria. *The ISME Journal*. 12, 1072–1083.

Venter, J.C., Remington, K. & Heidelberg, J.F. (2004) Environmental genome shotgun sequencing of the Sargasso Sea. *Science*. 304 (5667), 66–74.

Wang, D.Z., Xie, Z.X. & Zhang, S.F. (2014). Marine metaproteomics: Current status and future directions. *Journal of Proteomics*. 97, 27–35.

Weinstock, G.M. (2012) Genomic approaches to studying the human microbiota. *Nature*. 489 (7415), 250.

Wilmes, P. & Bond, P.L. (2004) The application of two-dimensional polyacrylamide gel electrophoresis and downstream analyses to a mixed community of prokaryotic microorganisms. *Environmental Microbiology*. 6, 911–920.

Xie, J., He, Z., Liu, X., Liu, X., Van Nostrand, J.D. & Deng, Y. (2011) GeoChip based analysis of the functional gene diversity and metabolic potential of microbial communities in acid mine drainage. *Applied and Environmental Microbiology*. 77, 991–999.

Yavitt, J.B., Roco, C.A., Debenport, S.J., Barnett, S.E. & Shapleigh, J.P. (2020) Community organization and metagenomics of bacterial assemblages across local scale pH gradients in Northern Forest Soils. *Microbial Ecology*. https://doi.org/10.1007/s00248-020-01613-7.

Zhu, W., Lomsadze, A. & Borodovsky, M. (2010) Ab initio gene identification in metagenomic sequences. *Nucleic Acids Research*. 38 (12), e132–e132.

2 Microbial Response to Lead Exposure

J. Anandkumar
National Institute of Technology Raipur

Jyoti Kant Choudhari
Maulana Azad National Institute of Technology

Jyotsna Choubey and Tanushree Chatterjee
Raipur Institute of Technology

Mukesh Kumar Verma and Biju Prava Sahariah
Chhattisgarh Swami Vivekanand Technical University

CONTENTS

2.1 INTRODUCTION

The world is providing habitat and nurturing its organisms; either tiny microbes or giant whales, each of them possess their significance for the continuous and balanced operation of the earth system. When these units are displaced, some disturbances are sure to occur, making the earth's condition unfavourable. The micro-units of the biosphere are termed as an individual ecosystem where abiotic and biotic components interact along with its different trophic statuses for circulation of essential nutrients/minerals and transfer of energy required for sustaining life. Establishment of ecosystem through succession towards stabilisation and execution of activities are enrolled on specific organisms. Microbes play various roles in governing the successful operation of an ecosystem. Some activities worth mentioning are detritus organisms for decomposition or mineralisation of compounds releasing locked components

DOI: 10.1201/9781003354147-2

from dead material to soil, conditioning and enriching the soil quality and making it available for plants, buffering the toxicity of elements, and so on. A huge number of microbial species are recorded with their efficiency for cost-effective and environment-friendly treatment of various pollutants. Similarly, microbes exhibit their functions in the production of industrial/medicinal/agricultural items, whereas some are recorded for causing negative impacts such as disease generation/contagious and degradation of item quality. Microbes are the active agents of an ecosystem that can positively or negatively affect the ecosystem. Similarly, though tiny, they are mostly living organisms and prone to the impacts of various environmental and anthropogenic activities. Microbes need positive environmental conditions such as pH, temperature, moisture and nutrients for their growth and survival.

2.1.1 Lead and Its Toxicity

Lead (atomic number: 82) a common heavy metal that the EPA enlisted as hazardous inorganic waste is naturally occurring in all parts of the environment, rarely beneficial to any organisms, but can cause negative impacts on microbes and higher organisms when entering the body. Anthropogenic activity such as the burning of fossil fuel is highly responsible for the release of lead into the environment in large amounts. Lead shows mutagenic and teratogenic effects on humans through damage to DNA, proteins, and lipids, and prone to cause poisonous effects on major body parts such as the nervous, renal and reproductive systems due to the replacement of essential metal ions (Davis et al., 2005; Oliveira et al., 1987). Children are reported to be more prone to lead effects. A diverse group of bacteria are identified to possess mechanisms and capability for removal of lead contamination and applied for the same. These microbes efficiently face the stress of lead exposure and remove the toxicity or bring the same to a less toxic level. This chapter aims to discuss the response of microbes to lead exposure at the genetic and molecular levels.

2.1.2 Mechanism of Lead Survival by Microbes

Metal pollutants are generally treated through a change in their state so as to reduce/remove their toxicity instead of complete elimination or mineralisation as in the case of organic matters. Although high toxicity is associated with various heavy metals, fortunately, many natural microbes can withstand their toxicity and finds their place for bioremediation of these pollutants. Four principal mechanisms, namely, biodegradation, bioaccumulation, bioleaching and biosorption govern the bioremediation of metal pollutants. Biological agents extract energy and nutrients for their survival while changing the state of pollutants. Bioaccumulation of metals (here lead) occurs via either extracellular adsorption (metabolism-independent) of metal ions at cell surface components due to various interactions, or intracellular accumulation (metabolism-dependent) governed by localisation of the metal within specific organelles, enzymatic detoxification and efflux (Velásquez and Dussan, 2009; Samsuri et al., 2019). Bioleaching occurs through direct incorporation of microbes with metal or an indirect mechanism (initiation of by-product generation that incorporates the metal) (Chen and Lin, 2001). Bioleaching is preferred for recovery of metals from low-grade/or concentrated ore where the conventional process is not economically viable (Borja et al., 2019). The biosorption phenomenon is mediated either metabolically or in physico-chemical pathways in certain biomass influenced by components of the cell and spatial orientation of the cell wall. Biomass uptake and accumulation of heavy metals from their sources passively concentrate in cellular structures, mostly on the surface of biomass (Xu et al., 2019; Tiquia-Arashiro, 2018; Shah, 2020).

Unlike other organic or inorganic pollutants, metal pollutants cannot be totally diminished from the environment. Protective mechanisms of tolerance and resistance to metal pollutants at the cellular/molecular level of microbes enable them to beat the toxicity of metals and remain unaffected (Schimel et al., 2007)

A few identified mechanisms associated with lead-resistant bacteria are mentioned below:

a. Efflux mechanism

   Efflux of metal ions from a cell is an energy-dependent activity. A group of transporter proteins present in microbes (predominantly in the membrane) control the intracellular level of heavy metal ions by transporting them outside the cell membrane, preventing excessive accumulation while maintaining heavy metal resistance. Protein P_{1B}-type ATPases are highly recognised for Pb (II) efflux in *E. coli* (encoded by gene zntA) and *Staphylococcus aureus* (gene cad A).

b. Intracellular bioaccumulation

   The resistance to toxic metals in microbes is largely accomplished by the presence of metallothioneins (MTs, low-molecular-weight proteins metal binding) and phytochelatins (PCs, polypeptide with a high number of gammaPCs, dipeptide residues) that enables accumulation, confiscation and immobilisation of the metals within the cell. The MTs and PCs are rich in amino acid cysteine (Cys) containing sulphur atoms that facilitate metal binding.

c. Extracellular sequestration

   Exopolysaccharides (EPSs, high-molecular-weight polyanionic polymers) secreted by microbes are key responsible factors for protection and resistance to harmful elements present in their environment. The presence of functional groups namely, hydroxyl, carboxyl, amides, phosphoryl, sulfhydryl, etc. in EPSs increases their chemical diversity and enhances affinity toward heavy metals. Lead sequestration in EPSs occurs due to electrostatic interaction between Pb^{2+} and the negatively charged functional groups present in EPSs.

d. Biosorption

   The presence of negatively charged chemical groups such as the phosphate group in Gram-negative bacteria as well as the carboxyl group in Gram-positive bacteria mediates surface biosorption through processes like ion exchange, chelation, adsorption and diffusion through the cell wall and cell membrane etc. Biosorption of lead is the non-enzymatic adsorption of metal ions to proteins and EPS/cell surface-associated polysaccharides. The biosorption efficiency is diverse with species and can be active (rapid, irreversible and metabolism-independent) or passive process (slow and metabolism-dependent).

e. Bioprecipitation

Bioprecipitation occurs when the negative functional groups such as phosphate (PO_4^{2-}) and sulphate (SO_4^{2-}) present in microbes react with Pb^{2+} and precipitate the same as insoluble lead compounds such as $Pb_9(PO_4^{2-})_6$, $PbHPO_4$, and PbS. This causes a significant reduction in the bioavailability of toxic Pb^{2+}. Many microbes release siderophore pigments that function as chelators also enhance the bioprecipitation of lead and produce multi-metal complex suitable for precipitation.

f. Enzyme detoxification

Enzymes present in microbes are involved in various cell activities and can cause the transformation of Pb^{2+} to other less toxic lead compounds by altering the chemical composition following precipitation and confiscation in cell mass. Production of organic acid and metal leaching sulphuric acid in heterotrophic and autotrophic bacteria, respectively, are associated with lead detoxification.

Biomethylation of lead results in the production of methyl lead and ethyl lead also frequent incidents encountered in microbial activity with lead waste.

2.1.3 Microbes Identified for Lead Resistance

Several microbes with lead-resistant genes present in their extrachromosomal circular DNA (example: plasmid) that ensure gain of resistance, which are identified to have tolerance and resistance toward the lead. In the presence of a certain level of Pb concentration, genes get induced and expressed for their function in regulation with the promoter and regulatory genes present in the bacterial operon. A few bacteria species and their lead-resistant genes are given in Table 2.1.

TABLE 2.1
Prominent Microbes with Pb resistant Gene for Treatment of Lead from the Environment

Organisms	Gene (Pb resistant)
Staphylococcus aureus *Staphylococcus saprophyticus* *Staphylococcus epidermidis*	cadC
Stenotrophomonas maltophilia	
Actinosynnema mirum DSM 43827	
Gordonia polyisoprenivorans VH2	
Gordonia sputi NBRC 100414	
Gordonia terrae NBRC 100016	
Kitasatospora setae KM-6054	

(Continued)

TABLE 2.1 (*Continued*)
Prominent Microbes with Pb resistant Gene for Treatment of Lead from the Environment

Organisms	Gene (Pb resistant)
Mycobacterium smegmatis str. MC2 155	
Mycobacterium tuberculosis	
Nocardia cyriacigeorgica GUH-2	
Nocardia farcinica IFM 10152	
Rhodococcus erythropolis SK121	
Rhodococcus pyridinivorans AK37	
Saccharomonospora azurea SZMC 14600	
Saccharomonospora glauca K62	
Saccharopolyspora erythraea NRRL 2338	
Saccharopolyspora spinosa NRRL 18395	
Streptomyces bingchenggensis BCW-1	
Streptomyces cattleya DSM 46488	
Streptomyces coelicoflavus ZG0656	
Streptomyces coelicolor A3(2)	nmtR
Streptomyces flavogriseus ATCC 33331	
Streptomyces ghanaensis ATCC 14672	
Streptomyces griseoaurantiacus M045	
Streptomyces griseoflavus Tu4000	
Streptomyces hygroscopicus ATCC 53653	
Streptomyces lividans TK24	
Streptomyces pristinaespiralis ATCC 25486	
Streptomyces roseosporus NRRL 15998	
Streptomyces scabiei 87.22	
Streptomyces venezuelae ATCC 10712	
Streptomyces violaceusniger Tu 4113	
Streptomyces viridochromogenes DSM 40736	
Streptomyces zinciresistens K42	
Ralstonia metallidurans CH34	pbrA
Cupriavidus metallidurans CH34	pbrD
Herbaspirillum seropedicae SmR1	pbrR
Cupriavidus metallidurans CH34	pbrD
Herbaspirillum seropedicae SmR1	pbrR
Achromobacter xylosoxidans A8, Acidovorax ebreus TPSY	pbrT
Arsenophonus nasoniae	
Brenneria sp. EniD312	
Citrobacter freundii 4_7_47CFAA	
Citrobacter koseri ATCC BAA-895	
Citrobacter rodentium ICC168	
Citrobacter youngae ATCC 29220	
Cronobacter sakazakii	
Cronobacter turicensis z3032	
Edwardsiella ictaluri 93–146	

(*Continued*)

TABLE 2.1 (*Continued*)
Prominent Microbes with Pb resistant Gene for Treatment of Lead from the Environment

Organisms	Gene (Pb resistant)
Edwardsiella tarda	
Enterobacter aerogenes KCTC 2190	
Enterobacter asburiae LF7a	
Enterobacter cancerogenus ATCC 35316	
Enterobacter cloacae	
Enterobacter hormaechei ATCC 49162	
Enterobacter mori LMG 25706	
Enterobacteriaceae bacterium 9_2_54FAA	
Erwinia amylovora	
Erwinia billingiae Eb661	
Erwinia pyrifoliae Ep1/96	
Erwinia tasmaniensis	
Escherichia albertii TW07627	
Escherichia coli	
Escherichia fergusonii B253	
Escherichia hermannii NBRC 105704	
Hafnia alvei ATCC 51873	
Klebsiella oxytoca	
Klebsiella pneumoniae subsp. pneumoniae MGH 78578	
Klebsiella variicola At-22	
Marinomonas sp. MWYL1	
Oryza sativa Indica Group	zntA/yhhO
Pantoea ananatis	
Pantoea sp. aB	
Pantoea sp. SL1_M5	
Pectobacterium atrosepticum SCRI1043	
Pectobacterium carotovorum subsp. brasiliensis PBR1692	
Pectobacterium carotovorum subsp. carotovorum PC1	
Pectobacterium carotovorum subsp. carotovorum WPP14	
Pectobacterium wasabiae WPP163	
Photobacterium damselae subsp. damselae CIP 102761	
Photorhabdus asymbiotica subsp. asymbiotica ATCC 43949	
Photorhabdus luminescens subsp. laumondii TTO1	
Proteus mirabilis	
Proteus penneri ATCC 35198	
Providencia alcalifaciens DSM 30120	
Providencia rettgeri DSM 1131	
Providencia rustigianii DSM 4541	
Providencia stuartii ATCC 25827	
Rahnella aquatilis CIP 78.65 = ATCC 33071	
Rahnella sp. Y9602	
Salmonella bongori NCTC 12419	
Salmonella enterica subsp. enterica serovar Baildon str. R6–199	
Serratia odorifera	

(*Continued*)

TABLE 2.1 (*Continued*)
Prominent Microbes with Pb resistant Gene for Treatment of Lead from the Environment

Organisms	Gene (Pb resistant)
Serratia proteamaculans 568	
Serratia symbiotica str. Tucson	
Shigella boydii CDC 3083–94	
Shigella dysenteriae CDC 74–1112	
Shigella flexneri 2a str. 301	
Shigella sonnei Ss046	
Vibrio alginolyticus	
Vibrio harveyi 1DA3	
Vibrio shilonii AK1	
Vibrio splendidus	
Vibrionales bacterium SWAT-3	
Xenorhabdus bovienii SS-2004	
Xenorhabdus nematophila ATCC 19061	
Yersinia aldovae ATCC 35236	
Yersinia bercovieri ATCC 43970	
Yersinia enterocolitica subsp. enterocolitica 8081	
Yersinia enterocolitica	
Yersinia frederiksenii ATCC 33641	
Yersinia intermedia ATCC 29909	
Yersinia kristensenii ATCC 33638	
Yersinia mollaretii ATCC 43969	
Yersinia pestis Antiqua	
Yersinia pestis KIM10+	
Yersinia pseudotuberculosis	
Yersinia rohdei ATCC 43380	
Yersinia ruckeri ATCC 29473	
Yokenella regensburgei ATCC 43003	
Citrobacter rodentium ICC168	
Citrobacter youngae ATCC 29220	
Enterobacter aerogenes KCTC 2190	
Enterobacter cloacae SCF1	
Escherichia albertii TW07627	
Escherichia coli	
Escherichia fergusonii ATCC 35469	
Klebsiella oxytoca	
Klebsiella pneumoniae 342	zraS/hydG
Klebsiella pneumoniae subsp. pneumoniae MGH 78578	
Klebsiella variicola At-22	
Salmonella bongori NCTC 12419	
Salmonella enterica	
Shigella boydii	
Shigella dysenteriae Sd197	
Shigella flexneri	
Shigella sonnei Ss046	

2.2 CELLULAR AND MOLECULAR RESPONSES

Metal ions such as Ni^{+}, Cd^{+}, and Ca^{2+} serve as micronutrients for microbes. These elements can enter organisms through osmotic regulation and are utilised to perform necessary activities when associated with specific organelles (Adekanmbi et al., 2019). Microbes hold inbuilt signalling mechanisms when exposed to toxic levels of metal ions, and in the presence of unfavourable conditions, microbes can develop various types of adaptive defence mechanisms. These defence mechanisms termed stress responses involve physiological as well as metabolic adaptations. Metabolic activities can induce stress responses which further influence normal physiological activities as well as lead to the synthesis of a specific protein required for survival. Gene expression changes also directly influence the sigma protein factor and RNA polymerase responsible for various protein synthesis.

Synthesis of adhesion protein facilitates aggregation of bacterial species, resulting in the formation of biofilm, where cells complete their lifecycle at different rates when the biofilm is exposed to metal ions. The cells present in the biofilm are especially accomplished for the synthesis of EPS, proteins, and nucleic acids favourable for biofilm formation, resulting in additional microbial resistance. Thus, biofilm formation, EPS or other biosorbent production, synthesis of secondary metabolites and similar activities are executed by microbes to encounter an environmental crisis. Due to toxicity from the heavy metals, the number of active microbes decreases, but the remaining active microbes go through required physiological adaptations to face the situation. Moreover, microbes possess quorum sensing capability to regulate the expression of the genes according to the fluctuations of the cell population density. Transcriptional regulators and signalling proteins play significant roles in adaptations as well as mutations under different conditions (Bleuven and Landry, 2016). Cadmium and lead can extremely cause stress to soil actinomycetes and cause little response toward bacterial growth, interfere with metabolic activity and hence influence microbial diversity (Xiao et al., 2020). The inhibition level of heavy metals on bacterial response is directly related to metal, bacterial species, metal species and bioavailable duration and quantity. In case of failure in metal toxicity reduction after interaction with various bacterial adaptation mechanisms bacterial biomass component and its various protein synthesis activities are greatly affected (Chen and Wang, 2007; Xiao et al., 2020; Shah, 2021). The geochemical gradients of soil and soil parameters directly influence the distribution of soil microbial diversity and hence can govern the pathway of the level of pollutant and microbial interactions against available pollutants for specific microbe (Fan et al., 2020; Šolić et al., 2010).

2.3 CASE STUDY

Three mining sites, namely, Nandini Limestone Mine (NM), Limestone Century Cement (LCC) Mine and Dalli Rajhara (DL) Iron Mine located in Chhattisgarh, India, are considered for mine tailing (MT) characterisation and identification of indigenous less resistant microbes. Significant concentrations of lead, iron and sulphate are present in MTs from all the sites. The acclimatised culture originated from MTs mixed with fresh MTs at four various ratios (3:1, 2:1, 1:1 and 1:2) are exposed to different

TABLE 2.2
Biomass at Various Ratios of Acclimatised and Fresh MTs

Sl. No.	Sample	Biomass (mg/g)			
		3:1	2:1	1:1	1:2
1	Anaerobic (NM)	74.26	69.26	71.22	73.62
2	Anoxic (NM)	92.32	94.26	94.58	98.7
3	Aerobic (NM)	52.42	58.96	76.22	61.34
4	Anaerobic (LCC)	36.68	36.32	44.86	52.14
5	Anoxic (LCC)	36.94	56.28	58.36	58.78
6	Aerobic (LCC)	44.9	58.36	58.36	61.28
7	Anaerobic (DL)	55.34	55.81	59.36	61.24
8	Anoxic (DL)	79	78.68	82.05	82.66
9	Aerobic (DL)	103.8	111.24	120.42	120.28

environments, namely, anaerobic, anoxic and aerobic, for determining the treatment potential. Lead- and iron-resistant species are identified in all the sites capable of MT treatment at the ratio of 3:1, 2:1 and 1:1 all the reactors provided similar performance, whereas there was a drop in performance efficiency of all the reactors at the ratio of 2:1. The biomass is originated from the mining sites and considered for MTs treatment at various ratios and environments to assess their efficiency. The biomass concentration is given in Table 2.2, where the volatile solids are the considered parameter to assess biomass growth in the reactors at various ratios. The dominant species identified in the MTs are *Bacillus cereus*, *Sinorhizobium meliloti* and *Bacillus thuringiensis* from Nandini mines; *Bacillus megaterium*, *Escherichia coli* and *Clostridium perfringens* from the LCC site; and *Bordetella ansorpi*, *S. meliloti*, *Escherichia coli* and *Bacillus amyloliquefaciens* from the DL site. *Bacillus thuringiensis*, *Clostridium perfringens*, *Escherichia coli* and *Bacillus amyloliquefaciens* are identified to have lead resistance capacity present in the MTs through molecular analysis.

2.4 FUTURE SCOPE

The complexity of pollutant interaction and properties can invariably influence microbial diversity and activity. With enduring pollutant and microbial interactions, microbes consistently act on their physiological and morphological attributes and evolutions occur in response to the removal of either pollutant or microbial species as well as resist in their habitat. Therefore, researchers are still working on various configurations of pollutants–species–environmental parameter interaction for identification of resistant mechanisms and pathways in active microbes. The system biology approach associated with various computational tools plays a great role in metagenome analysis for detection of suitable species for lead-species intensity from an environment safety point of view. There is much focus required to find out more suitable indigenous species capable to withstand and resolve stress from the web of toxic heavy metals such as lead.

2.5 CONCLUSIONS

Lead is a pervasive heavy metal with various health impacts on humans or other organisms when it comes in contact and demands efficient treatment to detoxify the same. Microbes, mostly bacteria, are highly utilised for lead waste treatment. Bacteria possess unique defence mechanisms that enable them to resist or tolerate the toxicity originating from heavy metals while detoxifying the pollutant into less toxic or inert compounds. In the presence of definite pollutant levels, bacteria possess an inbuilt capacity for gene expression that enhance defence mechanisms together with the synthesis of various protein required for EPS generation, biosorption, efflux pump as well as detoxification of enzymes.

REFERENCES

Adekanmbi, A.O., Adelowo, O.O., Okoh, A.I., Fagade, O.E. 2019. Metal-resistance encoding gene-fingerprints in some bacteria isolated from wastewaters of selected printeries in Ibadan, South-western Nigeria. *Journal of Taibah University for Science*, 13(1), 266–273.

Bleuven, C., Landry, C.R. 2016. Molecular and cellular bases of adaptation to a changing environment in microorganisms. *Proceedings of the Royal Society B: Biological Sciences*, 283. https://doi.org/10.1098/RSPB.2016.1458

Borja, D., Nguyen, K.A., Silva, R.A., Ngoma, E., Petersen, J., Harrison, S.T.L., Park, J.H., Kim, H. 2019. Continuous bioleaching of arsenopyrite from mine tailings using an adapted mesophilic microbial culture. *Hydrometallurgy*, 187, 187–194. https://doi.org/10.1016/j.hydromet.2019.05.022.

Chen, S.Y., Lin, J.G. 2001. Bioleaching of heavy metals from sediment: Significance of pH. *Chemosphere*, 44, 1093–1102. https://doi.org/10.1016/S0045-6535(00)00334-9.

Chen, C., Wang, J. 2007. Response of Saccharomyces cerevisiae to lead ion stress. *Applied Microbiology and Biotechnology*, 74, 683–687. https://doi.org/10.1007/S00253-006-0678-X.

Davis, K.E.R., Joseph, S.J., Janssen, P.H. 2005. Effects of growth medium, inoculum size, and incubation time on culturability and isolation of soil bacteria. *Applied and Environmental Microbiology*, 71, 826. https://doi.org/10.1128/AEM.71.2.826-834.2005.

Fan, R., Ma, W., Zhang, H. 2020. Microbial community responses to soil parameters and their effects on petroleum degradation during bio-electrokinetic remediation. *Science of the Total Environment*, 748, 142463. https://doi.org/10.1016/j.scitotenv.2020.142463.

Oliveira, J.S., Rodrigues, A., Mendes, B., Dias, M. 1987. Toxicity of lead on microorganisms: Interaction with protein content of culture medium. *Toxic Assessement*, 2, 175–191. https://doi.org/10.1002/TOX.2540020206.

Samsuri, A.W., Tariq, F.S., Karam, D.S., Aris, A.Z., Jamilu, G. 2019. The effects of rice husk ashes and inorganic fertilizers application rates on the phytoremediation of gold mine tailings by vetiver grass. *Applied Geochemistry*, 108, 104366. https://doi.org/10.1016/j.apgeochem.2019.104366.

Schimel, J., Balser, T.C., Wallenstein, M. 2007. Microbial stress-response physiology and its implications for ecosystem function. *Ecology*, 88, 1386–1394. https://doi.org/10.1890/06-0219.

Shah, M.P. 2020. *Advanced Oxidation Processes for Effluent Treatment Plants*. Elsevier.

Shah, M.P. 2021. *Removal of Emerging Contaminants through Microbial Processes*. Springer.

Šolić, M., Krstulović, N., Kušpilić, G., Ninčević Gladan, Ž., Bojanić, N., Šestanović, S., Šantić, D., Ordulj, M. 2010. Changes in microbial food web structure in response to changed environmental trophic status: A case study of the Vranjic Basin (Adriatic Sea). *Marine Environmental Research*, 70, 239–249. https://doi.org/10.1016/j.marenvres.2010.05.007.

Tiquia-Arashiro, S.M. 2018. Lead absorption mechanisms in bacteria as strategies for lead bioremediation. *Applied Microbiology and Biotechnology*, 102(13), 5437–5444.

Velásquez, L., Dussan, J. 2009. Biosorption and bioaccumulation of heavy metals on dead and living biomass of Bacillus sphaericus. *Journal of Hazardous Materials*, 167, 713–716. https://doi.org/10.1016/j.jhazmat.2009.01.044.

Xiao, L., Yu, Z., Liu, H., Tan, T., Yao, J., Zhang, Y., Wu, J. 2020. Effects of Cd and Pb on diversity of microbial community and enzyme activity in soil. *Ecotoxicology*, 29, 551–558. https://doi.org/10.1007/S10646-020-02205-4.

Xu, K., Lee, Y.S., Li, J., Li, C. 2019. Resistance mechanisms and reprogramming of microorganisms for efficient biorefinery under multiple environmental stresses. *Synthetic and Systems Biotechnology*, 4(2), 92–98.

3 Metagenomics and Metatranscriptomic Analysis of Wastewater

Jyotsna Choubey
Raipur Institute of Technology

Jyoti Kant Choudhari
Maulana Azad National Institute of Technology

Mukesh Kumar Verma
Chhattisgarh Swami Vivekanand Technical University

Tanushree Chatterjee
Raipur Institute of Technology

Biju Prava Sahariah
Chhattisgarh Swami Vivekanand Technical University

CONTENTS

DOI: 10.1201/9781003354147-3

3.1 INTRODUCTION

Water resources provide valuable food through aquatic life and irrigation for agriculture production along with the clean water required for establishment and maintenance of various human activities. However, throughout the world, liquid and solid wastes produced by human settlements and industrial activities pollute most water sources. Rapid industrialisation and urbanisation around the world add high pressure to water resources, and demands lead to the recognition and proper understanding of the relationship between environmental contamination and public health. Industries are inevitable for the economy of a country and also contribute as a major polluter of the environment if processes and waste are not handled properly. Compared to other environmental pollution elements, such as domestic/municipal and agriculture, industrial waste discharge is a major source of environmental pollution, where industries use various chemicals for processing raw materials. In general, utilisation of cheap and poor or non-biodegradable chemicals and ignorance of their toxicity results in havoc on many occasions. The presence of a variety of highly toxic/hazardous chemicals in industrial wastewater is confirmed by the majority of literature reports available in the public domain.

The utilisation of microbes for wastewater treatment is highly recognised with microbes enriched with great catabolic potential for transformation of wastewater contaminants to less toxic simple compounds or eliminating the same from the vicinity. In-depth knowledge of microbial activities related to catabolism of various organic and inorganic compounds (held as pollutants) enlightens the path of wastewater treatment using various microbes to great extents (Daims et al. 2006; He and McMahon 2011; Jenkins 2008). Microbes are capable to detoxify/remove pollutants in various circumstances using available pollutants as their substrate for growth and maintenance of their life, such as in the absence of molecular oxygen (anaerobic), in the presence of oxygen (aerobic) or in the presence of the electron acceptor other than oxygen (anoxic). Several biological reactor designs are currently applied for growth and activity of microbes. From the widely used conventional activated sludge (AS) system to more advanced systems with improved sludge retention, such as membrane reactors, biofilm reactors, and granular sludge reactors. Therefore, a huge number of microbes are identified with their potential and still large number remain undetected. Laboratory cultivation methods provide significant insights regarding the physiological properties of microorganisms involved in pollutant removal mechanisms such as nitrification, denitrification, phosphorus removal, sulphate reduction, methanogenesis, and xenobiotic remediation, whereas whole genome sequencing of microbes results in extraction of information about functional genes and biochemical pathways of microorganisms involved in wastewater treatment. However, most microorganisms present in wastewater treatment systems still remain non-cultivated, due to either laboratory biases or interspecies metabolic dependence. To encompass the whole microbial communities along with their performance might provide an imperative breakthrough for bioengineers and microbiologists. This is feasible to a great extent with advanced bioinformatic techniques, namely, metagenomics (the study of all genes in a microbial community), metatranscriptomics (the study of gene expression in a microbial community), metaproteomics (the study of proteins from a microbial community), and metabolomics (metabolite profiling and analysis of metabolic fluxes). These practices empower cultivation-independent analysis of metagenomes under specific environmental conditions to derive various vital information that includes ecosystem (genetic and physiological diversity), and microbial identification and its potential metabolic capabilities (Handelsman et al. 1998). The extracted information builds the bridge for a knowledge group of uncultured microbes with wastewater treatment efficiency imperative for efficient bioreactor design parameters. Also, metagenomic approach encompass various mathematical model derivations through meta-omics data integration to predict environmental or engineered systems to external perturbation response. Next-generation sequencing (NGS) technologies provide reliable information about the key enzymes/genes involved in the degradation and detoxification of environmental pollutants. The objective of this chapter is to provide the basic knowledge of metagenomic approaches and their applications to better understand the microbial community structure and functions during bioremediation of environmental pollutants in a contaminated matrix. This chapter highlights the potential of genomic and meta-omic approaches to understand the structure, function and interactions of microbial communities of wastewater treatment systems and discusses how these genomic and meta-omics results influence the development and monitoring of these biotechnologies.

3.2 OPPORTUNITIES AND CHALLENGES WITH BIOLOGICAL TREATMENT OF WASTEWATER

Strict legislation for effluent control in addition to regional characteristics and socio-economic conditions demands suitable efficient treatment systems. A variety of physicochemical and biological treatment plants are in use for various kinds of wastewater in the current scenario. Highly efficient advanced physicochemical process generates successfully treated effluent but these processes are often associated with limitations such as secondary waste generation and high costs. With high simplicity, cost-effectiveness and high efficiency involving diverse microbes, biological treatment process for wastewater still occupies top priority among environmental scientists. Microbes possess the ability to gradually acclimatise towards exposed pollutants when suitable environmental factors such as pH, temperature and nutrients are provided. Mixed microbial cultures with a diverse group of microbes with enormous enzymatic/catabolic activities take part in the treatment of a variety of pollutants whether organic/inorganic present in wastewater. The mechanisms involve biodegradation (mineralisation of pollutants) and bioaccumulation, and biotransformation or biosorption (where recovery of the pollutants, mostly metals) takes place (Bader et al. 2018; Kolhe et al. 2018; Shah 2021). With time, currently enormous complex pollutants are getting released to the environment from various industries such as pesticides, fertilisers (agricultural), dyes (textiles), heavy metals and high inorganic/organics load (tanneries, petrochemicals, and powerplants). The pollutants when present in combination can inhibit the performance of microbes and hence influence reactor performance. Selection of suitable microbial culture to tackle the complex pollutants in single or combined is high responsibility to achieve simultaneous removal of these pollutants from wastewater.

Detailed information regarding microbial response in terms of functional genes, as well as various pathways towards pollutant–microbe interactions and reactor parameters (temperature, mixing/oxygenation and retention time, etc.), is precious to overcome pollution problems caused by a variety of pollutants. This information can provide a base for an effective approach to bioreactor composition and performance. It can also minimise unavoidable conditions such as the generation of new secondary potentially harmful contaminants, e.g., iodinated organic compounds in the effluent of a BAF due to iodide-oxidising bacteria activity (Almaraz et al. 2020).

3.3 METAGENOMICS: A TECHNOLOGICAL DRIFT

Metagenomics was first coined by Handelsman et al. in 1998 that encompasses information on entire microbial community composition and function widening the area of genomics where only the genetic material is studied. A few prior studies, like Pace et al. (1985), about phylogenetic analyses of environmental microbial communities are also reported. In the process of metagenomics study, genomic DNA from all organisms in a community (metagenome) is extracted for its fragmentation, cloning, transformation, and subsequent screening of the constructed metagenomic library. Initially, the primary target of metagenomics was limited to screening environmental communities for a specific biological activity and identifying the related genomics (Yun and Ryu 2005). Furthermore, the development of high-throughput NGS technologies [such as 454 pyrosequencing, Illumina (Solexa) sequencing, sequencing by oligonucleotide ligation and detection (SOLiD) sequencing] opens the approach to sequence-based screening. Yun and Ryu (2005) and Venter et al. (2004) first demonstrated environmental genome shotgun sequencing focussing on the vast phylogenetic and metabolic diversity of microbial communities of the Sargasso Sea.

3.3.1 Metagenomic Approaches

Metagenomic studies can be tackled from two different approaches, namely, the targeted metagenomic approach and the shotgun metagenomic approach with fundamental differences based on methodology and objective. In targeted metagenomics, a gene or a few genes are sequenced and used primarily to carry out phylogenetic-type studies, while in shotgun metagenomics, all present DNA is sequenced and used in functional gene analysis assays (Morgan et al. 2013). This process usually involves NGS after the DNA is extracted from the samples and results in a huge amount of data in the form of short reads.

Metagenomic analyses many times include high-throughput microarrays such as GeoChip and PhyloChip microarrays for analysing microbial communities and monitoring environmental biogeochemical processes. GeoChip microarrays are mostly used to study the structure, dynamics, and potential metabolic activity of microbial communities and their variations depending on different stimuli (Uhlik et al. 2013). PhyloChip is used for microbial ecology (phylogenetic analyses) and contaminant biodegradation (DeAngelis et al. 2011; Brodie et al. 2006). Various metagenomic fundamental approaches are described in the following section.

3.3.1.1 Metatranscriptomics

Based on gene expression and total mRNA of a given sample at the experimental moment, the functional profile of a microbial community is derived in metatranscriptomics as a snapshot (Moran 2009). Recovery of high-quality mRNA from environmental samples, short half-lives of mRNA species, separation of mRNA from other RNA species, etc. cause limitations to metatranscriptomics (Simon and Daniel 2011). However, direct cDNA sequencing employing NGS technologies can overcome these limitations of metatranscriptome (Simon and Daniel 2011).

3.3.1.2 Metaproteomics

Metaproteomics adopts direct cDNA sequencing through NGS and identifies the key enzymes of significant metabolic pathways and associates them with expression and genomic profiles of microbial populations (Chistoserdova 2010; Roh et al. 2010). Metaproteomics analyses are restricted by uneven species distribution, the broad range of protein expression levels within microorganisms and the large genetic heterogeneity within microbial communities. However, its importance remains unaffected due to its potential to discover key biochemical metabolic pathways, enzymatic markers, the protein expression patterns in the microbial communities during microbe–pollutant interaction.

3.3.1.3 Metabolomics

Metabolomics includes wide-ranging analysis, identification and quantification of metabolites of a sample (Fiehn 2002). Diversion in specific metabolite generation indicates deviation in metabolic activity and pathways (Krumsiek et al. 2015).

3.3.1.4 Fluxomics

Similar to metabolomics, fluxomics identify ways to determine the rates of metabolic reactions within a biological entity. The set of metabolic fluxes represents a dynamic picture of the phenotype, in as much as it captures the metabolome in its functional interactions with the environment and the genome.

3.4 IMPORTANCE OF METAGENOMICS, TRANSCRIPTOMICS, PROTEOMICS AND METABOLOMICS IN OPTIMISING WASTEWATER TREATMENT

Generally, biological wastewater treatment methods are established based on analytical reports which are very limited compared to a vast diversity of wastewater and microbes available with high treatment potential. The metagenomic approaches possess a high prospect to identify pollutant–microbial (genetic-metabolic pathway/mechanisms) interactions. This will help in unfolding unidentified diverse microbes in environmental samples with relevant and potential genes enzymes for specific wastewater treatment.

3.5 APPLICATION OF THE OMICS APPROACH IN WASTEWATER TREATMENT

Metagenomics helps in information extraction and selection of suitable species for target pollutants. Different groups can be compared in terms of microbial functional diversity and provided environment while targeting specific pollutants. In-depth knowledge about phylogenetic composition and functional diversity of microbial communities can be achieved through advanced technologies such as NGS in an unbiased view in a short span of time (Zwolinski 2007).

Tyson et al. (2005) isolated *Leptospirillum ferrodiazotrophum* from an acid mine drainage biofilm after discovering that a minor member of the community had a single nitrogen fixation operon (nifHD-KENX). Using meta-transcriptomics, Bomar et al. (2011) claimed that an uncultured *Rikenella*-like leech gut symbiont was able to forage on host mucin glycans.

Meta-omic approaches are also powerful tools for the development of specific biomarkers or the analysis of protein expression patterns of microbial communities that could be used as predictive indicators of process performance, aiding in the monitoring and control of wastewater treatment processes.

3.5.1 Computational and Bioinformatics Tools for Metagenomic Data Analysis

For interpretation and information extraction, numerous in silico software, pipelines, web resources, and algorithms are available (Table 3.1). Resources of bioinformatics in the field of wastewater treatment still remain less utilised. The University of Minnesota Biocatalysis/Biodegradation Database has been the pioneer and most prominent web resource in the field of bioremediation available online at the link (http://umbbd.msi.umn.edu/).

The first step in the metagenomic analysis is always the comparison of the sequences with databases for taxonomy, functional annotation, binning of sequences, phylogenomic profiling, and metabolic reconstruction. A free server that can be used for this purpose is the MG-RAST—rapid annotation using a subsystem technology server, which processes and integrates all metagenomic data. Raw reads can be uploaded in the FASTA format on the server; then, the server normalises the sequences and automatically generates a summary of information. Additionally, users can compare their data with other metagenomes or complete genomes through the SEED environment (Meyer et al. 2008).

3.5.2 Advantages and Limitations of the Omics Approach in Wastewater Treatment

The metagenomic analysis allows the researchers to better understand microbial populations in their natural environment and also to identify the genes of interest. An important feature of metagenomics techniques is that it is performed using metagenomic libraries constructed from total DNA isolated from a particular niche,

TABLE 3.1
Computational and Bioinformatics Software and Database Used for Metagenomic Data Analysis

S.No.	Name of Software	Web Server	Application
1	FOAM	http://portal.nersc.gov/project/m1317/ FOAM/	It was developed to screen the environmental metagenomic sequence datasets and provides a new functional ontology dedicated to classify gene functions relevant to environmental microorganisms based on HMMs
2	JANE	http://jane.bioapps.biozentrum.uniwuerzburg.de	Mapping of ESTs and variable length prokaryotic genome sequence reads on related template genomes
3	SmashCommunity	http://www.bork.embl.de/software/ smash	A metagenomic annotation and analysis tool
4	UniFrac	http://bmf.colorado.edu/unifrac/	An online tool for comparing microbial diversity
6	MetaVelvet	http://metavelvet.dna.bio.keio.ac.jp	
7	MetAMOS	https://github.com/treangen/MetAMOS	A tool for assembly and analysis of metagenomes that draws a community taxonomic profile, performs gene prediction, and identifies potential genomic variants. The software package is a collection of public tools for genome assembly and analysis, concatenated by the Ruffus system
8	MetaQUAST	http://bioinf.spbau.ru/metaquast	Tool for checking the quality of genome assembly which can detect assembly errors based on alignments against reference genomes. The software reports and plots statistics related to contigs, such as the N50, and has advantages of accessing an unlimited number of reference genomes, automating the detection of species, and detecting the formation of chimeric contigs
9	Pfam	http://pfam.sanger.ac.uk/ http://pfam. janelia.org/ http://pfam.sbc.su.se/ http://pfam.jouy.inra.fr/ http://pfam.ccbb.re.kr/	A comprehensive collection of protein domains and families, represented as multiple sequence alignments and as profile HMMs
10	WebCARMA	http://webcarma.cebitec.unibilefeld.de	A web application for the functional and taxonomical classification of unassembled metagenomic reads

(*Continued*)

TABLE 3.1 (*Continued*)
Computational and Bioinformatics Software and Database Used for Metagenomic Data Analysis

S.No.	Name of Software	Web Server	Application
11	COGNIZER	http://metagenomics.atc.tcs.com/cognizer, https://metagenomics.atc.tcs.com/function/cognizer	A framework for functional annotation of metagenomic datasets
12	Prodigal	http://compbio.ornl.gov/prodial/	Prokaryotic gene recognition and translation initiation site identification
13	RDP	http://rdp.cme.msu.edu/	Database and tools for high-throughput rRNA analysis
14	MEGAN	www.ab.informatik.unituebingen.de/software/megan	Illumina sequencing metagenome reads analysis tool
15	myPhyloDB	http://www.ars.usda.gov/services/software/download.htm?softwareid 5 472 http://www.myphylodb.org	A local web server for the storage and analysis of metagenomic data
16	MG-RAST	http://metagenomics.anl.gov/	Metagenome annotation server
17	WebMGA	http://weizhongli-lab.org/	Metagenomic analysis: A customisable web server for fast metagenomic sequence analysis
18	CAMERA	http://camera.calit2.net	Metagenomic database server which contains sequences from environmental samples collected during the GOS
19	UCHIME	http://drive5.com/uchime	For improved sensitivity and speedy chimaera detection
20	envDB	http://metagenomics.uv.es/envDB/	Database and tool server for environmental distribution of prokaryotic taxa
21	UCLUST	http://www.drive5.com/	usearch: A new clustering method that exploits USEARCH to assign sequences to clusters
22	FUNGIpath	http://www.fungipath.upsud.fr	Database and tool serve for fungal orthology and metagenomics
23	GreenGene	http://greengenes.lbl.gov	A chimaera-checked 16S rRNA gene database and workbench compatible with ARB
24	METAREP	http://www.jcvi.org/metarep	Annotate the data with GO and KEGG

(*Continued*)

TABLE 3.1 (*Continued*)
Computational and Bioinformatics Software and Database Used for Metagenomic Data Analysis

S.No.	Name of Software	Web Server	Application
26	Human Metabolome Database	https://hmdb.ca/	A cross-referenced database about the small metabolites found in the human body
27	BioMagResBank	www.bmrb.wisc.edu	Works as a central repository for experimental NMR data including both small metabolites and macromolecules
28	Madison-Qingdao Metabolomics Consortium Database	https://www.allacronyms.com/MMCD/Madison_Metabolomics_Consortium_Database	Includes both NMR and MS data thoroughly annotated and collected from other databases and literature
29	MassBank	https://massbank.eu//MassBank/	Merges spectral data from different collision-induced dissociation conditions to improve the precision in the identification of compounds
30	Golm	http://gmd.mpimp-golm.mpg.de/	Stores spectral data with retention indexes, useful for automated identification of compounds analysed with GC–MS
31	METLIN	http://metlin.scripps.edu.	Contains curated spectral information of biological metabolites without information on the environmental context from which the samples were obtained. Each of them differs slightly in functionality but pursues similar goals, serving as repositories of spectral data and offering links to their biological interpretation

instead of laboratory culture. Therefore, these techniques allow direct access to all the genetic resources present in a particular environment, irrespective of whether or not the microorganisms can be cultured in the laboratory. The major limitation of metagenomics is the inefficient expression of some metagenomic genes in the host bacteria used for screening (Terrón-González et al. 2014; Shah 2020). Metagenomic approaches describe only the presence of microorganisms or genes in an environment but are not able to clarify their activity. To overcome these drawbacks, microbiome scientists started employing metatranscriptomics approaches. Metaproteomics can approximate the activity of microorganisms by investigating the protein content of a sample. A limitation of metatranscriptomics analysis is the lack of adequate reference genome which results in an incomplete functional and taxonomical characterisation of reads in any dataset (Shakya et al. 2019). Though metaproteomics is a very powerful method, bioinformatics evaluation of its data hinders its success.

In particular, the construction of databases for protein identification, grouping of redundant proteins as well as taxonomic and functional annotation pose big challenges. Moreover, increasing amounts of data within a metaproteomics study require dedicated algorithms and software (Heyer et al. 2017).

3.6 CONCLUSION

In the 20th century, water pollution has dramatically increased; therefore, wastewater treatment attracts increasing interest to the environmental biologist. The process is carried out in specific plants which follow the incomplete or complete removal of the organic load through different physical, chemical, biological, and oxidative treatments. Before the elimination of all these organic and inorganic contaminants, the first step is to monitor them. After the different treatments, the values of the measurements made to these contaminants have to conform to the parameters established by the legislation. With the parameterisation of the microbiological values, the situation is complicated. First, it is because 99% of the microorganisms present in the wastewater are not cultivable in the laboratory and therefore have not been traditionally done. Application metagenome in the bioremediation of water contamination is one of the best ways to reduce water contamination. Recently various case studies suggest that metagenomic applications have been widely used for the identification and treatment of pollutants and contaminations in the sea, groundwater, and drinking water. Metagenomics is a faster, more accurate, and highly efficient genomic approach that overcomes the limitations of conventional molecular techniques. This gives a glimpse into the microbial community's view of "Uncultured Microbiota". With the rapidly decreasing sequencing prices, the NGS technologies are also becoming popular among molecular biologists/environmental microbiologists as these high-throughput techniques are easy to handle, scalable, and speedy.

REFERENCES

Almaraz, N., Regnery, J., Vanzin, G.F., Riley, S.M., Ahoor, D.C., Cath, T.Y. (2020). Emergence and fate of volatile iodinated organic compounds during biological treatment of oil and gas produced water. *Sci Total Environ* 699:134202. doi: 10.1016/j.scitotenv.2019.134202.

Bader, M., Müller, K., Foerstendorf, H., et al. (2018). Comparative analysis of uranium bioassociation with halophilic bacteria and archaea. *PLoS One* 13:e0190953. doi: 10.1371/journal.pone.0190953.

Bomar, L., Maltz, M., Colston, S., Graf, J. (2011). Directed culturing of microorganisms using metatranscriptomics. *mBio* 2:e00012-00011. doi: 10.1128/mBio.00012–11.

Brodie, E.L., Desantis, T.Z., Joyner, D.C., et al. (2006). Application of a high-density oligonucleotide microarray approach to study bacterial population dynamics during uranium reduction and reoxidation. *Appl Environ Microbiol* 72:6288–6298.

Chistoserdova, L. (2010). Recent progress and new challenges in metagenomics for biotechnology. *Biotechnol Lett* 32:1351–1359. doi: 10.1007/s10529-010-0306–9.

Daims, H., Taylor, M.W., Wagner, M. (2006). Wastewater treatment: a model system for microbial ecology. *Trends Biotechnol* 24:483–489. doi: 10.1016/j.tibtech.2006.09.002.

DeAngelis, K.M., Wu, C.H., Beller, H.R., et al. (2011). PCR amplification-independent methods for detection of microbial communities by the high-density microarray PhyloChip. *Appl Environ Microbiol* 77:6313–6322.

Fiehn, O. (2002). Metabolomics – the link between genotypes and phenotypes. *Plant Mol Biol* 48(1–2):155–171.

Handelsman, J., Rondon, M.R., Brady, S.F., Clardy, J., Goodman, R.M. (1998). Molecular biological access to the chemistry of unknown soil microbes: a new frontier for natural products. *Chem Biol* 5:R245–R249.

He, S., McMahon, K.D. (2011). 'Candidatus Accumulibacter' gene expression in response to dynamic EBPR conditions. *ISME J*, 5(2):329–340.

Heyer, R., Schallert, K., Zoun, R., Becher, B., Saake, G., Benndorf, D. (2017). Challenges and perspectives of metaproteomic data analysis. *J Biotechnol* 261. doi: 10.1016/j.jbiotec.2017.06.1201.

Jenkins, D. (2008). From total suspended solids to molecular biology tools—a personal view of biological wastewater treatment process population dynamics. *Water Environ Res* 80:677–687.

Kolhe, N., Zinjarde, S., Acharya, C. (2018). Responses exhibited by various microbial groups relevant to uranium exposure. *Biotechnol Adv* 36:1828–1846. doi: 10.1016/j.biotechadv.2018.07.002.

Krumsiek, J., Mittelstrass, K., Do, K.T., et al. (2015). Gender-specific pathway differences in the human serum metabolome. *Metabolomics* 11(6):1815–1833.

Meyer, F., Paarman, D., D'Souza, M., Olson, R., Glass, E.M., Kubal, M., Paczian, T., Rodrigues, A., Stevens, R., Wilke, A., Wilkening, J., Edwards, R.A. (2008). The metagenomics RAST server- a public resource for the automatic phylogenetic and functional analysis of metagenomes. *BMC Bioinf* 9:386. https://doi.org/10.1186/1471–2105–9–386.

Moran, M.A. (2009). Metatranscriptomics: eavesdropping on complex microbial communities. *Microbiome* 4(7):329–334.

Morgan, X.C., Segata, N., Huttenhower, C. (2013). Biodiversity and functional genomics in the human microbiome. *Trends Genet* 29:51–58.

Pace, N.R., Stahl, D.A., Lane, D.J., Olsen, G.J. (1985). Analyzing natural microbial populations by rRNA sequences. *ASM News* 51:4–12.

Roh, S.W., Abell, G.C., Kim, K.H., Nam, Y.D., Bae, J.W. (2010). Comparing microarrays and next-generation sequencing technologies for microbial ecology research. *Trends Biotechnol* 28:291–299. doi: 10.1016/j.tibtech.2010.03.001.

Shah, M.P. (2020). *Advanced Oxidation Processes for Effluent Treatment Plants*. Elsevier.

Shah, M.P. (2021). *Removal of Emerging Contaminants through Microbial Processes*. Springer.

Shakya, M., Lo, C-C., Chain, P. (2019). Advances and challenges in metatranscriptomic analysis. *Front Genet* 1.

Simon, C., Daniel, R. (2011). Metagenomic analyses: past and future trends. *Appl Environ Microbiol* 77:41153–41161.

Terrón-González, L., Genilloud, O., Santero, E. (2014). Potential and limitations of metagenomic functional analyses.

Tyson, G.W., Lo, I., Baker, B.J., Allen, E.E., Hugenholtz, P., Banfield, J.F. (2005). Genome-directed isolation of the key nitrogen fixer Leptospirillum ferrodiazotrophum sp. nov. from an acidophilic microbial community. *Appl Environ Microbiol* 71(10):6319–6324.

Uhlik, O., Leewis, M.-C., Strejcek, M., et al. (2013). Stable isotope probing in the metagenomics era: a bridge towards improved bioremediation. *Biotechnol Adv* 31(2):154–165.

Venter, J.C., Remington, K., Heidelberg, J.F., et al. (2004). Environmental genome shotgun sequencing of the Sargasso Sea. *Science* 304:66–74.

Yun, J., Ryu, S. (2005). Screening for novel enzymes from metagenome and SIGEX, as a way to improve it. *Microb Cell Fact* 4:8.

Zwolinski, M.D. (2007). DNA sequencing: strategies for soil microbiology. *Soil Sci Soc Am J* 71:592–600.

4 Environmental Metaproteomics
Tools to Study Microbial Communities

Hiren K. Patel and Priyanka D. Sheladiya
P. P. Savani University

Rishee K. Kalaria
Navsari Agricultural University

CONTENTS

DOI: 10.1201/9781003354147-4

4.1 INTRODUCTION

4.1.1 Microbial Ecology

The structure and function of the microbial population that exists in the environment are being studied on a global scale. They play important roles in the biogeochemical cycles, and they can decompose almost all natural compounds, so they exert a permanent effect on the biosphere and environment. Around 20 years before, ecologists understood that microbial activity and physiology in a defiant environment are powerfully reliant on the arrangement of the current community and the connection of the community members during predation, cellular signalling and nutrient competition (Brock, 1987).

However, more than 90% of the microbial community in a given environment is not readily cultured applying the standard procedures (Amann et al., 1995). The disadvantaged investigations targeting a deeper understanding of the structure and function of the biological organisation for a long time and separately contributions of the various species to a definite environment remained basically unknown. The current improvement of the various molecular tools that avoid essentially isolating and culturing individual microbial species has afforded a capable new understanding of microbial ecology that potentially revolutionises our thought of microbial variety and physiology within complex groups and activities to whole ecosystems: (i) 16S rRNA sequencing method provides significant information about the species arrangement and development (Schloss & Handelsman, 2006); (ii) new shotgun sequencing and pyrosequencing methods are enabled to the mapping of entire metagenomes (Vieites et al., 2009; Cardenas & Tiedje, 2008; Singh et al., 2009), for example, the study of the transcriptional profiles of the microbial groups; and (iii) proteomics techniques allowed qualitative and quantitative valuation of the protein supplement in a given environment (Keller & Hettich, 2009; Wilmes & Bond, 2009; VerBerkmoes et al., 2009a; Lacerda & Reardon, 2009; Lopez-Barea & Gomez-Ariza, 2006; Schulze, 2004). In certain, metagenomics has developed as a powerful tool to examine structural, evolutionary, and metabolic properties of the multifaceted microbial communities. Venter (Venter J.C. 2004) creation of the Sargasso Sea sequencing in 2004 was a key turning point in the history of metagenome investigations. In the interim, the metagenomes from the various habitats, like soil, global ocean, human gut, and faeces, are completely sequenced (Markowitz et al., 2008). Recognition of the current development in sequencing technologies, microbial genomics and metagenomics sequencing information will remain to develop exponentially and will consequently suggest a solid basis for several postgenomic research (Pignatelli et al., 2008; Tringe & Rubin, 2005; Tringe et al., 2005). Proteome attention and resolving power of the environmental proteomics analyses powerfully depend upon the size and quality of the reference protein database toward the MS and MS/MS data which are to be searched. As shown in various currently published environmental proteomics studies, like strain determined communal proteomics of acid mine drainage biofilms (Denef et al., 2009; Goltsman et al., 2009; Shah, 2021), activated sludge (Wilmes & Bond, 2009) and communal proteomics of the leaf theorist (Delmotte et al., 2009), the arrangement of metagenomics and metaproteomics can deliver appreciably understanding into structure and physiology of diverse phylogenetic groups which are present in a specific environment.

4.1.2 Historical Retrospective of "Omics" Technologies

Proteomes, transcriptomes, and even metabolomes of microorganisms were restricted to species amenable to isolated cultivation before the most recent age of widespread investigations of their microbial genomes. Additionally, rapid improvements in "omics" technologies have completed it probably to study not only previously uncultivable species but multifaceted microbial communities and straight to entire ecosystems. At the opportunity of the millennium, novel shotgun DNA sequencing methods like 454 pyrosequencing (Ahmadian et al., 2006) attached with significant cost drops gave a marvellous improvement to culture-independent metagenomics investigation and started to reveal the variety and circulation of the local microbial communities in natural environments (Tringe et al., 2005). Metagenomic approaches unaided cannot clarify the functionality of the microbes present in the particular ecosystem. Furthermore, a huge number of the newly recognised ORFs through no homology to well-categorised genes still expect functional assignment. These constraints have ignited an increase in the study of environmental transcriptomes, but the metatranscriptome studies still have significant drawbacks, which include the frequently low correlation between transcription levels and definite protein expression, the extraction procedures are complicated by the interference of organic and inorganic molecules, and the short half-lives of the mRNA particles. Progressively, proteomics has developed as a capable technique to describe microbial activity at a molecular level. Proteomics is generally defined as "the study of the protein which is expressed by an organism" (Wilkins et al., 1995), in progress to the development in the 1960s when protein outlines of single organisms were analysed by 2-DE. At that time, the identification of the protein was time-consuming and cost-intensive due to the lack of the genome sequence available and advanced protein sequence analyses. Meanwhile, the 1990s proteomics has become much more extensive, viable, and reliable, thanks to the three mechanical revolutions: (i) marvellous progress in understanding and accuracy of mass spectrometers enable a correct, high-throughput protein identification, comparative and absolute quantification of proteins and the purpose of post-translation modification; (ii) the huge amount of increase of genomic and metagenomics data delivers a solid basis for the identification of the protein; and (iii) tough developments in calculating the power and bioinformatics permit the processing and assessments of substantial datasets. Worldwide investigation of proteins elaborated in biotransformation, like enzymes, finally permits a complete characterisation of microbial metabolic subtitles and sheds light on the parameter of the metabolomes, a whole set of metabolic intermediates, signalling particles, and secondary metabolites originating within the biological sample (Bochner, 2009).

4.1.3 Terminology of Environmental Proteomics

More than 5 years ago, Wilmes et al. (2008) well defined metaproteomics as "the important characterisation of the whole protein supplement of the environmental microbiota at a specified point in a time"; temporarily, quick development and multifold submission of high-throughput "omics" knowledge have directed to many new values including environmental proteomics, metaproteomics, community proteomics, or community proteogenomics. Environmental proteomics must be started as a generic term basically describing the proteome investigation of environmental samples; metaproteomics includes studies of extremely complex biological arrangements which do not permit transmission of a large number of proteins to exact species within the phylotypes. In difference, the word community proteomics indicates that most of the recognised proteins can be associated with definite members of the community.

Low- or medium-complexity environments. The word proteogenomics, which was primarily used to define the application of proteomics for the improvements of gene annotations, currently also describes the valuation of the strain or species variations and the evolutionary improvement of the genomics makeup to certain environments (VerBerkmoes et al., 2009a; Shah, 2020). Additionally, proteogenomics contributes to the identification of the real gene function by linking recognised protein information to the DNA level.

4.1.4 Potential Applications of Environmental Proteomics

In their normal habitat, microbes are frequently facing quick and punitive changes in environmental limits such as temperature, humidity, predators and nutrient availability. A mutual approach of microorganisms to overcome these experiments is alteration of their protein appearance profiles. Accordingly, a simple study of the separate genes and their regulation is not enough to fully comprehend microbial adaptation approaches, and post-genomics analyses with transcriptomics and proteomics are immediately needed for investigation of the physiology of the complex microbial association at a molecular level. Environmental proteomics previously comprises a real "treasure chest" of knowledge ranging from the simple protein cataloguing to relative and quantitative proteomics, analyses of protein localisations, identification of the post-translational modifications which might affect the functionality of the protein, study of the protein–protein interactions, to determination of amino acid sequences and their genotypes. Therefore, probably application of the upstairs technologies in microbial ecology is abundant and contains the account of new functional genes, the documentation of the entirely new catalytic enzymes or complete metabolic pathways, and the account of the functional bioindicator to screen dynamics and sustainability of the environmental class (Maron et al., 2007). Additionally, development and intensive usage of the whole set of "omics" technologies will permit us to reexamine microbial ecology ideas by linking genetic and functional verities in microbial communities and involving the taxonomic and functional diversity to the ecosystem constancy.

4.1.5 Does Microbial Composition Affect Ecosystem Processes?

A microbial species' direct understanding and explanation, the finding of a substantial nearby gene transfer, in these environmental challenges, the occurrence of this marvelous variety in the mixing, and the development of microbes globally are all examples of this.. It has been suggested that "there is a steadiness of energy fluctuation and informational transfer from the genome up through cells, community, virosphere and environment," but it is not clear whether genomes are separate or if they are modified and assimilate to the needs and pressures communicated with a particular environment (Goldenfeld & Woese, 2007). An increasing body of evidence indicates that microorganism's arrangement also affects ecosystem processes, which include CO_2 respiration and de-composition, nitrogen cycling (Naeem & Li, 1997), and autotrophic and heterotrophic production (Horz et al., 2004). Before few years, Baas-Becking and Beijerink assumed that microbial taxa have preferred environments: "everything is everywhere, but the environment is selected." This suggestion indicates that microorganisms are often dispersed globally and that they are then selected through the environments in which they exist on the basis of their functional ability. Currently, communities would thus continuously be challenged by intruders from the non-specialist taxa that may irregularly survive simply through chance, obtaining the essential functionality through horizontal gene transfer (von Mering et al., 2007). The paper by von Mering et al. supports this hypothesis; it shows strong environmental partiality along lineages but through a time-dependent degeneration. They perceived a remarkable time-dependent constancy of habitats and showed that for any two microbial isolates, the comparison of their interpreted habitats is toughly connected to their evolutionary understanding. Even strain associated only at the level of taxonomic order is still meaningfully more frequently originated in the same environment than a causal pair of isolates. Therefore, furthermost microbial families remain related to a certain environment for lengthy time eras, and effective competition in a new environment seems to be an occasional event, requiring more than just the gaining of a few vital functions. Consequently, a query for future research is as follows.

What are the characteristics that lead to the extensive variety of colonisation, divergence and extinction taxes in microbes?

4.1.6 Proteomics in the Postgenomic Era

The marvellous growth in genome sequencing volume, coupled with important cost drops, has led to a new tendency of metagenomics in which entire-community of DNA shotgun sequencing can be led to describe at least the leading associates of microbial group, thus avoiding the essential to isolate and culture separately microbial species. However, recent shotgun sequencing and in silico assembly procedures are defined in the assembly and assignment of sequences to exact species, strains, or ecotypes through a characteristic interspecies genetic difficulty (Gill et al., 2006). The task of the function and the metabolic influence of exact microbial ecotypes or species to the examined environment make the discarding hard to forecast, particularly when understated species are of attention. General, despite all the wonderful advantages of shotgun sequencing techniques, the post-genomic period also showed the limitations of nucleic acid-based tools for disseminating information about the in situ functional interaction between members of a microbial community (Maron et al., 2007). This is the motivation for the arena of proteomics, in that this investigational method is designed to deliver complete qualitative and quantitative dimensions of the final gene products like protein, as biomarkers of the metabolic movement happening in microbial communities. It is serious to understand that the colossal 16S rRNA work and deep metagenomes annotation are important basic needs for the achievement of the proteomic amounts. The 16S rRNA data deliver energetic information around the species association of a sample, which is an essential contribution to the meaningful proteome evaluations of environmental microbial groups. The metagenome sequence gives a list of every gene product that might exist, and as a result, it supplies the generic catalogue from which proteome identification is derived.

4.2 WHAT IS PROTEOMICS?

Proteins, as opposed to lipids and nucleic acids, are able to serve as alternative biomarkers; in addition, they reflect the precise activity with regard to metabolic reactions and regulatory cascades and provide more direct information about microbial activity than functional genes and even their consistency mRNAs (Wilmes & Bond, 2006). The question might arise "why not simply use the well-built method of transcriptomics to outline the gene expression and thus avoid the need for the proteomics capacity?" To handle this situation, it is important to understand that proteomics not only describes the final gene products but also includes information about protein abundance, stabilities, turnover rates, post-translational modification, and protein–protein interaction, all of which provide important metabolic activity data at the genome and transcriptome levels. The use of proteins as substitute markers led to the formation of proteomics as a research part. Proteomics is distinct as the whole protein supplements of the expressed genome and includes the various techniques that deliver a macroscopic opinion of what is communicated or expressed genome and present under various growth conditions, thus improving more positive targeted investigation (Graves & Haystead, 2002). Initial criticism of the developing field of proteomics was a concept that this quantity would offer the information on the most abundant organised protein and thus be of very limited importance for microbiology research. However, the arrival of more classy and high-throughput chromatographic–mass spectrometric instrumentation has significantly progressed the depth of proteome attention, at least for the microbial classes (Figure 4.1). It is readily probable to use proteomics methods to classify at least 50%–70% of the foretold proteome for most bacteria grown below a single growth aliment. Like, 2,000–2,500 proteins can be recognised for bacteria through a genome of 4,000 open analysis frames. Although there is a significant assumption about what portion of a bacterial genome is essentially expressed under a single growth condition, approximation suggests that most of the bacteria might be occupied only 50%–80% of their foretold genes under a single growth condition. Therefore, the recent level of proteome amount is previously fairly deep into the active range of protein expressed. Obviously, low-copy-number proteins like transcription factors are still problematic to classify, but this investigational method has established remarkable success at penetrating well beneath "only the most abundant organising proteins." The application of single-organism proteomics to entire bacterial groups and environments is a fair creation.

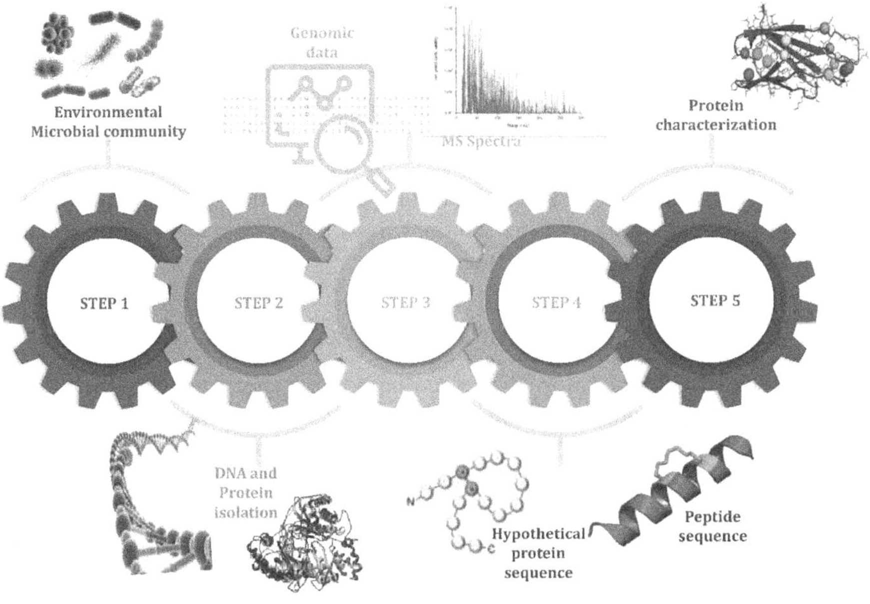

FIGURE 4.1 Overview of environmental proteomics analysis for the microbial community.

4.3 ENVIRONMENTAL PROTEOMICS

4.3.1 Optimisation of Sample Preparation Protocols

Although it may seem a little too dull, sample preparation is one of the important aspects of environmental proteomics and has long been a development in this field's Achilles' heel. As exposed to environmental genomic library assembly, important method improvement was essential to develop procedures prominent to representative environmental genomic libraries from soil and supplementary environments (Green & Keller, 2006). Emerging procedures to achieve an environmental illustrative protein extract will be interesting and will vary for the two main proteomics methods (LC versus 2D gel separation). Ogunseitan established and assessed two approaches for extracting proteins from water, sediments, and soil samples (Ogunseitan, 1993, 1997). In the first process, microbial proteins were extracted from 1 g soil or sediment by boiling the samples in a solution, although in the second method, similar amounts of environmental samples were incubated for 1 hour at 0°C in a solution monitored by four 10-minute freeze–thaw rotations. The boiling process improved high concentrations of proteins from wastewater but not from soils and sediments. The freeze–thaw technique completed well for soils and sediments (Ogunseitan, 1993, 1997). Singleton et al. observed a variety of approaches to extract total soil protein. A simple snap-freeze protein extraction method by liquid nitrogen was originated to extract the most protein from soil samples associated with a bead-beating technique used normally for DNA extraction from soil (Singleton et al., 2003). Schulze et al. established a procedure to study soil proteins isolated from thawed organic matter, directing extracellular microbial enzymes that are perhaps significant in the carbon cycle. Water-soluble proteins were removed from the soil by dissolving soil minerals with hydrofluoric acid, and the mined proteins were recognised through MS coupled to liquid chromatography (LC) (Schulze et al., 2005). Wilmes and Bond reported a method that allowed the effective extraction and purification of the whole proteome from a laboratory-scale simulated sludge system improved for enhanced biological phosphorus elimination (EBPR) followed through separation by 2D-GE. The procedure includes numerous washing solutions, lysis buffers, French press lyses, and precipitation processes (Wilmes & Bond, 2004).

4.3.2 Community Proteomics of Marine Symbionts of *R. pachyptila*

The deep-sea cylinder worm *R. pachyptila* harbours a specific organ, the trophosome, occupied by sulphide-oxidising endosymbiotic bacteria that deliver the larva with carbon, nitrogen, and other nutrients. An intracellular and membrane protein orientation map founded on metagenomic data of the endosymbionts (Robidart et al., 2008) was produced through 2-D polyacrylamide gel electrophoresis (PAGE) coupled with matrix-assisted laser desorption ionisation time-of-flight—mass spectrometry (MALDI-TOF-MS) and 1-D PAGE combined with 2-D LC-MS/MS (Markert et al., 2007). It showed that the bacteria concurrently express enzymes of the Calvin cycle and the reductive tricarboxylic acid (TCA) cycle to fix CO_2. Furthermore, the association of protein profiles derivative from sulphide-rich and sulphide-depleted environments designated that the Riftia endosymbionts contain the appearance of energetically costly sulphide-oxidation-related enzymes and the key Calvin cycle enzyme RUBISCo in favour of fewer ATP-consuming TCA cycle enzymes when H_2S is limited (Markert et al., 2007).

4.3.3 Proteome Studies of WasteWater Management Plants and Activated Sludge

Lacerda et al. (2007) examined the reaction of a natural communal in a continuous-flow wastewater management bioreactor to an inhibitory level of cadmium through 2-D PAGE shared by MALDI-TOF/TOF-MS then *de novo* sequencing. The authors detected an important change in the communal proteome after cadmium shock, as specified through the differential expression of more than 100 proteins with ATPases, oxidoreductases, and transport proteins. Park et al. (2008) analysed the protein accompaniment of extracellular polymeric substances of initiated sludge flocs by 1-D PAGE shared with LC-MS/MS and recognised a limited number of bacterial but also human polypeptides, amongst them proteins related to bacterial defence, cell attachments, outer membrane proteins and a human elastase. In 2004, Wilmes and Bond (2004) deliberate the molecular apparatuses of improved biological phosphorus removal (EBPR) by a relative metaproteome analysis of two laboratory wastewater sludge microbial groups by and without EBPR performance by 2-D PAGE combined to MALDI-TOF-MS. Foremost alterations in protein expression profiles among the two apparatuses were detected. A short time later, more than 2,300 proteins were acknowledged by 2-D LC-MS/MS examinations of activated sludge (Wilmes et al., 2008; Wilmes & Bond, 2004), assisted by reference metagenomic data from studies of EBPR sludge (Garcia Martin et al., 2006). The achieved data specified that the uncultured polyphosphate-accumulating bacterium "*Candidatus accumulibacter phosphatis*" controls the microbial community of the EBPR reactor and further enables an extensive investigation of metabolic pathways, e.g., denitrification, fatty acid cycling, and glyoxylate bypass, all central to EBPR.

4.3.4 Community Proteogenomics of Phyllosphere Bacteria

Very recently, Delmotte et al. (2009) shared a culture-independent metagenome and metaproteome methodology to study the microbiota related to leaves of soybean, clover and Arabidopsis thaliana plants. Phyllosphere bacteria were washed away from the leaves and DNA and proteins were removed and analysed by pyrosequencing and 1-D PAGE LC-MS/MS resulting in the documentation of 2,883 proteins. The majority of the proteins were connected to *Methylobacterium*, *Sphingomonas*, and *Pseudomonas*, representative of the predominance of these genera within the phyllosphere communal. Functional assignments of the proteins recommended that phyllosphere Methylobacteria are capable to exploit methanol as a carbon and energy source and that Sphingomonads possess a mainly large substrate spectrum on plant leaves.

4.3.5 Community Proteomics of Animal Intestinal Tracts

Warnecke et al. (2007) employed a mutual genomics and multidimensional-LC-MS/MS proteomics method to examine the microbial community present in the hindgut of higher wood-feeding termites. For peptide separation, they used a 2-D method containing three steps: RP LC followed through a strong cation-exchange (SCX) chromatography and an extra RP LC step. This three-step system has been used positively to increase the resolving power of LC foremost to an increased number of recognised proteins in a complex sample (Wei et al., 2005). The authors described the occurrence of a large set of bacterial enzymes elaborate in the degradation of cellulose and xylan and other significant symbiotic functions such as H_2 metabolism, CO_2- reductive acetogenesis, and N_2 fixation. In a more current study, Toyoda et al. (2009) used 1-D PAGE attached to MS/MS to classify cellulose-binding proteins derived from sheep rumen microorganisms; amongst these proteins were endoglucanase F of the cellulolytic bacterium *Fibrobacter succinogenes* and exoglucanase Cel6A of the fungus *Piromyces equi*.

4.3.6 Community Proteomics of Human Intestinal Tracts

Klaassens et al. (2007) deliberate the functionality of the uncultured microbiota of human infant stool samples by 2-D PAGE mutual with MALDI-TOF-MS. The authors detected time-dependent variations in the gut metaproteome but were not capable to classify more than one protein exhibiting high comparison to a bifidobacterial transaldolase due to at that time limited microbiome sequence info. Finally, Verberkmoes et al. (2009b) recognised several thousand proteins existing in two female twin faecal samples by an extensive semi-quantitative shotgun proteome examination; amongst them, bacterial proteins elaborate in well-known but also undescribed microbial pathways and human antimicrobial peptides.

4.3.7 Metaproteome Analyses of Ocean Water

One of the very first metaproteome examines was existing by Kan et al. (2005), who associated protein profiles from different sample origins of the Chesapeake Bay by 2-D PAGE and tried to classify protein spots removed from the gels by an LC-MS/MS method; though, the acquired information was rather limited, as a considerable DNA sequence background was still lacking. Recently, Sowell et al. (2009) distributed a complete study of the Sargasso Sea surface metaproteome. The authors used 2-D LC coupled to MS/MS and recognised over 1,000 proteins, among them an irresistible number of SAR11 periplasmic substrate-binding proteins as well as *Prochlorococcus* and *Synechococcus* proteins complicated in photosynthesis and carbon fixation. High profusion of SAR11 transporters as determined through spectral counting proposes that cells endeavour to exploit nutrient acceptance activity and therefore improve a competitive benefit in nutrient-depleted environments.

4.3.8 Metaproteome Studies of Highly Complex Groundwater and Soil Environments

A fascinating functional insight of the complex microbial groups found in dissolved organic matter from lake water and seepage water after soil micro-particles was obtained by Schulze et al. (2005).Though the number of proteins recognised by 2-D LC-MS/MS was relatively low, the authors were capable of allocating functional proteins to broad taxonomic groups and experiential rather unexpected seasonal

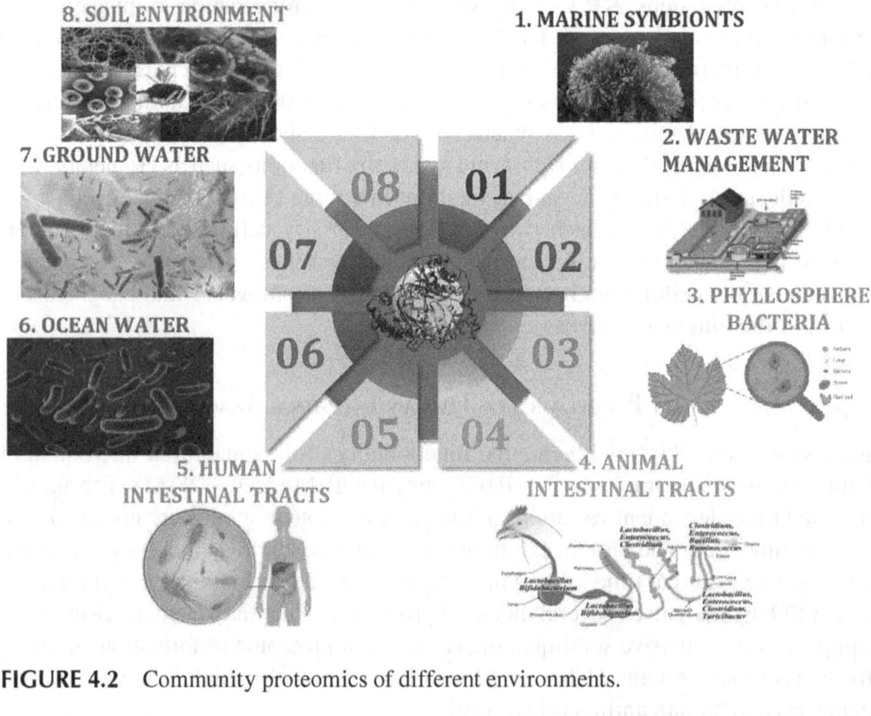

FIGURE 4.2 Community proteomics of different environments.

differences of the protein accompaniment. Notably, decomposing enzymes only originated amongst proteins removed from soil particles, thereby representing that the degradation of soil organic matter generally takes place in biofilm-associated groups. More recently, Benndorf et al. (2007) published a metaproteome examination of protein sources from contaminated soil and groundwater paying either 1-D or 2-D PAGE combined with LC and MS/MS. Only 59 proteins could be identified despite the authors' multi-step purification protocol, which included phenol extraction and NaOH treatment. This is because soil proteome analyses are primarily hampered by numerous inorganic and organic contaminants, which make it difficult to separate and document proteins (Figure 4.2).

4.4 FUTURE PERSPECTIVES

When viewed in relation to its huge possibility, the real output of environmental proteomics seems so far to be insufficiently limited. Present studies have generally concentrated on microbial communities with a comparatively low diversity or conquered by a particular phylogenetic group. The main problems toward complete metaproteome attention seem to be (i) the unbalanced distribution of species within microbial communities, (ii) within microbes, there is a wide range of protein expression levels, and (iii) microbial communities have enormous genetic heterogeneity (Wilmes & Bond, 2009). It is inspiring to note, though, that continually improving extraction procedures alongside developments in downstream MS technology and a gradually growing pool of bioinformatics data might soon help to overcome the present challenges and limitations of metaproteomics research.

4.4.1 Improvements in Mass Spectrometer Sensitivity and Accuracy

A positive environmental proteomics experiment involves the dependable identification not only of the predominant but also of low-abundance proteins, which closely depends on ultrasensitive and highly accurate mass spectrometers. An important and predictable development in MS performance will allow us to (i) classify low-abundance but important gene products, e.g., proteins complicated in transcriptional and translational regulation, (ii) evaluate the 3-D delivery of proteins within multifaceted habitats, and (iii) investigate proteins on such a fine scale that we might even visualise single-cell proteome examines.

4.4.2 Quantitative Environmental Proteomics

Another significant future line of research will take benefit from the increasing power of metaproteomics tools to quantitatively associate protein expression rates in environmental samples. Despite its well-known disadvantages, 2-D PAGE has controlled quantitative protein expression studies until lately; furthermore, quantitative proteomics has been limited to biological arrangements of low or limited complexity. Nowadays, 2-D gel-free LC-MS-based technologies have developed as powerful tools for comparative or quantitative proteome studies and might be functional for in-depth, quantitative proteome profiling of complex environments. Newly, label-free techniques (Gilchrist et al., 2006), which are constructed by counting fragment ranges of peptides used to classify a certain protein, have been active for quantitative environmental proteome examines. An improvement of label-free methods is their comparably large dynamic range, which is of particular significance when multifaceted and large protein variations within altered samples have to be anticipated (Sowell et al., 2009). Because samples have to be analysed distinctly and are so accountable to experimental differences, it is essential that sample preparation and examination become highly consistent and reproducible (Bantscheff et al., 2007). A revolutionary progress in the field of measurable proteomics was the outline of isotope- or isobar-tag-based technologies, e.g., ICAT, iTRAQ, and ANIBAL, which allow the examination of changed samples in a single MS measurement. Though, so far none of these methods was used to measure protein appearance in complex environmental samples, which might be due to the detail that these methods are comparatively costly, can only be useful to a limited number of samples and are essential for substantial post-processing of the original samples. Moreover, the improvement of corresponding hardware and, even more prominently, software tools for label-based measurable proteomics lag behind the developments in MS (Bantscheff et al., 2007).

4.5 CONCLUSION

Proteomics is one of today's wildest emerging research parts and has contributed considerably to our understanding of separate organisms at the cellular level. Its draw stems from the existing ability to investigate many protein functions and replies simultaneously and seems also preferably suitable to increase our information on the complex interaction between the composition of habitat, diversity, and architecture of microbial groups and ecosystem functioning. Recently, an incomplete number of studies relating to large-scale proteome investigations of environmental samples have established the enormous possibility of metaproteomics to reveal the molecular devices complicated in function, interaction, physiology, and evolution of microbial groups. Furthermore, the quickly growing number of genomic and metagenomic sequences composed with innovative advances in protein study and bioinformatics have opened up an entirely new range of uses, e.g., studying the influence of environmental changes upon protein expression profiles of whole microbial communities or measuring low-level protein appearance differences in order to resolve the functional implication of three-dimensional protein distribution within an assumed environment. In conclusion, the complete information increased by the concentrated application of system-level methods such as genomics, transcriptomics, proteomics, and metabolomics will importantly advance our thoughtful biogeochemical cycles and will simplify the biotechnological attaching of microbial groups or sterile organisms.

REFERENCES

Ahmadian, A., Ehn, M., Hober, S., Pyrosequencing: history, biochemistry and future. *Clin. Chim. Acta* 2006, 363, 83–94.

Amann, R. I., Ludwig, W., Schleifer, K. H., Phylogenetic identification and in situ detection of individual microbial cells without cultivation. *Microbiol. Rev.* 1995, 59, 143–169.

Bantscheff, M., Schirle, M., Sweetman, G., Rick, J., Kuster, B., Quantitative mass spectrometry in proteomics: a critical review. *Anal. Bioanal. Chem.* 2007, 389, 1017–1031.

Benndorf, D., Balcke, G. U., Harms, H., von Bergen, M., Functional metaproteome analysis of protein extracts from contaminated soil and groundwater. *ISME J.* 2007, 1, 224–234.

Bochner, B. R., Global phenotypic characterization of bacteria. *FEMS Microbiol. Rev.* 2009, 33, 191–205.

Brock, T. D., The study of microorganisms in situ: progress and problems. *Symp. Soc. Gen. Microbiol.* 1987, 41, 1–17.

Cardenas, E., Tiedje, J. M., New tools for discovering and characterizing microbial diversity. *Curr. Opin. Biotechnol.* 2008, 19, 544–549.

Delmotte, N., Knief, C., Chaffron, S., Innerebner, G. et al., Community proteogenomics reveals insights into the physiology of phyllosphere bacteria. *Proc. Natl. Acad. Sci. USA* 2009, 106, 16428–16433.

Denef, V. J., VerBerkmoes, N. C., Shah, M. B., Abraham, P. et al., Proteomics-inferred genome typing (PIGT) demonstrates interpopulation recombination as a strategy for environmental adaptation. *Environ. Microbiol.* 2009, 11, 313–325.

Garcia Martin, H., Ivanova, N., Kunin, V., Warnecke, F. et al., Metagenomic analysis of two enhanced biological phosphorus removal (EBPR) sludge communities. *Nat. Biotechnol.* 2006, 24, 1263–1269.

Gilchrist, A., Au, C. E., Hiding, J., Bell, A. W. et al., Quantitative proteomics analysis of the secretory pathway. *Cell* 2006, 127, 1265–1281.

Gill, S. R., Pop, M., DeBoy, R. T., Eckburg, P. B., Turnbaugh, P. J., Samuel, B. S., Gordon, J. I., Relman, D. A., Fraser-Liggett, C. M., Nelson, K. E., Metagenomic analysis of the human distal gut microbiome. *Science* 2006, 312, 1355–1359.

Goldenfeld, N., Woese, C., Biology's next revolution. *Nature* 2007, 445, 369.

Goltsman, D. S. A., Denef, V. J., Singer, S. W., VerBerkmoes, N. C. et al., Community genomic and proteomic analysis of chemoautotrophic, iron-oxidizing "Leptospirillum rubarum" (Group II) and Leptospirillum ferrodiazotrophum (Group III) in acid mine drainage biofilms. *Appl. Environ. Microbiol.* 2009, 75, 4599–4615.

Graves, P. R., Haystead, T. A. J., Molecular biologist's guide to proteomics. *Microbiol. Mol. Biol. Rev.* 2002, 66, 39–63.

Green, B. D., Keller, M., Capturing the uncultivated majority. *Curr. Opin. Biotechnol.* 2006, 17, 236–240.

Horz, H. P., Barbrook, A., Field, C. B., Bohannan, B. J. M., Ammonia-oxidizing bacteria respond to multifactorial global change. *Proc. Natl. Acad. Sci. USA* 2004, 101, 15136–15141.

Kan, J., Hanson, T. E., Ginter, J. M., Wang, K., Chen, F., Metaproteomic analysis of Chesapeake Bay microbial communities. *Saline Syst.* 2005, 1, 7.

Keller, M., Hettich, R., Environmental proteomics: a paradigm shift in characterizing microbial activities at the molecular level. *Microbiol. Mol. Biol. Rev.* 2009, 73, 62–70.

Klaassens, E. S., de Vos, W. M., Vaughan, E. E., Metaproteomics approach to study the functionality of the microbiota in the human infant gastrointestinal tract. *Appl. Environ. Microbiol.* 2007, 73, 1388–1392.

Lacerda, C. M., Choe, L. H., Reardon, K. F., Metaproteomic analysis of a bacterial community response to cadmium exposure. *J. Proteome Res.* 2007, 6, 1145–1152.

Lacerda, C. M., Reardon, K. F., Environmental proteomics: applications of proteome profiling in environmental microbiology and biotechnology. *Brief. Funct. Genomic. Proteomic.* 2009, 8, 75–87.

Lopez-Barea, J., Gomez-Ariza, J. L., Environmental proteomics and metallomics. *Proteomics* 2006, 6, 51–62.

Markert, S., Arndt, C., Felbeck, H., Becher, D. et al., Physiological proteomics of the uncultured endosymbiont of Riftia pachyptila. *Science* 2007, 315, 247–250.

Markowitz, V. M., Ivanova, N. N., Szeto, E., Palaniappan, K. et al., IMG/M: a data management and analysis system for metagenomes. *Nucleic Acids Res.* 2008, 36, D534–D538.

Maron, P. A., Ranjard, L., Mougel, C., Lemanceau, P., Metaproteomics: a new approach for studying functional microbial ecology. *Microb. Ecol.* 2007, 53, 486–493.

Naeem, S., Li, S. B., Biodiversity enhances ecosystem reliability. *Nature* 1997, 390, 507–509.

Ogunseitan, O. A., Direct extraction of proteins from environmentalsamples. *J. Microbiol. Methods* 1993, 17, 273–281.

Ogunseitan, O. A., Direct extraction of catalytic proteins from natural microbial communities. *J. Microbiol. Methods* 1997, 28, 55–63.

Park, C., Helm, R. F., Novak, J. T., Investigating the fate of activated sludge extracellular proteins in sludge digestion using sodium dodecyl sulfate polyacrylamide gel electrophoresis. *Water Environ. Res.* 2008, 80, 2219–2227.

Pignatelli, M., Aparicio, G., Blanquer, I., Hernandez, V. et al., Metagenomics reveals our incomplete knowledge of global diversity. *Bioinformatics* 2008, 24, 2124–2125.

Robidart, J. C., Bench, S. R., Feldman, R. A., Novoradovsky, A. et al., Metabolic versatility of the Riftia pachyptila endosymbiont revealed through metagenomics. *Environ. Microbiol.* 2008, 10, 727–737.

Schloss, P. D., Handelsman, J., Toward a census of bacteria in soil. *PLoS Comput. Biol.* 2006, 2, e92.

Schulze, W. X., Environmental proteomics – what proteins from soil and surface water can tell us: a perspective. *Biogeosci. Discuss.* 2004, 1, 195–218.

Schulze, W. X., Gleixner, G., Kaiser, K., Guggenberger, G., Mann, M., Schulze, E. D., A proteomic fingerprint of dissolved organic carbon and of soil particles. *Oecologia* 2005, 142, 335–343.

Shah, M. P., *Advanced Oxidation Processes for Effluent Treatment Plants.* Elsevier, 2020.

Shah, M. P., *Removal of Emerging Contaminants through Microbial Processes.* Springer, 2021.

Singh, J., Behal, A., Singla, N., Joshi, A. et al., Metagenomics: concept, methodology, ecological inference and recent advances. *Biotechnol. J.* 2009, 4, 480–494.

Singleton, I., Merrington, G., Colvan, S., Delahunty, J. S., The potential of soil protein-based methods to indicate metal contamination. *Appl. Soil Ecol.* 2003, 23, 25–32.

Sowell, S. M., Wilhelm, L. J., Norbeck, A. D., Lipton, M. S. et al., Transport functions dominate the SAR11 metaproteome at low-nutrient extremes in the Sargasso Sea. *ISME J.* 2009, 3, 93–105.

Toyoda, A., Iio, W., Mitsumori, M., Minato, H., Isolation and identification of cellulose-binding proteins from sheep rumen contents. *Appl. Environ. Microbiol.* 2009, 75, 1667–1673.

Tringe, S. G., Rubin, E. M., Metagenomics: DNA sequencing of environmental samples. *Nat. Rev.* 2005, 6, 805–814.

Tringe, S. G., von Mering, C., Kobayashi, A., Salamov, A. A. et al., Comparative metagenomics of microbial communities. *Science* 2005, 308, 554–557.

Venter, J. C., Remington, K., Heidelberg, J. F., Halpern, A. L. et al., Environmental genome shotgun sequencing of the Sargasso Sea. *Science* 2004, 304, 66–74.

VerBerkmoes, N. C., Denef, V. J., Hettich, R. L., Banfield, J. F., Systems biology: functional analysis of natural microbial consortia using community proteomics. *Nat. Rev. Microbiol.* 2009a, 7, 196–205.

Verberkmoes, N. C., Russell, A. L., Shah, M., Godzik, A. et al., Shotgun metaproteomics of the human distal gut microbiota. *ISME J.* 2009b, 3, 179–189.

Vieites, J. M., Guazzaroni, M. E., Beloqui, A., Golyshin, P. N., Ferrer, M., Metagenomics approaches in systems microbiology. *FEMS Microbiol. Rev.* 2009, 33, 236–255.

von Mering, C., Hugenholtz, P., Raes, J., Tringe, S. G., Doerks, T., Jensen, L. J., Ward, N., Bork, P., Quantitative phylogenetic assessment of microbial communities in diverse environments. *Science* 2007, 315, 1126–1130.

Warnecke, F., Luginbuhl, P., Ivanova, N., Ghassemian, M. et al., Metagenomic and functional analysis of hindgut microbiota of a wood-feeding higher termite. *Nature* 2007, 450, 560–565.

Wei, J., Sun, J., Yu, W., Jones, A. et al., Global proteome discovery using an online three-dimensional LC-MS/MS. *J. Proteome Res.* 2005, 4, 801–808.

Wilkins, M. R., Sanchez, J. C., Gooley, A. A., Appel, R. D. et al., Progress with proteome projects: why all proteins expressed by a genome should be identified and how to do it. *Biotechnol. Genet. Eng. Rev.* 1995, 13, 19–50.

Wilmes, P., Andersson, A. F., Lefsrud, M. G., Wexler, M. et al., Community proteogenomics highlights microbial strain-variant protein expression within activated sludge performing enhanced biological phosphorus removal. *ISME J.* 2008, 2, 853–864.

Wilmes, P., Bond, P. L., The application of two-dimensional polyacrylamide gel electrophoresis and downstream analyses to a mixed community of prokaryotic microorganisms. *Environ. Microbiol.* 2004, 6, 911–920.

Wilmes, P., Bond, P. L., Metaproteomics: studying functional gene expression in microbial ecosystems. *Trends Microbiol.* 2006, 14, 92–97.

Wilmes, P., Bond, P. L., Microbial community proteomics: elucidating the catalysts and metabolic mechanisms that drive the Earth's biogeochemical cycles. *Curr. Opin. Microbiol.* 2009, 12, 310–317.

5 Events and Hazards in Biotransformation of Contaminants

Pratik Jagtap
Dr. Homi Bhabha State University

Aniket K. Gade
Sant Gadge Baba Amravati University
Institute of Chemical Technology, Matunga

Rajesh W. Raut
Dr. Homi Bhabha State University

CONTENTS

5.1 INTRODUCTION

Recently, environmental sustainability is considered an accountable concern with the environment to avoid damage to natural resources. The implementation of environmental sustainability practices necessitates that vulnerable natural resources should be protected for future generations to meet their basic needs. The

DOI: 10.1201/9781003354147-5

environment has a remarkable ability to protect itself from anthropogenic factors around the globe. Humans are continuously exploiting natural resources resulting in their drastic depletion, thus compromising their enduring feasibility. In these circumstances, sustainable water supply systems can supply ample quantities of water and its eminence standard for the basic needs of mankind. For the same purpose, consideration of environmental integrity and economic feasibility is a concern for its utilisation. Water is the ultimate origin of every form of life on the earth and also plays a vital role in the constancy of life. It acts as an essential component of cells. All the life forms on the earth require water for survival, and some of them can live with a modest amount of water. Life is made up of minute cells acting as structural and functional units of organisms from simple prokaryotes to more complex eukaryotes. These life forms need biochemical metabolism to obtain energy, grow and dispose of the waste generated. Thus, water mediates the entire metabolism through it by serving as a universal solvent for the functioning of life (Karigar and Rao 2011).

However, at present access to safe potable water has already become a crucial issue in developing countries characterised by various issues like industrialisation, population explosion and untreated wastewater disposal. The chief sources of concern for this cause drastic water pollution by sewage water, textile industry effluents, chemical waste, domestic waste, pesticides and fertilisers, pharmaceuticals, xenobiotics, etc. All these pollutants contribute biological, mutagenic, organic and inorganic substances that cause water to be unhealthy for any use. They become a life threat to all animals and plants, as well as humans (Karigar and Rao 2011).

The vital role played by microorganisms in the treatment of wastewater has already been reported. In both naturally occurring and genetically engineered microorganisms, alga-, bacterium-, and fungus-based treatment processes play a crucial role in the conversion of various wastes to more stable and less polluting substances. Three major processes that can be employed for wastewater treatment using microorganisms are aerobic, anaerobic, or photosynthetic processes. These are distinguished on the basis of degradation or transformation of wastes in the contaminated water. In the aerobic treatment, microorganisms utilise oxygen for the oxidation purpose, while under the anaerobic treatment, microorganisms utilise nitrates and sulphates as a source of energy during the degradation of waste. Meanwhile, photosynthetic microorganisms utilise the carbon compounds from the waste for carbon fixation and naturally degrade them by converting them into simpler non-hazardous compounds (Hussaini et al. 2013).

In this chapter, we emphasise the understanding of microorganisms involved in wastewater treatment using biodegradation and biotransformation capabilities to degrade or transform a wide variety of pollutants. We focus on the accountability of microorganisms and their ability to degrade different pollutants from wastewater. Herein, biodegradation and biotransformation methods are reported that attempt to connect naturally occurring organisms and their catabolic activities to degrade and transform a wide variety of compounds including petroleum waste, aromatic and aliphatic hydrocarbons, crude oils, textile dyes, azo-dye effluents, agricultural pesticides and fertilisers, heavy metals, etc.

5.2 TYPES OF WATER CONTAMINANTS

Water contaminants are the substances produced by many geogenic and anthropogenic activities which can ultimately have an adverse effect on the water reservoirs. It is a very conspicuous problem throughout the world. Water is the most important resource that exists on the earth essential for the survival of all the life forms present on it. Rapid industrialisation is the chief cause of uncontrolled discharge of contaminants in the reservoirs. A wide variety of both organic and inorganic contaminants are continuously discharged from breweries, tanneries, dyeing, textiles, paper industries, steel manufacturers, and mining operations. The contaminants that are discharged from the industrial effluents range from crude oil, hydrocarbons, PAHs, PCBs, pesticides, fertilisers, aliphatic hydrocarbons, heavy metals, etc.

Organic contaminants such as various dyes, aromatic hydrocarbons (phenols), petroleum, surfactants, fertilisers, pesticides/herbicides, and pharmaceuticals have important concerns during wastewater treatment. These should be removed before discharge. Inorganic contaminants of aquatic environments are caused by naturally occurring substances (fluoride, arsenic and boron) and industrial waste (heavy metals like mercury, cadmium, chromium, cyanide, etc.). Naturally occurring inorganic materials mainly contaminate groundwater and industrial and agricultural waste, mainly surface water such as rivers, lakes, ponds and pipes of distribution systems, mainly tap water.

5.2.1 Organic Contaminants

Organic contaminants are chemicals that are carbon-based molecules such as organic solvents, crude oils, petroleum-based wastes, incomplete burned woods, gases and volatile organic compounds. These contaminants also include industrial dyes, humic acid phenolic compounds, petroleum surfactants and pharmaceuticals are the major sources of organic pollutants in wastewater. The occurrence of organic contaminants can cause toxic effects and can degrade the potable water quality. The majority of organic contaminants can easily enter the water reservoirs through various anthropogenic factors like the use of synthetic pesticides, detergents, surfactants, pharmaceutical products, paints, pigment manufacturers, etc.

Organic contaminants are generally intact in the environment for longer durations and are reported to accumulate/magnify in the local ecosystem and can ultimately affect humans. The organic contaminants can effortlessly pollute the drinking water supplies and can cause extreme long-term health problems for humans. These organic pollutants may include a very large number of different organic substances like organic solvents, pesticides, polychlorinated biphenyls, furans, and some nitrogenous derivatives. The most usual anthropogenic sources of the pollutants include the utilisation of wood preservatives, antifreeze, dry-cleaning materials, cleansers, etc. The two main significant sources of organic contaminants are inappropriate disposal of industrial wastewater and the leaching of pesticides/fertilisers through the runoff towards the water reservoirs that cause hazardous health effects. This form of organic compound causes several health problems such as skin irritations, inflammations, carcinogenic effects such as leukaemia, thyroid glands, kidney lymphoma and tumours in the testicles (Table 5.1) (Borah et al. 2020).

TABLE 5.1
List of Organic Contaminants Found in Water Reservoirs, Its Potential Sources and Hazardous Effects

Sr. No.	Organic Pollutants	Sources	Health/ Environmental Hazards	References
1	PAHs	Residential heating, coal gasification, coal tar pitch and asphalt production, aluminium production, petroleum refineries and motor vehicle exhaust.	Cataracts, kidney and liver damage, and jaundice	Abdel-Shafy and Mansour (2016)
2	PCBs	Ballast from lighting, old electric appliances, caulking, elastic sealants	Effects on the immune system, reproductive system, nervous system, and endocrine system	United States Environmental Protection Agency (2017)
3	Phenol	Coal tar and creosote, combustion of wood and auto exhaust	Anorexia, progressive weight loss, diarrhoea, vertigo, salivation, dark colouration of the urine, and blood and liver effects	United States Environmental Protection Agency (2017)
4	Benzene	Part of crude oil, gasoline, and cigarette smoke	Leukaemia, cancer of the blood-forming organs	CDC
5	Xylene	Airplane fuel, gasoline and cigarette smoke	Depression of the central nervous system, headache, dizziness, nausea and vomiting	United States Environmental Protection Agency (2017)
6	Phenanthrene	Dyes, plastics and pesticides, explosives and drugs	Carcinogenic effects	United States Environmental Protection Agency (2017)
7	Benzopyrene	Burning plants, wood, coal, and operating cars, trucks and other vehicles	Increased risk of skin, lung, bladder, and gastrointestinal cancers	United States Environmental Protection Agency (2009, 2017)
8	Naphthalene	Metal industries, biomass burning, gasoline and oil combustion, tobacco smoking, the use of mothballs, fumigants and deodorisers	Haemolytic anaemia, damage to the liver, and neurological damage	United States Environmental Protection Agency (2009)

(Continued)

TABLE 5.1 (*Continued*)
List of Organic Contaminants Found in Water Reservoirs, Its Potential Sources and Hazardous Effects

Sr. No.	Organic Pollutants	Sources	Health/ Environmental Hazards	References
9	Crude oil	Accidental crude oil spills, oil vessel leak; crude oil extraction platforms, and oil pipelines	Skin and eye irritation, neurologic and breathing problems	http://oils.gpa.unep.org/facts/sources.htm
10	Azo-dyes	Textile dye industries	Cause bladder and liver cancers	Sarkar et al. (2017)
11	Industrial dyes	Textile dye industries, paint industries, pigment production	Carcinogenic, mutagenic and/or toxic to life, cause skin, kidney, urinary bladder and liver cancer	United States Environmental Protection Agency (2009)
12	Textile dyes	Textile dye industries, tanning industries	Skin and eye irritation, neurologic and breathing problems	United States Environmental Protection Agency (2009)
13	Oil paints	Paint industries, pigment production	Cause allergic reactions, gastrointestinal symptoms, difficulty in breathing, rapid heartbeat	https://www.livestrong.com/article/125804-health-effects-oil-based-paint/

5.2.2 Inorganic Pollutants

The potential inorganic contaminant sources range from all possible geogenic and anthropogenic factors. Recognition and deliberation of these factors are essential to design a technology that can solve the present pollution scenario. Recently, extensive developments are going on, but still we seek eco-friendly and economical methods for water treatment.

Due to the drastic industrialisation, mismanaged agricultural practices and loads of domestic wastewater are continuously adding toxic pollutants (e.g., heavy metals, pesticides/herbicides, fertiliser, acidic contents) in the water reservoirs (Velizarov et al. 2004). The potable water quality is incessantly disgracing caused by commonly found inorganic contaminants such as heavy metals (arsenic, lead, chromium, mercury, etc.), fluorides, nitrates, and other metalloids. Among all these contaminants, arsenic, fluoride, and iron have a geogenic origin, while nitrates, phosphates; heavy metals are added by anthropogenic sources such as poor sewage systems, poor agricultural practices, and industrial discharges. Transformation of colonisation, drastic industrial growth, urban development/expansion, and population explosion are the major causes of water quality degradation (Ahluwalia and Goyal 2007). Majorly, arsenic, copper, chromium, lead, mercury, nickel, and zinc are the frequently reported heavy metals in wastewater originating from various industrial sources (Fu and Wang 2011).

Inorganic contaminants in the water do not degrade easily and can persist for a longer time, ultimately causing deterioration. Many parts of the world have already reported fluorides and nitrates with their toxicological effects. High concentrations of heavy metals cause severe damage to the local ecosystem and mankind as well (Table 5.2) (Mehndiratta et al. 2013, Shah 2020).

TABLE 5.2
List of Inorganic Contaminants Found in Water Reservoirs, Their Potential Source and Hazardous Effects

Sr. No.	Inorganic Pollutants	Sources	Health/Environmental Hazards	References
1	Arsenic (As)	Fossil fuel combustion and leaching of disposed fly ash and coal cleaning waste	Ulceration and chronic diseases such as bladder/ kidney, lung, and skin cancer	Herawati et al. (2000), Finkelman et al. (2002)
2	Zinc (Zn)	Metal corrosive resistant coatings, dry cell batteries, paints, plastics, wood preservatives, rubber, cosmetics	Nausea, stomach cramps, lungs and the body's temperature control system	Finkelman et al. (2002)
3	Cadmium (Cd)	Metal smelter, mines	Cardiovascular disease, fibrosis of the lung, renal injury	Finkelman et al. (2002)
4	Chromium (Cr)	Battery, electronics, and pigment production, electroplating processes	Toxic and carcinogenic effects, disorders of the liver and spleen	Herawati et al. (2000), Finkelman et al. (2002)
5	Hydrogen sulphide	Oil or gas extraction and sewage treatment plants	Haemorrhage, headache, coma, blurred vision, dizziness, and nausea	Finkelman et al. (2002)
6	Fluorine (F)	Industries, brickwork, electronics and dyes	Fluorosis, severe bone deformation, osteosclerosis, knock-knees and spinal curvature	Finkelman et al. (2002)
7	Manganese (Mn)	Combustion of fossils and leaching of ash	Respiratory problems	Finkelman et al. (2002), Geiger and Cooper (2010)

(Continued)

TABLE 5.2 (*Continued*)
List of Inorganic Contaminants Found in Water Reservoirs, Their Potential Source and Hazardous Effects

Sr. No.	Inorganic Pollutants	Sources	Health/Environmental Hazards	References
8	Lead (Pb)	Oil/petroleum spills, Batteries, car exhaust, components of domestic appliances, dyes, pigments	Damages kidney, liver, and nervous system, mental retardation, behavioural disorders, high blood pressure and heart disease	Geiger and Cooper (2010)
9	Selenium (Se)	Carbonaceous shales (stone coals) for home heating and cooking purposes	Gastrointestinal disturbance, anaemia, liver, and spleen damage	Geiger and Cooper (2010)
10	Mercury (Hg)	Gold mining, production of nonferrous metals and fossil fuel combustion, cement production, electronic waste	Toxic and carcinogenic effects	Finkelman et al. (2002)
11	Nitrates (NO_3^-, NO_2^-)	Inorganic fertilisers during agriculture	Methemoglobinemia among infants, gastric and intestinal carcinogenicity	Bhatnagar and Sillanpää (2011)
12	Fluorides	Natural rock weathering, inorganic fertilisers during agriculture	Teeth decay, crippling, as well as skeletal fluorosis	Sayato (1989)
13	Pesticides/herbicides	Inorganic pesticides/ herbicides during agriculture	Toxic and carcinogenic effects on various body parts	Mohamed et al. (2011)

5.3 BIODEGRADATION

Biodegradation is a naturally occurring process that constantly recycles the biologically essential macro- and micronutrients by microbe-mediated activities. This is exclusively initiated through the enzymatic action of microorganisms for the breakdown of complex compounds into simpler end products all the way through the series of intermediate compound metabolism. Thus, conversion of toxic chemicals to completely or partially oxidised end products has less or negligible environmental effects. Biodegradation is natural way of recycling the major components of dead organisms, particularly carbon and nitrogen, and also all other elements that participate in the functioning of living systems. In natural conditions, biodegradable components are generally organic materials of various biotic components. Anthropogenic factors have been added much more to natural biodegradation by synthesising xenobiotic compounds which have never occurred geogenically. Among these, some are biodegradable and some are not readily biodegradable depending upon the structure and composition of synthetic organic components. Microorganisms with their amazing metabolic diversity and abilities evolve to have, in particular, an incredible catabolic capacity that enables them to degrade or transform a vast range of compounds including natural materials such as plant residues (e.g., cellulose and lignin), hydrocarbons (e.g., crude oil), and radionuclides or heavy metals (e.g., uranium and mercury), and also xenobiotic compounds such as polychlorinated biphenyls (PCBs), pesticides (e.g., pentachlorophenol (PCP)), and even novel pharmaceutical substances never before seen in nature (e.g., novel antibiotics and other drugs). Researchers are constantly focusing on the improving biodegradation process for more than few decades. As it was recognised, biodegradation takes place through the enzymatic processes produced by microorganisms through discrete stepwise catabolic pathways emphasised for biodegradability of organic contaminants (Figure 5.1).

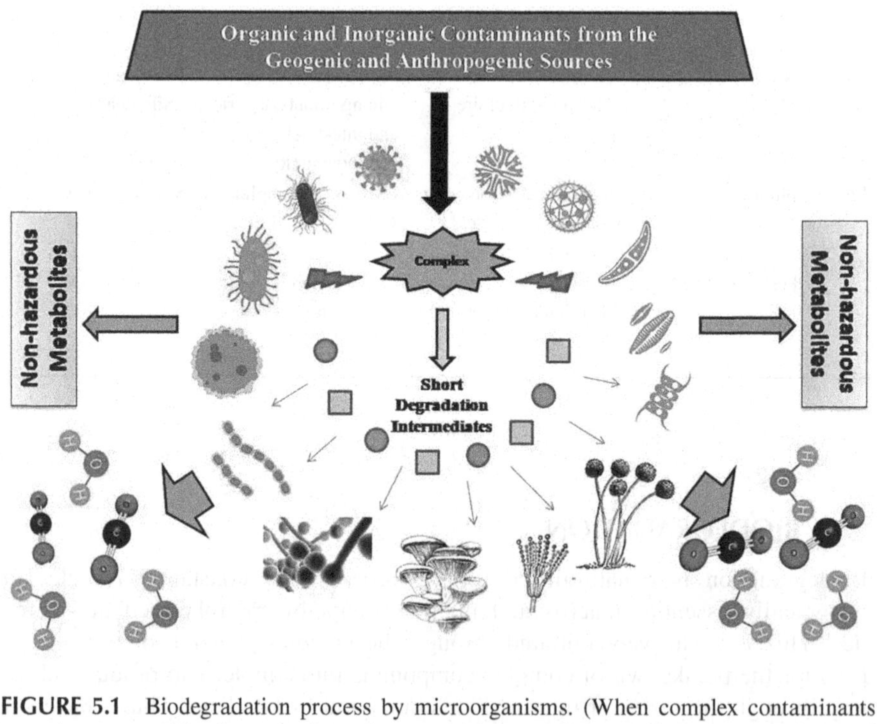

FIGURE 5.1 Biodegradation process by microorganisms. (When complex contaminants come in contact with microorganisms, it secrets the extracellular catabolic enzymes that lead to a short degradation reaction). The intermediate products are then diffused in the cells by the microorganisms as the energy source and finally oxidised into CO_2, H_2O and non-hazardous end products.

5.4 BIOTRANSFORMATION

Biotransformation is another approach that can be employed to obtain cleaner water through the utilisation of microorganisms. In this context, toxic and wastewater pollutants are being biologically transformed using enzymes or whole microbial organisms in order to achieve the conversion of toxic compounds into non-toxic compounds. Biotransformation also serves as an important mechanism that converts the toxic xenobiotic compounds into less harmful substances. For example, metals cannot be degraded, but they can be transformed by oxidation or reduction from one form to another (e.g., FeII to FeIII) (Basniwal et al. 2017).

Biotransformation plays a promising role in the toxicity reduction of hazardous chemical contaminants. The biotransformation can be exploited using a variety of transforming microorganisms like bacteria, fungi, and their respective extracellular enzymes which catalysed the catabolic activity of the water contaminants. Naturally occurring transformation is quite a slow process, and it may be non-specific and less productive in condition. Biotransformation succeeded by assembling the sequence of oxidation, reduction, hydrolysis, isomerisation, condensation, formation of altered carbon bonds, and introduction of functional groups that make non-hazardous by-products (Basniwal et al. 2017).

Microorganisms have been broadly applied for steroid biotransformation to prepare specific chemical derivatives and the production of specific compounds by manual methods. Thus, biotransformation leads to an excellent way of handling environmental water pollution hazards in an efficient manner. Therefore, microbial biotransformation is a boon for the current world scenario for water pollution (Figure 5.2).

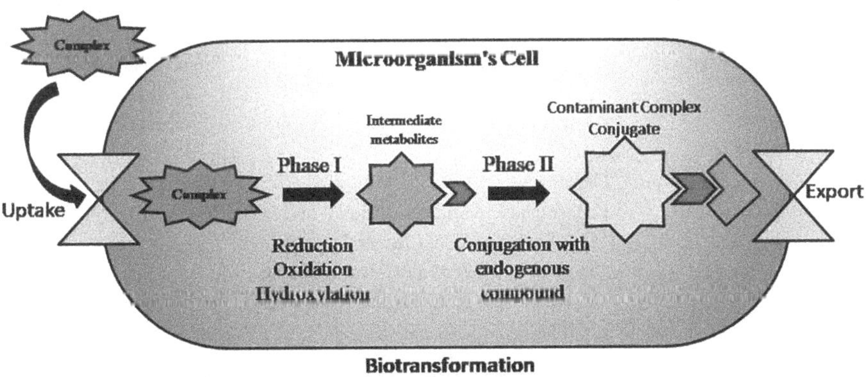

FIGURE 5.2 Schematic representation of biotransformation of organic and inorganic contaminants. It shows phase I (reduction, oxidation and hydroxylation processes that are responsible for formation of intermediate compounds) and then, phase II (conjugation with the contaminant intermediates obtain transformed non-hazardous end products that are exported out of the cell by transporter proteins on the membrane).

5.5 MICROORGANISM FLORA INVOLVED IN THE BIODEGRADATION OF ORGANIC AND INORGANIC POLLUTANTS

Microbial degradation is a highly promising, eco-friendly and economically important contaminant remediating technique. It can be employed through in situ and ex situ applications on water reservoirs. Drastic discharge of this contaminant can deteriorate the water quality. In this chapter, it has been attempted to showcase the ability of different microorganisms and their potential actions on water contaminants (Table 5.3).

TABLE 5.3
List of Microorganisms That Are Reported for Biodegradation of Organic and Inorganic Water Contaminants

Sr. No.	Organism	Organic and/or Inorganic Contaminants	Reference
		Bacteria	
1	*Achromobacter*	Petrol and diesel	Hesham et al. (2012)
2	*Acinetobacter* spp.	Chlorpyrifos and methyl parathion, petroleum compounds, pesticides (Ridomil, Fitoraz, Cleaner, Score EC, Zero 5 EC, Bravo SC, Meltatox, Mirage CE and TeldorCombi)	Pérez et al. (2016)
3	*Acinetobacter baumanii*	Petrol and diesel	Hesham et al. (2012), Lin et al. (2010)
4	*Acinetobacter radioresistens*	Chlorpyrifos	Hussaini et al. (2013)
5	*Aeromonas* spp.	U, Cu, Ni, Cr, Pb, Cd	Sinha and Biswas (2014)
6	*Alcaligenes seutrophus*	2,4-D, 1,4DCB	Basniwal et al. (2017)
7	*Alcaligenes sodorans*	Phenol	Basniwal et al. (2017)
8	*Arthrobacter* spp.	Endosulphate, pesticides (Ridomil MZ 68 MG, Fitoraz WP 76, Cleaner, Decis 2.5 CE, Score 250 EC, Zero 5 EC, Bravo 720 SC, Meltatox, Mirage 45 CE and TeldorCombi), 3CBA	Pérez et al. (2016), Basniwal et al. (2017)
9	*Bacillus* spp.	Endosulfan, petroleum compounds, azo-dye industrial effluent	Basniwal et al. (2017)
10	*Bacillus cereus*	Diesel, azo-dye industrial effluent	Maliji et al. (2013)
11	*Bacillus coagulans*	Diesel, crude oil	Fagbemi and Sanusi (2016)

(*Continued*)

TABLE 5.3 (*Continued*)
List of Microorganisms That Are Reported for Biodegradation of Organic and Inorganic Water Contaminants

Sr. No.	Organism	Organic and/or Inorganic Contaminants	Reference
		Bacteria	
12	*Bacillus fusiformis*	Naphthalene, diesel, crude oil	Simarro et al. (2013), Fagbemi and Sanusi (2016)
13	*Bacillus licheniformis*	Crude oil	El-Borai et al. (2016)
14	*Bacillus pumilus*	Commercial textile dyes such as remazol black, sulfonated di-azo-dyes Red HE8B, and RNB dye. Malathion and dimethoate	Patel and Gupte (2016), Hussaini et al. (2013)
15	*Bacillus safensis*	Cadmium	Priyalaxmi et al. (2014)
16	*Bacillus sphaericus*	Crude oil	El-Borai et al. (2016)
17	*Bacillus subtilis*	Oil-based paints	Phulpoto et al. (2016)
18	*Brevibacillus brevis*	Crude oil	El-Borai et al. (2016)
19	*Burkholderia cepacia*	Phthalate	Basniwal et al. (2017)
20	*Burkholderia gladioli*	Dimethoate	Hussaini et al. (2013)
21	*Citrobacter koseri*	Diesel, crude oil	Fagbemi and Sanusi (2016)
22	*Coprinellus radians*	PAHs, methylnaphthalenes, and dibenzofurans	Karigar and Rao (2011)
23	*Corynebacterium propinquum*	Phenol	Pradeep et al. (2015)
24	*Dechloromonas* spp.	Benzene	Basniwal et al. (2017)
25	*Dehalococcoides* spp.	Dioxins	
26	*Desulfovibrio* spp.	RDX	
27	*Dehalococcoides* spp.	Vinyl chloride	
28	*Dehalococcoides ethenogenes*	PCE	
29	*Dehalospirilum multivorans*	DDT	
30	*Enterobacter spp.*	Chlorpyrifos and methyl parathion	Niti et al. (2014)
31	*Exiguobacterium indicum*	Azo-dye industrial effluent	Kumar et al. (2016)
32	*Exiguobacterium aurantiacums*	Azo-dye industrial effluent	
33	*Flavobacterium* spp.	Petrol and diesel	Sivakumar et al. (2012)
34	*Geobacter* spp	Fe (III), U (VI)	Mirlahiji and Eisazadeh (2014)
35	*Listeria denitrificans*	Textile azo-dyes (Novacron dye, viz orange W3R, red FNR, yellow FN2R, blue FNR or navy WB)	Hassan et al. (2013)

(*Continued*)

TABLE 5.3 (*Continued*)
List of Microorganisms That Are Reported for Biodegradation of Organic and Inorganic Water Contaminants

Sr. No.	Organism	Organic and/or Inorganic Contaminants	Reference
		Bacteria	
36	*Lysinibacillus sphaericus*	Cobalt, copper, chromium and lead	Infante (2014)
37	*Microbacterium profundi*	Fe	
38	*Micrococcus luteus*	Textile azo-dyes (Novacron dye, viz orange W3R, red FNR, yellow FN2R, blue FNR or navy WB)	Hassan et al. (2013)
39	*Mycobacterium* spp.	Endosulfan, pyrene, petroleum compounds	Hussaini et al. (2013), Basniwal et al. (2017)
40	*Nocardia atlantica*	Petroleum compounds, textile azo-dyes (Novacron dye, viz orange W3R, red FNR, yellow FN2R, blue FNR or navy WB)	Hassan et al. (2013), Basniwal et al. (2017)
41	*Penicillium chrysogenum* fungi	Xenobiotics (aromatic hydrocarbons, benzene, toluene, ethyl benzene, xylene, phenolic compounds, etc.)	Singh (2017)
42	*Phanerochaete chrysosporium*	Biphenyl and triphenylmethane	Flayyih and Jawhari (2014)
43	*Photobacterium* spp.	Chlorpyrifos and methyl parathion	Kumar et al. (2015)
44	*Pseudomonas* spp.	Chlorpyrifos and methyl parathion, atrazine, PCP, 2,3,4-chloroaniline and 2,4,5-T, azo-dye industrial effluent	Kumar et al. (2015), Basniwal et al. (2017)
45	*Pseudomonas aeruginosa*	Phenol, crude oil, Fe2+, Zn2+, Pb2+, Mn2+, Cu2+, U, Cu, Ni, Cr, Commercial textile dyes such as remazol black, sulfonated di-azo-dyes Red HE8B, and RNB dye	Singh (2017), Sukumar and Nirmala (2016), De et al. (2008), Sinha and Biswas (2014), Kumar et al. (2016)
46	*Pseudomonas alcaligenes*	Petrol and diesel	Safiyanu et al. (2015)
47	*Pseudomonas cepacia*	Diesel, crude oil, fluoranthene	Fagbemi and Sanusi (2016), Basniwal et al. (2017)
48	*Pseudomonas fluorescens*	Fe^{2+}, Zn^{2+}, Pb^{2+}, Mn^{2+}, Cu^{2+}	De et al. (2008)
49	*Pseudomonas frederiksbergensis*	Endosulfan, chlorpyrifos and malathion	Hussaini et al. (2013)
50	*Pseudomonas mendocina*	Petrol and diesel	Safiyanu et al. (2015)

(*Continued*)

TABLE 5.3 (*Continued*)
List of Microorganisms That Are Reported for Biodegradation of Organic and Inorganic Water Contaminants

Sr. No.	Organism	Organic and/or Inorganic Contaminants	Reference
		Bacteria	
51	*Pseudomonas putida*	Petrol and diesel, benzene and xylene, naphthalene, pesticides (Ridomil, Fitoraz, Cleaner, Score EC, Zero 5 EC, Bravo SC, Meltatox, Mirage CE and TeldorCombi)	Pérez et al. (2016), Kumar et al. (2015)
52	*Pseudomonas spinosa*	Endosulfan	Mohamed et al. (2011)
53	*Pseudomonas veronii*	Petrol and diesel	Karigar and Rao (2011)
54	*Rhodococcus* spp.	PCB	Basniwal et al. (2017)
55	*Rhodopseudomonas palustris*	Pb, Cr, Cd	Mohamed et al. (2011)
56	*Serratia ficaria*	Diesel, crude oil	Fagbemi and Sanusi (2016)
57	*Serratia liquefaciens*	Diazinon and malathion	Hussaini et al. (2013)
58	*Serratia marcescens*	Diazinon, chlorpyrifos, Igepal CO-210 and methyl parathion	Hussaini et al. (2013)
59	*Sphingomonas* spp.	Azo-dye industrial effluent	Basniwal et al. (2017)
60	*Sphingomonas spaucimobilis*	Pyrene	Basniwal et al. (2017)
61	*Staphylococcus*	Endosulfan	Mohamed et al. (2011)
62	*Xanthomonas* spp.	Azo-dye industrial effluent	Basniwal et al. (2017)
	Fungi		
63	*Aspergillus fumigatus*	Naphthalene, PAHs, paraffin, cyclane, cadmium	Aranda et al. (2010), Fazli et al. (2015)
64	*Aspergillus niger*	Crude oil, naphthalene, PAHs, paraffin, cyclane	Aranda et al. (2010)
65	*Aspergillus solani*	Naphthalene, PAHs, paraffin, cyclane	Aranda et al. (2010)
66	*Aspergillus versicolor*	Cadmium	Fazli et al. (2015)
67	*Candida glabrata*	Crude oil	Burghal et al. (2016)
68	*Candida krusei*	Crude oil	Burghal et al. (2016)
69	*Candida viswanathii*	Phenanthrene, benzopyrene	Sivakumar et al. (2012)
70	Cladosporium spp.	Cadmium	Fazli et al. (2015)
71	Cunninghamella elegans	Dyes, salts, heavy metals and surfactants	Adebajo et al. (2017)
72	*Fusarium* spp.	Aliphatic hydrocarbon	(Hesham et al., 2012)
73	*Gloeophyllum striatum*	Pyrene, anthracene, dibenzothiophene lignin peroxidase etc.	Yadav et al. (2011)
74	*Gloeophyllum trabeum*	Hydrocarbons	Hesham et al. (2012)

(*Continued*)

TABLE 5.3 (*Continued*)
List of Microorganisms That Are Reported for Biodegradation of Organic and Inorganic Water Contaminants

Sr. No.	Organism	Organic and/or Inorganic Contaminants	Reference
		Bacteria	
75	*Microsporum* spp.	Cadmium	Fazli et al. (2015)
76	*Myrothecium roridum*	Malachite green (industrial dye)	Jasińska et al. (2015)
77	*Paecilomyces* spp.	Cadmium	Fazli et al. (2015)
78	*Penicillium funiculosum*	Naphthalene, PAHs, paraffin, cyclane	Aranda et al. (2010)
79	*Penicillium ochrochloron*	Malachite green (industrial dye)	Jasińska et al. (2015)
80	*Phanerochaete chrysosporium*	Malachite green (industrial dye)	
81	*Pycnoporus sanguineous*	Malachite green (industrial dye)	
82	*Saccharomyces cerevisiae*	Crude oil, heavy metals, lead, mercury and nickel	Infante (2014), Burghal et al. (2016)
83	*Trichoderma spp.*	Cadmium	Fazli et al. (2015)
84	*Trametes versicolor*	Hydrocarbons	Hesham et al. (2012)
85	*Trametes trogii*	Malachite green (industrial dye)	Jasińska et al. (2015)
86	*Tyromyces palustris*	Hydrocarbons	Hesham et al. (2012)
		Algae	
87	*Acrosiphonia coalita*	TNT	Cruz-Uribe et al. (2007)
88	*Chlamydomonas* spp.	Naphthalene	Sample et al. (1999)
89	*Chlorella* spp.	Chlordimeform	
90	*Chlorococcum* spp.	α-Endosulfan	Sethunathan et al. (2004)
91	*Desmodesmus* spp.	Bisphenol A	Wang et al. (2017)
92	*Dunaliella* spp.	Naphthalene	Sample et al. (1999)
93	*Euglena gracilis*	Naphthalene, phenol	
94	*Navicula pelliculosa*	Naphthenic acids	Mahdavi et al. (2015)
95	*Nitzschia* spp.	Phenanthrene	Hong et al. (2008)
96	*Portieria hornemannii*	TNT	Cruz-Uribe et al. (2007)
97	Scenedesmus spp.	α-Endosulfan	Sethunathan et al. (2004)
98	*Scenedesmus acutus*	Benzopyrene	Sample et al. (1999)
99	*Scenedesmus obliquus*	Sulfonic acid, naphthalene	
100	*Selenastrum capricornutum*	Atrazine, benzopyrene, benzo[a]pyrene	Sample et al. (1999)

5.6 MICROORGANISMS INVOLVED IN BIOTRANSFORMATION

TABLE 5.4
List of Microorganisms That Are Reported for Biotransformation of Organic and Inorganic Water Contaminants

Sr. No.	Organism	Contaminants	References
		Bacteria	
1	*Bacillus* spp.	Xenobiotic [acetone, cyclohexane, styrene, benzene, ethylbenzene, propylbenzene, dioxane, and 1,2-dichloroethylene]	Chowdhury et al. (2008)
2	*Desulfotomaculum* spp.		
3	*Desulfovibrio* spp.		
4	*Escherichia* spp.		
5	*Gordonia* spp.		
6	*Methanosaeta* spp.		
7	*Methanospirillum* spp.		
8	*Micrococcus* spp.		
9	*Moraxella* spp.		
10	*Pelatomaculum* spp.		
11	*Pseudomonas* spp.		
12	*Rhodococcus* spp.		
13	*Syntrophobacter* spp.		
14	*Syntrophus* spp.		
15	*Acetobacter* spp.	Azo-dyes	Gioia et al. (2008)
16	*Klebsiella* spp.		
17	*Pseudomonas* spp.		
18	*Cunninghamella elegans*	Pentafluorosulfanyl (SF5-)	Kavanagh et al. (2014)
19	*Pseudomonas knackmussii*		
20	*Pseudomonas pseudoalcaligenes*		
		Fungi	
21	Aspergillus niger	Endosulfan	Supreeth and Raju (2017)
22	*Aspergillus sydowii*		
23	*Aspergillus terreus*		
24	*Aspergillus terricola*		
25	*Chaetosartoryas tromatoides*		
26	*Pleurotus ostreatus*		
27	*Trametes hirsuta*		
28	*Trametes versicolor*		
		Algae	
29	*Chlamydomonas* spp.	Lindane, naphthalene, phenol	Sample et al. (1999)
30	*Chlorella* spp.	Lindane, chlordimeform	
31	*Dunaliella* spp.	DDT, naphthalene	
32	*Euglena gracilis*	Phenol	
33	*Phanerochaete chrysosporium*	TNT	
34	*Scenedesmus obliquus*	Naphthalene, sulfonic acid	
35	*Selenastrum capricornutum*	Benzo[a]pyrene	

(Continued)

TABLE 5.4 (*Continued*)
List of Microorganisms That Are Reported for Biotransformation of Organic and Inorganic Water Contaminants

Sr. No.	Organism	Contaminants	References
36	*Chlorella pyrenoidosa*	Heavy metals (e.g., Cu, Pb, Ni, Se, Cr, Cd, Hg, As, etc.)	Yao et al. (2012)
37	*Chlorococcum* spp.		
38	*Gloeocapsa* spp.		
39	*Lyngbya heironymusii*		
40	*Lyngbya spiralis*		
41	*Nostoc* spp.		
42	*Oscillatoria quadripunctulata*		
43	*Phormidium molle*		
44	*Scenedesmus acutus*		
45	*Scenedesmus bijuga*		
46	*Scenedesmus obliquus*		
47	*Spirogyra hyaline*		
48	*Spirulina platensis*		
49	*Stigonema* spp.		
50	*Tolypothrix tenuis*		

5.7 CONCLUSION

In recent years, water reservoirs in the environment are getting highly polluted with toxic organic and inorganic contaminants that have been released from numerous geogenic and anthropogenic sources over an extensive duration of time. Petroleum fuels, oil spills, PCBs, polycyclic aromatic hydrocarbons (PAHs), pesticides, fertilisers, combustion engines, azo-dyes, and industrial dye compounds are continuously discharged into water reservoirs.

Biodegradation is an exceptionally productive alternative to remediation, cleansing, and managing effective recovery techniques for solving contamination in water reservoirs through microbial activity. Microbial activities can degrade or transform organic and inorganic compounds like hydrocarbons and heavy metals that might have toxicological effects on the ecosystem and mankind. Many researchers have reported results on the degradation or transformation of environmental pollutants by the potential capabilities of different microorganisms. Dynamics of aerobic and anaerobic bacteria have been extensively used in biodegradation owing to adapting to any environmental condition and utilisation of pollutants as energy sources. Ultimately, the contaminants are converted into non-hazardous end products. The extracellular enzymatic action tendency of bacteria is employed here to reduce/remediate the drastic effect of contaminants by converting them into oxidised products. In other contexts, fungi are a vital piece of degrading microorganisms as they metabolise a wide range of organic and inorganic pollutants. The network of mycelia provides not only an entrapment point for contaminants but also effectively reduces the contaminant levels from water. Along with these, algal degradations also play a dual function to attend contaminant remediation and energy generation. Algal degradation of contaminants is again a natural approach which ensures the lowering of toxicological impacts. As an advantage, algal-based degradation has a greater value in biomass production along with detoxifying or degrading xenobiotics and organic and inorganic pollutants.

Biotransformation is another approach to solving the contaminants problems in water reservoirs. Microbial biotransformation is broadly utilised in the transformation of various organic and inorganic pollutants including hydrocarbons, xenobiotics, PAHs, PCBs, petroleum, heavy metals, etc. From ancient times, biotransformation has been demonstrated as a crucial tool for preparation of specific derivatives in traditional synthetic methods. Thus, it can also play a vital role to handle water pollution problems like transforming hazardous xenobiotics, hydrocarbons and heavy metals into non-hazardous chemical compounds.

It is seen that some aerobic and anaerobic bacterial genera have been associated with the biodegradation and biotransformation of a wide variety of organic and inorganic contaminants in the water reservoirs. Biodegradation and biotransformation chiefly employ aerobic microorganisms of genera such as *Bacillus*, *Pseudomonas*, *Escherichia*, *Rhodococcus*, *Gordonia*, *Moraxella*, and *Micrococcus* and anaerobic microorganisms of genera such as *Methanospirillum*, *Pelatomaculum*, *Syntrophobacter*, *Desulfotomaculum*, *Syntrophus*, *Desulfovibrio*, and *Methanosaeta*. Similarly, fungal genera, namely, *Aspergillus*, *Penicillium*, *Candia*, *Fusarium*, *Saccharomyces*, *Pleurotus*, and *Cephalosporium* have been noted to play important roles in biodegradation and biotransformation of organic and inorganic contaminants. In addition to the above, the potential microalgal genera utilised for biodegradation and biotransformation commonly include *Chlorella*, *Scenedesmus*, *Phormidium*, *Botryococcus*, *Chlamydomonas*, *Spirulina*, *Oscillatoria*, *Desmodesmus*, *Arthrospira*, *Nodularia*, and *Cyanothece*. Biodegradation and biotransformation technologies are becoming recognised as eco-friendly and cost-effective methods for remediating sites contaminated with toxic organic and inorganic contaminants at a fraction of the cost of conventional technologies.

REFERENCES

Abdel-Shafy, HI, and MSM Mansour. 2016. A Review on Polycyclic Aromatic Hydrocarbons: Source, Environmental Impact, Effect on Human Health and Remediation. *Egyptian Journal of Petroleum* 25 (1): 107–123. https://doi.org/10.1016/j.ejpe.2015.03.011.

Adebajo, S, S Balogun, and A Akintokun. 2017. Decolourization of Vat Dyes by Bacterial Isolates Recovered from Local Textile Mills in Southwest, Nigeria. *Microbiology Research Journal International* 18 (1): 1–8. https://doi.org/10.9734/mrji/2017/29656.

Ahluwalia, SS, and D Goyal. 2007. Microbial and Plant Derived Biomass for Removal of Heavy Metals from Wastewater. *Bioresource Technology* 98 (12): 2243–2257. https://doi.org/10.1016/j.biortech.2005.12.006.

Aranda, E, R Ullrich, and M Hofrichter. 2010. Conversion of Polycyclic Aromatic Hydrocarbons, Methyl Naphthalenes and Dibenzofuran by Two Fungal Peroxygenases. *Biodegradation* 21 (2): 267–281. https://doi.org/10.1007/s10532-009-9299–2.

Basniwal, RK, NJ Singh, M Kumar, V Kumar, N Tuteja, A Varma, and P Goyal. 2017. Biotransformation of Xenobiotic Compounds: Microbial Approach: 335–345. https://doi.org/10.1007/978-3-319-47744-2_22.

Bhatnagar, A, and M Sillanpää. 2011. A Review of Emerging Adsorbents for Nitrate Removal from Water. *Chemical Engineering Journal* 168 (2): 493–504. https://doi.org/10.1016/j.cej.2011.01.103.

Borah, P, M Kumar, and P Devi. 2020. Types of Inorganic Pollutants: Metals/Metalloids, Acids, and Organic Forms. *Inorganic Pollutants in Water.* INC. https://doi.org/10.1016/b978-0-12-818965-8.00002–0.

Burghal, AA, N Abu-Mejdad, and W Al-Tamimi. 2016. Mycodegradation of Crude Oil by Fungal Species Isolated from Petroleum Contaminated Soil. *International Journal of Innovative Research in Science* 5 (2): 1517–1524. https://doi.org/10.15680/IJIRSET.2016.0502052.

Chowdhury, A, S Pradhan, M Saha, and N Sanyal. 2008. Impact of Pesticides on Soil Microbiological Parameters and Possible Bioremediation Strategies. *Indian Journal of Microbiology* 48 (1): 114–127. https://doi.org/10.1007/s12088-008-0011–8.

Cruz-Uribe, O, DP Cheney, and GL Rorrer. 2007. Comparison of TNT Removal from Seawater by Three Marine Macroalgae. *Chemosphere* 67 (8): 1469–1476. https://doi.org/10.1016/j.chemosphere.2007.01.001.

De, J, N Ramaiah, and L Vardanyan. 2008. Detoxification of Toxic Heavy Metals by Marine Bacteria Highly Resistant to Mercury. *Marine Biotechnology* 10 (4): 471–477. https://doi.org/10.1007/s10126-008-9083–z.

El-Borai, AM, KM Eltayeb, AR Mostafa, and SA El-Assar. 2016. Biodegradation of Industrial Oil-Polluted Wastewater in Egypt by Bacterial Consortium Immobilized in Different Types of Carriers. *Polish Journal of Environmental Studies* 25 (5): 1901–1909. https://doi.org/10.15244/pjoes/62301.

Fagbemi, OK, and AI Sanusi. 2016. Effectiveness of Augmented Consortia of Bacillus Coagulans, Citrobacter Koseri and Serratia Ficaria in the Degradation of Diesel Polluted Soil Supplemented with Pig Dung. *African Journal of Microbiology Research* 10 (39): 1637–1644. https://doi.org/10.5897/ajmr2016.8249.

Fazli, MM, N Soleimani, M Mehrasbi, S Darabian, J Mohammadi, and A Ramazani. 2015. Highly Cadmium Tolerant Fungi: Their Tolerance and Removal Potential. *Journal of Environmental Health Science and Engineering* 13 (1): 1–9. https://doi.org/10.1186/s40201-015-0176–0.

Finkelman, RB, W Orem, V Castranova, CA Tatu, HE Belkin, B Zheng, HE Lerch, SV Maharaj, and AL Bates. 2002. Health Impacts of Coal and Coal Use: Possible Solutions. *International Journal of Coal Geology* 50 (1–4): 425–443. https://doi.org/10.1016/S0166–5162(02)00125–8.

Flayyih, I, and H Ai-Jawhari. 2014. Ability of Some Soil Fungi in Biodegradation of Petroleum Hydrocarbons. *Journal of Applied & Environmental Microbiology* 2 (2): 46–52. https://doi.org/10.12691/jaem-2-2–3.

Fu, F, and Q Wang. 2011. Removal of Heavy Metal Ions from Wastewaters: A Review. *Journal of Environmental Management* 92 (3): 407–418. https://doi.org/10.1016/j.jenvman.2010.11.011.

Geiger, A, and J Cooper. 2010. Overview of Airborne Metals Regulations, Exposure Limits, Health Effects, and Contemporary Research. *Draft Environmental Analysis* 3 (9): 1–50.

Gioia, DD, A Michelles, M Pierini, S Bogialli, F Fava, and C Barberio. 2008. Selection and Characterization of Aerobic Bacteria Capable of Degrading Commercial Mixtures of Low-Ethoxylated Nonylphenols. *Journal of Applied Microbiology* 104 (1): 231–242. https://doi.org/10.1111/j.1365-2672.2007.03541.x.

Hassan, MM, MZ Alam, MN Anwar. 2013. Biodegradation of Textile Azo Dyes by Bacteria Isolated from Dyeing Industry Effluent. *International Research Journal of Biological Sciences* 2 (8), 27–31.

Herawati, N, S Suzuki, K Hayashi, IF Rivai, and H Koyama. 2000. Cadmium, Copper, and Zinc Levels in Rice and Soil of Japan, Indonesia, and China by Soil Type. *Bulletin of Environmental Contamination and Toxicology* 64 (1): 33–39. https://doi.org/10.1007/s001289910006.

Hesham, AEL, S Khan, Y Tao, D Li, Y Zhang, and M Yang. 2012. Biodegradation of High Molecular Weight PAHs Using Isolated Yeast Mixtures: Application of Meta-Genomic Methods for Community Structure Analyses. *Environmental Science and Pollution Research* 19 (8): 3568–3578. https://doi.org/10.1007/s11356-012-0919–8.

Hong, YW, DX Yuan, QM Lin, and TL Yang. 2008. Accumulation and Biodegradation of Phenanthrene and Fluoranthene by the Algae Enriched from a Mangrove Aquatic Ecosystem. *Marine Pollution Bulletin* 56 (8): 1400–1405. https://doi.org/10.1016/j.marpolbul.2008.05.003.

Hussaini, SZ, M Shaker, MA Iqbal, and KSK College Beed. 2013. Isolation of Fungal Isolates for Degradation of Selected Pesticides 2 (March): 50–53.

Infante J, C. 2014. Removal of Lead, Mercury and Nickel Using the Yeast Saccharomyces Cerevisiae. *Revista MVZ Córdoba* 19 (2): 4141–4149. https://doi.org/10.21897/rmvz.107.

Islam, M. 2008. Development of Adsorption Media for Removal of Lead and Nitrate from Water 9 (17 2).

Jasińska, Anna, K Paraszkiewicz, A Sip, and J Długoński. 2015. Malachite Green Decolorization by the Filamentous Fungus Myrothecium Roridum – Mechanistic Study and Process Optimization. *Bioresource Technology* 194: 43–48. https://doi.org/10.1016/j.biortech.2015.07.008.

Karigar, CS, and SS Rao. 2011. Role of Microbial Enzymes in the Bioremediation of Pollutants: A Review. *Enzyme Research* 2011 (1). https://doi.org/10.4061/2011/805187.

Kavanagh, E, M Winn, CN Gabhann, NK O'Connor, P Beier, and CD Murphy. 2014. Microbial Biotransformation of Aryl Sulfanylpentafluorides. *Environmental Science and Pollution Research* 21 (1): 753–758. https://doi.org/10.1007/s11356-013-1985–2.

Kumar, S, P Chaurasia, and A Kumar. 2016. Isolation and Characterization of Microbial Strains from Textile Industry Effluents of Bhilwara, India: Analysis with Bioremediation. *Journal of Chemical and Pharmaceutical Research* 8 (4): 143–150.

Lin, C, L Gan, and ZL Chen. 2010. Biodegradation of Naphthalene by Strain Bacillus Fusiformis (BFN). *Journal of Hazardous Materials* 182 (1–3): 771–777. https://doi.org/10.1016/j.jhazmat.2010.06.101.

Mahdavi, H, V Prasad, Y Liu, and AC Ulrich. 2015. In Situ Biodegradation of Naphthenic Acids in Oil Sands Tailings Pond Water Using Indigenous Algae-Bacteria Consortium. *Bioresource Technology* 187: 97–105. https://doi.org/10.1016/j.biortech.2015.03.091.

Maliji, D, Z Olama, and H Holail. 2013. Environmental Studies on the Microbial Degradation of Oil Hydrocarbons and Its Application in Lebanese Oil Polluted Coastal and Marine Ecosystem. *International Journal of Current Microbiology and Applied Sciences* 2 (6): 1–18.

Mehndiratta, P, A Jain, S Srivastava, and N Gupta. 2013. Environmental Pollution and Nanotechnology. *Environment and Pollution* 2 (2). https://doi.org/10.5539/ep.v2n2p49.

Mirlahiji, SG, and K Eisazadeh. 2014. Bioremediation of Uranium Via Geobacter Spp. *Journal of Research and Development* 1 (12): 52–58.

Mohamed, AT, AA El Hussein, MA El Siddig, and AG Osman. 2011. Degradation of Oxyfluorfen Herbicide by Soil Microorganisms Biodegradation of Herbicides. *Biotechnology*. https://doi.org/10.3923/biotech.2011.274.279.

Niti, C, S Sunita, K Kamlesh, and K Rakesh. 2014. Bioremediation : An emerging technology for remediation of pesticides.

Patel, Y, and A Gupte. 2016. Evaluation of Bioremediation Potential of Isolated Bacterial Culture YPAG-9 (Pseudomonas Aeruginosa) for Decolorization of Sulfonated Di-Azodye Reactive Red HE8B under Optimized Culture Conditions. *International Journal of Current Microbiology and Applied Sciences* 5 (8): 258–272. https://doi.org/10.20546/ijcmas.2016.508.027.

Pérez, M, OD Rueda, M Bangeppagari, JZ Johana, D Ríos, BB Rueda, IM Sikandar, and RM Naga. 2016. Evaluation of Various Pesticides-Degrading Pure Bacterial Cultures Isolated from Pesticide-Contaminated Soils in Ecuador. *African Journal of Biotechnology* 15 (40): 2224–2233. https://doi.org/10.5897/ajb2016.15418.

Phulpoto, AH, MA Qazi, S Mangi, S Ahmed, and NA Kanhar. 2016. Biodegradation of Oil-Based Paint by Bacillus Species Monocultures Isolated from the Paint Warehouses. *International Journal of Environmental Science and Technology* 13 (1): 125–134. https://doi.org/10.1007/s13762-015-0851-9.

Pradeep, NV, S Anupama, K Navya, HN Shalini, M Idris, and US Hampannavar. 2015. Biological Removal of Phenol from Wastewaters: A Mini Review. *Applied Water Science* 5 (2): 105–112. https://doi.org/10.1007/s13201-014-0176-8.

Priyalaxmi, R, A Murugan, P Raja, and KD Raj. 2014. Bioremediation of Cadmium by Bacillus Safensis (JX126862), a Marine Bacterium Isolated from Mangrove Sediments. *International Journal of Current Microbiology and Applied Sciences* 3 (12): 326–335. http://www.ijcmas.com.Journal.

Safiyanu, I, A Abdulwahid Isah, US Abubakar, and M Rita Singh. 2015. Review on Comparative Study on Bioremediation for Oil Spills Using Microbes. *Research Journal of Pharmaceutical, Biological and Chemical Sciences* 6 (6): 783–790.

Sample, KT, RB Cain, and S Schmidt. 1999. Biodegradation of Aromatic Compounds by Microalgae. *FEMS Microbiology Letters* 170 (2): 291–300. https://doi.org/10.1016/S0378-1097(98)00544-8.

Sarkar, S, A Banerjee, U Halder, R Biswas, and R Bandopadhyay. 2017. Degradation of Synthetic Azo Dyes of Textile Industry: A Sustainable Approach Using Microbial Enzymes. *Water Conservation Science and Engineering* 2 (4): 121–131. https://doi.org/10.1007/s41101-017-0031-5.

Sayato, Y. 1989. WHO Guidelines for Drinking-Water Quality. *Eisei Kagaku* 35 (5): 307–312. https://doi.org/10.1248/jhs1956.35.307.

Sethunathan, N, M Megharaj, ZL Chen, BD Williams, G Lewis, and R Naidu. 2004. Algal Degradation of a Known Endocrine Disrupting Insecticide, α-Endosulfan, and Its Metabolite, Endosulfan Sulfate, in Liquid Medium and Soil. *Journal of Agricultural and Food Chemistry* 52 (10): 3030–3035. https://doi.org/10.1021/jf035173x.

Shah, MP. 2020. *Advanced Oxidation Processes for Effluent Treatment Plants*. Elsevier.

Shah, MP. 2021. *Removal of Emerging Contaminants through Microbial Processes*. Springer.

Simarro, R, N González, LF Bautista, and MC Molina. 2013. Assessment of the Efficiency of in Situ Bioremediation Techniques in a Creosote Polluted Soil: Change in Bacterial Community. *Journal of Hazardous Materials* 262: 158–167. https://doi.org/10.1016/j.jhazmat.2013.08.025.

Singh, R. 2017. Microbial Biotransformation: A Process for Chemical Alterations. *Journal of Bacteriology & Mycology: Open Access* 4 (2): 47–51. https://doi.org/10.15406/jbmoa.2017.04.00085.

Sinha, SN, and K Biswas. 2014. Bioremediation of Lead from River Water through Lead-Resistant Purple-Nonsulfur Bacteria. *International Research Publication House* 2 (1): 11–14.

Sivakumar, G, J Xu, RW Thompson, Y Yang, P Randol-Smith, and PJ Weathers. 2012. Integrated Green Algal Technology for Bioremediation and Biofuel. *Bioresource Technology* 107: 1–9. https://doi.org/10.1016/j.biortech.2011.12.091.

Sukumar, S, and P Nirmala. 2016. Screening of Diesel Oil Degrading Bacteria from Petroleum Hydrocarbon Contaminated Soil. *International Journal of Advanced Research in Biological Sciences* 3 (8): 18–22. http://s-o-i.org/1.15/ijarbs-2016-3-8-4.

Supreeth, M, and Ns Raju. 2017. Biotransformation of Chlorpyrifos and Endosulfan by Bacteria and Fungi. *Applied Microbiology and Biotechnology* 101 (15): 5961–5971. https://doi.org/10.1007/s00253-017-8401–7.

United States Environmental Protection Agency. 2009. National Water Quality Inventory: Report to Congress. *Water*, (January): 43. http://water.epa.gov/lawsregs/guidance/cwa/305b/upload/2009_01_22_305b_2004report_2004_305Breport.pdf.

United States Environmental Protection Agency. 2017. National Water Quality Inventory: Report to Congress, EPA 841-R-16-011, no. August. http://www.ncbi.nlm.nih.gov/pubmed/2347274.

Velizarov, S, JG Crespo, and MA Reis. 2004. Removal of Inorganic Anions from Drinking Water Supplies by Membrane Bio/Processes. *Reviews in Environmental Science and Biotechnology* 3 (4): 361–380. https://doi.org/10.1007/s11157-004-4627–9.

Wang, R, S Wang, Y Tai, R Tao, Y Dai, J Guo, Y Yang, and S Duan. 2017. Biogenic Manganese Oxides Generated by Green Algae Desmodesmus Sp. WR1 to Improve Bisphenol A Removal. *Journal of Hazardous Materials* 339: 310–319. https://doi.org/10.1016/j.jhazmat.2017.06.026.

Yadav, KK, N Gupta, V Kumar, and JK Singh. 2017. Bioremediation of Heavy Metals from Contaminated Sites Using Potential Species: A Review. *Indian Journal of Environmental Protection* 37 (1): 65–84.

Yao, J, W Li, F Xia, Y Zheng, C Fang, and D Shen. 2012. Heavy Metals and PCDD/Fs in Solid Waste Incinerator Fly Ash in Zhejiang Province, China: Chemical and Bio-Analytical Characterization. *Environmental Monitoring and Assessment* 184 (6): 3711–3720. https://doi.org/10.1007/s10661-011-2218-0.

6 Microbial Community Analysis of Contaminated Soils

Charles Oluwaseun Adetunji
Edo State University Uzairue

Ruth Ebunoluwa Bodunrinde
Federal University of Technology Akure

Abel Inobeme
Edo State University Uzairue

Kshitij RB Singh
Banaras Hindu University

John Tsado Mathew
Ibrahim Badamasi Babangida University

Olugbemi T. Olaniyan
Rhema University

Ogundolie Frank Abimbola
Baze University

Jay Singh
Banaras Hindu University

Vanya Nayak and Ravindra Pratap Singh
Indira Gandhi National Tribal University

CONTENTS

DOI: 10.1201/9781003354147-6

6.1 INTRODUCTION

Soil pollution also known as soil contamination is characterised by the presence of various contaminants which include toxic compounds, benzene, radioactive wastes, heavy metals, agricultural wastes, chemicals, biological pathogens and petroleum products, accumulating in the soil (Fergusson, 1990; Duffus, 2002; Bradl, 2002, He et al., 2005).

Petroleum products such as crude oil derivatives (diesel, kerosene, petrol, jet fuel) fertilisers, insecticides, asphalt, bitumen, oil, and petrochemical plastics are one of the leading soil pollutants in the world. These pollutants contain a reasonable quantity of heavy metals which include vanadium, iron (Fe), chromium (Cr), antimony (Sb), thallium (Ti), copper (Cu), arsenic (As), zinc (Zn), mercury (Hg), beryllium (Be), cadmium (Cd), manganese (Mn), molybdenum (Mo), lead (Pb), cobalt (Co), selenium (Se), and nickel (Ni). Though metals such as copper (Co), selenium (Se), and zinc (Zn) at lower concentrations are important to sustain the metabolic function of the body, at higher concentrations, they become toxic (Pacyna, 1996; WHO, 1996).

Microbial population of soil contaminated with petroleum hydrocarbons through their metabolic and physiological activities has been involved in the breaking down of hydrocarbons and other toxic end products through a process called biodegradation. However, the heavy metals contained in this contaminated soil can be absorbed in a process called bioaugmentation (Kapahi and Sachdeva, 2019). The microbes can be biostimulated to aid the process of bioremediation, and also, plants could be utilised for the purpose of clean-up of the heavy metals present through the process of phytoremediation.

Microbes present in this environment that aids these processes can be identified based on their conserved regions, functional genes and traditionally morphological and biochemical characteristics. Improvement in technology has resulted in a number of new methods over the years to address the demerits of the traditional (morphological and biochemical characteristics) methods. The use of polymerase chain reaction (PCR)-based identification such as 16S RNA sequencing through Sanger sequencing, and next-generation sequencing (Clarridge, 2004).

In addition to these, several new approaches have been adopted for analysing microbial community, techniques such as metatranscriptomics, genome binning, high-throughput image analysis, single-cell genome sequencing, and metagenomics (Kunin et al., 2008; Shakya et al., 2019). Other approaches including cell sorting, metaproteomics, nanoSIMS (nanoscale mass spectrometry), flow cytometry, and metabolomics are used for accessing microbial activity at single-cell level or bulk and also for understanding their origin through their phylogeny (Hugerth and Andersson, 2017). Hence, this chapter elaborates on microbial community analysis of soils contaminated with nickel, arsenic, chromium, and petroleum hydrocarbon.

6.2 APPLICATION OF NOVEL TECHNIQUES FOR EVALUATION OF TAXONOMIC AND FUNCTIONAL PROPERTIES THAT COULD LEAD TO REMEDIATION OF HEAVY METAL-POLLUTED SOIL

As a result of anthropogenic (industrial) activity as well as natural processes, heavy-metal buildup in soil has rapidly grown. Since heavy metals are non-biodegradable, possibly they can survive in the environment cause food poisoning via crop plants, and finally deposit in the human body by biomagnification. Contamination by means of heavy metals has posed a severe threat to the environment as well as human health as a result of its poisonous nature. As a result, land contamination must be remedied as soon as possible.

Phytoremediation is an environmentally benign strategy which can be a cost-effective solution for the effective management of heavy metal-polluted soil. However, a deeper knowledge of the mechanisms behind heavy-metal tolerance along with accumulation in plants is required to advance phytoremediation effectiveness. However, the devices by which heavy metals are taken up are translocation as well as detoxification in plants. The study concentrated on ways to improve phytostabilisation and phytoextraction efficiencies, such as microbe-assisted, chelate-assisted techniques, and genetic engineering (Yan et al., 2020; Shah, 2021).

Soil is a nonrenewable natural resource which must be managed carefully to meet the Sustainable Development Goals situated by means of the United Nations. On the other hand, agricultural as well as industrial activity is repeatedly hazardous to soil health and could spread heavy metal through to the soil surroundings, causing harm to humans and ecosystems. This review looks at methods that can lead to heavy-metal pollution of farming land, ranging from mine tailings runoff entering local irrigation channels to the incinerator and coal-fired power-plant emissions being deposited in the atmosphere. The discussion was made about how heavy-metal bio-geochemical changes in the soil affect their bioavailability. Microbial bioremediation and plant-based remediation are two biological solutions for polluted agricultural land repair that offer cost-efficient and long-term options to standard chemical as well as physical clean-up procedures.

Soil pollution as a result of human activity is a growing concern in today's globe. Several contaminants from various businesses are emitted into the surrounding soils. Heavy metals are non-biodegradable, poisonous, and persistent pollutants that have a negative impact on the ecological niche of all life forms, including humans. The harmful effects of these heavy metals on living beings can be attributed to various cellular and metabolic processes. These are known to induce a variety of physiological problems in humans, including renal, respiratory, and gastrointestinal issues. Heavy metals' bio-toxicity is determined by their concentration, bioavailability and chemical forms (Brevik et al., 2020; Shah, 2020).

Contamination through organic pollutants, hazardous waste, and heavy metals has had a negative effect on the natural environment, which has harmed humans. These pollutants are produced by both human-made and natural calamities like volcanic eruptions and hurricanes. Toxic metals in agricultural soils can build up and enter the food chain, posing a danger to food security. Traditional and physical approaches are costly and ineffective in locations where metal toxicity is minimal. Bioremediation is thus an environmentally benign and effective approach to reclaiming heavy metal-contaminated ecosystems by utilising the biological mechanisms of microbes and plants to eliminate harmful toxins. The hazardous impacts of heavy-metal contamination are discussed as the techniques used by bacteria and plants to remediate the environment. It also highlighted the importance of nanofibers methods and practices in enhancing microbial enzymes' ability to successfully degrade heavy metals at a quicker pace and current progress in microbial phytoremediation as well as bioremediation for heavy-metal removal from the ecosystem, along with limitations and prospects. On the other hand, to ensure environmental safety, stringent adherence to biosafety rules must be maintained while using biotechnological procedures (Ojuederie and Babalola, 2017).

Scientists have traditionally paid close attention to environmental issues. Pollution is a major environmental issue that has caused considerable health risks to humans. It has been discovered that sythentic pesticides have been discovered that they could pose a threat to the ecosystem. The harmful effects of heavy metals [cadmium (Cd), lead (Pb), copper (Cu) as well as zinc (Zn)], as well as pesticides (herbicides, fungicides and insecticides), on the agricultural environment (soil and plant) and human health, are the subject of this review. Additionally, disorders induced by heavy metals and pesticides have been recorded in humans. Both heavy metals and insecticides have bioaccumulation, mode of action, and transmission channels that are highlighted. In addition, the bioavailability of these pollutants in soil and their uptake by plants have been taken into account. Previous relevant studies are presented to cover all the topics discussed in this chapter elaborately. The information in this review goes into great detail about environmental toxicants and their potentially dangerous effects (Delgado-Baquerizo et al., 2020).

6.3 THE PROCESS INVOLVED IN THE MICROBIAL COMMUNITY ANALYSIS OF CONTAMINATED SOILS

Igiri et al. (2018), in their review, documented the need to safeguard the decreasing population of microbial communities in the area contaminated by various environmental organic and inorganic pollutants. They focused on recent advances in microbial-based remediation of these contaminants such as heavy metals, especially cadmium, chromium, lead, and mercury, which do not have any known function in biological systems but are toxic even at low doses. They also discussed at length strategies for evaluating the bioremediation potentials of microorganisms. Similarly, Cheema et al. (2015) evaluated the variations in the community of microorganisms in agricultural soils that have been contaminated with hydrocarbons through the use of an independent culture approach of 16S rRNA sequencing gene analyses. They reported varying species of bacteria in the soil samples investigated, which included *Firmicutes*, *Bacteriodetes*, *Planctomycetes*, *Proteobacteria*, *Armaticmonadetes* and *Verrrucomicrobia*. The most dominant group was observed to be in the family *Proteobacteria*.

Mejeha (2016) investigated the effects of lead, nickel and cadmium pollution on the degradation of petroleum hydrocarbons biologically in the complex system of soil through the use of the microbial method combined with a geochemical approach. The findings from their study showed that nickel does not affect the degradation of the hydrocarbons in the soil environment. However, there was a stimulatory impact in nickel-porphyrin-polluted soils, which decreased with a rise in nickel concentration. The extent of diversity of the microorganisms within the community showed the dominant species of microorganism *Rhodococcus*. There were two dominating species in the lead-contaminated soil analysed, both of which are strains of *Bacillales*.

Azarbad et al. (2016) opined that several studies on microbial ecological assessment had shown remarkable variations in the compositions of microorganisms brought about by environmental contaminations by heavy metals and organic contaminants. Furthermore, various groups of microorganisms have also been shown to display different kinds of adaptations to survive such conditions. They focused on studies carried out in areas with prolonged pollution records. They concluded that the history of various environmental conditions plays a vital role in microbial communities' adaptive responses towards sensors.

Sarkar et al. (2016) characterised communities of microorganisms around refinery wastes and evaluated the biological stimulation based on *in situ* bioremediation. The samples around the lagoon containing wastewater from the refinery showed the presence of a high concentration of petroleum hydrocarbons. They also observed an enhancement in the intrinsic biodegradation potentials of the microorganisms found around. Findings from intensive sequencing of the 16S rRNA gene revealed the native community was made up mainly of methanogenic, hydrocarbon-degrading, nitrate-reducing and syntropic bacteria.

Ventorino et al. (2018) reported the capacity of inorganic and organic xenobiotic agents to affect the soil's ecological functions, hence affecting biodiversity. Their study utilised a sequencing approach in the investigation of the fungal and bacteria community structure and contaminants on the richness and diversity of polluted and non-polluted soils. They observed a significant shift in the microbial community in the contaminated soil. Furthermore, the statistical analysis revealed that the indigenous microbial communities were affected by the pollution and not the origin of the site.

Joynt et al. (2006) investigated the impact of prolonged exposure of the environment to organic contaminants and heavy metals on microorganisms' community. They collected soil samples in an area polluted with paints and solvents for paint production. Analysis was done, which indicated the presence of chromium and lead together with some organic compounds, especially toluene. Furthermore, they observed a weak association between the organic carbon content of the soil and the biomass of microorganisms. The result of the 16S rRNA gene confirmed the presence of various groups of bacteria such as beta, alpha and delta Proteobacteria. In a related study, Salam et al. (2017) analysed the microbial community's structural composition in an area contaminated with used engine oil. They set up microcosms of soil for a proper understanding of the effect of the spent engine. The taxonomic and functional parameters were assessed for metagenomic DNA extracts. The results showed the dominance of some groups of microorganisms, including *Pseudomonas*, *Burkholderia*, *Proteobacteria*, *Geodermatophilus* and *Actinobacteria*. Further analysis revealed the adaptation of the microbial community to varying stressors within the environment such as heavy metals, hydrocarbons, starvation in nutrients and oxidative stress. There was a significant decline in the population of microbiota that are nonhydrocarbon degrading in nature.

Zainab et al. (2015) showed remediation of heavy metal into three groups: gentle *in situ* remediation, harsh *in situ* soil restrictive measure, and *in situ* or *ex situ* harsh soil destructive measure. The main principle of the two hash methods can be hazardous either to man, animals or plants. The gentle measure does restoration of polluted soil. It has been validated that toxic materials could cause environmental pollution when they are released into the environment via anthropogenic activities The ecosystem's main sources of these heavy metals are during the production of energy, discharge from vehicles, industrial wastes, agrochemicals and particulates. Zainab et al. (2015) concluded that remediation of heavy metals is usually carried out by physical elimination. They added that the main purpose of this technique was to lower bioavailability and discharge of chemicals and hazardous materials by the production of advanced eco-friendly materials.

6.3.1 Phytoremediation

This technology utilises green plants, microorganisms and other soil amendments to clean up or reduce or render a toxic environment harmless, as illustrated in Figure 6.1 (Das et al., 2018; Feng et al., 2017). The technique of phytoremediation is cost-effective when compared with the traditional technique of *in situ* and *ex situ*. This technique gives a chance for the recovery of material and an opportunity for re-use of those materials. Due to phytoremediation, topsoil is usually not affected, thereby maintaining soil fertility (Ali et al., 2013). This technique was also reported to reduce erosion and leaching of metal in the soil. Othman and Leskovar (2018) reported that the phytoremediation technique helps increase soil health, improve crop yields, and preserve phytochemicals inherent in the soil. Schnoor et al. (1995) reported the effectiveness of this technique in areas with shallow contaminants. Chien et al. (2015) and Quinn et al. (2001) used this technique to treat wastewater from swine forms by the application of *Eichhornia crassipes*, and also Doni et al. (2015) used *Paspalum vaginatum* and other plant species to perform compost modification and phytoremediation, which degrades marine sediment of metallic pollutants.

Since phytoremediation has been found to be surface-limited and the depth occupied by roots is also limited, this technique is not self-sufficient in the complete protection against the leaching of contaminants into groundwater. The percentage of plant survival on this technique is meagre as plants could be affected by the toxicity of the contaminated soil and the land condition. When there is bioaccumulation of toxic metals/contaminants in the soil, it could hamper the plants' yields, and the plant's consumption could lead to serious economic/health challenges. Extraction of plants in these areas could also prove abortive when the plants have already been exposed to heavy metals. These metals could become attached to the organic matter of the soil. Farraji et al. (2016) reported that there could be accumulation of pollutants in fruits and other parts of the plant, crops and vegetables that could be edible. Due to reduced biomass, there is a need to decontaminate frequently by planting and harvesting (Pilipovic et al., 2008). Chelate-enhanced phytoremediation was found to cause environmental pollution (Römkens, 2002; Melo et al., 2008; Nam et al., 2008). Chintakovid et al. (2008) reported the slow nature and seasonal application of the method. When contaminants are dissolved in groundwater, it renders the water bodies unsuitable for the aquatic population due to pollution (Van Den Bos, 2002).

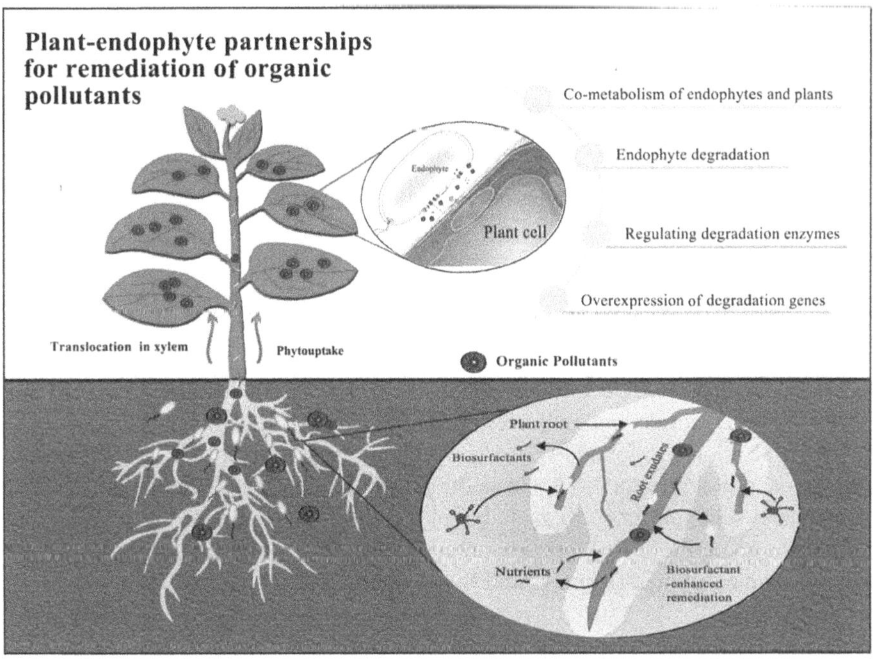

FIGURE 6.1 Communication between plant and endophyte for phytoremediation of soils contaminated with various organic pollutants. (Reprinted with permission from Feng, N. X., Yu, J., Zhao, H. M., Cheng, Y. T., Mo, C. H., Cai, Q. Y.,… & Wong, M. H. (2017). *Efficient phytoremediation of organic contaminants in soils using plant–endophyte partnerships. Science of the Total Environment*, *583*, 352–368.)

6.3.2 Phytoextraction

Phytoextraction is defined as the use of plants to decrease the amount of pollutants in the soil. These contaminants are concentrated in a smaller volume at the disposal period than the initial volume of the pollutants. Although Pilon-Smits (2005) reported that there was a lower degree of contaminants remaining in the soil after harvest, then they stated that this technique could only be effective if the process of planting/harvest is done in the cycle to have a significant clean-up. If this is not done, the inherent pollutants could kill plants, e.g., chromium in the soil was toxic to higher plants at 100 μm concentration (Shanker et al., 2005).

6.3.3 Phytostabilisation

The reduction of the mobility of the heavy metals in the soil is known as phytostabilisation, which is accomplished by reducing wind-blown dust and decreasing soil erosion. Phytostabilisation was found to limit the leaching of the soil substance. Plants has potential to prevent contaminants from binding pollutants to soil particles for less available uptake by the plants or human beings (Lone et al., 2008). However, this technique only focuses on the sequestration of soil pollutants but not plant tissues.

6.3.4 Phytostimulation

It refers to the breaking of organic contaminants in the soil through microbial activity in the rhizosphere; therefore, it is also known as enhanced rhizosphere biodegradation or plant-assisted biodegradation. Pilon-Smits (2005) reported phytostimulation to degrade hydrocarbons from PAHs and PCBs effectively. It was also found to improve microorganisms' populations that degrade in aquatic plants (Rupassara et al., 2002).

6.3.5 Phytovolatilisation

Phytovolatilisation is the uptake and transpiration of different contaminants through the plant, further released into the atmosphere. Phytovolatilisation exists in two different forms: direct and indirect phytovolatilisation, as shown in Figure 6.2. Phytovolatilisation was found to be effective via the removal of toxic soil substances with the release of air. Plants also have the potential to bioremediate the presence of mercury and selenium that are available in the soil. The limitation of this was that plants take up some of these toxic contaminants, thereby leading to plant/human toxicity when consumed (Limmer and Burken, 2016).

6.3.6 Immobilisation Technique

This technique is known to decrease metal mobility by introducing immobilising agents to contaminated soils. This technique was used by Shahid et al. (2014), and Ashraf et al. (2016) by organic and inorganic amendments to the soils. This technique immobilises heavy metals by forming stable compounds (Sabir et al.,

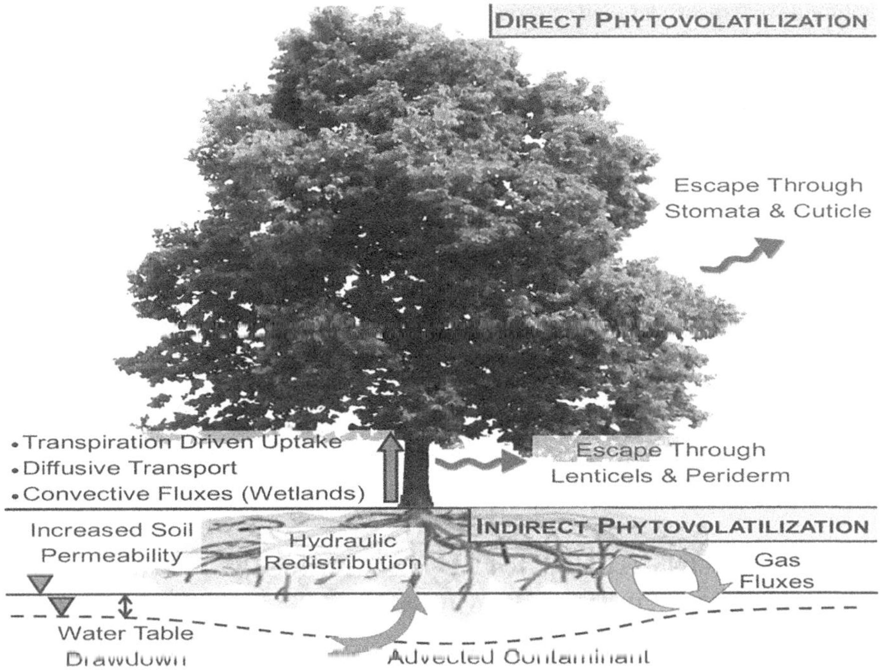

FIGURE 6.2 Schematic illustration of the direct and indirect photovolatisation procedures. (Reprinted with permission from Limmer, M., & Burken, J. (2016). *Phytovolatilisation of organic contaminants. Environmental science & technology*, *50*(13), 6632–6643.)

2015). Venegas et al. (2015) and Shakoor et al. (2015) reported this technique's importance as an adsorbent for solubilising heavy metals in the soil. Bolan et al. (2014) utilised manure from farmland to immobilise Fe, Ni, Cr, Pb and Mn and found di-ammonium phosphate to be more efficient for stabilising Zn, Cd and Cu in the soil. They reported that this process helps to improve soil organic content. It is fast and easy to apply. It is relatively low cost and easy to operate (Wuana and Okieimen, 2011). Moreover, Cele and Maboeta (2016) reported the setback of this method as the biosolid being a source of heavy-metal contamination of soil. It is highly invasive to the environment. There must be a fit land site for byproducts to be stored. Pollutants' activation could occur when the physicochemical feature of soil changes; for this process to be effective, there is a need for permanent monitoring (Martin and Ruby, 2004; USEPA, 1997).

6.3.7 Soil Washing

Research stated that this technique removes heavy metals from the soil using several reagents and extracts, thereby helping to percolate the heavy metals from the soil. This method is advantageous in the sense that it removes metals completely from the soil. Park and Son (2016) and Khalid et al. (2017) added that soil washing is a quick technique that meets detailed criteria without any delay in action. It is also considered a cost-efficient method of bioremediation technique. Ethylenediaminetetraacetic acid (EDTA) was found to be the most efficient conventional chelating agent suitable for soil washing (Udovic and Lestan, 2010). Wood (1997) reported the *ex situ* application of the soil washing technique to remove contamination completely. The technique does quick clean up and produces recyclable materials or energy. It is completely non-toxic and can be utilised at ambient temperature. The efficiency of soil washing is dependent on the kind of heavy metal and site for the remediation (Liao et al., 2015). When an extractant does not possess a high degree of extracting properties, it would not be suitable for soil cleaning. The contaminants are not located, and there is a need for the contaminants to move to another location; for proper monitoring, it could be hazardous to spread soil that is contaminated, dust particles during removal and contamination of contaminated soil. When there is a need to remove or purify large quantities of soil, excavation could be expensive (Liao et al., 2015). The effectiveness is site-dependent.

6.4 NUMEROUS TECHNIQUES FOR TAXONOMIC PROFILING OF THE SOIL MICROCOSMS

Meyer et al. (2008) used MG-RAST to determine functional and taxonomical relationships between mutagenic arrangements. These techniques utilise the SEED algorithm for analysing the standard pipeline of metagenomic deoxyribonucleic acid (DNA) sequence. Peng et al. (2012) used MG-RAST in the taxonomic and functional comments by observing the similarities within the MNR database during their work. The protein M5NR database that MG-RAST utilised was composed of non-reducing agent protein and ribosomal RNA sequence initiating from several databases. Fang et al. (2013) used a biodegradation gene database in probing for genes that could have the ability to degrade biologically. This technique (BDG) has 50,000 non-reducing agent protein sequences alongside their NCBI accessors, making it effective. The BDG comprises some gene families such as *alkb*, *hpb*, *bphA2*, *dxnA*, *glx*, *npah*, and *ppo*. They reported using the KEGE pathway, which reveals the required enzymes for the metabolism of catechol and benzoate via hydroxylation. They found some enzymes with matching genes MG-RAST ID 4514941 and other enzymes that originated in Illumina read MG-RAST ID 449432.3. All the enzymes needed for the conversion were inherent. MG-RAST was utilised to determine taxonomical and functional association for a sequence in a database from Illumina sequencing and assembled contigs. These sequences produce increased confidence hits when compared with the raw Illumina reads. They reported that this could be because assembly produces expanded sequences longer than 100 nbp that comprise raw reads and that there is a possible reason for procuring longer than a database entry. Taxonomic clusters that were identified via the use of MG-RAST were consistent between contigs (Fang et al., 2013).

Proteobacteria sp. were the most prevalent microbial group created from the shotgun sequence with close to 90% of arrangement from raw reads and also assembly reads. *Pseudomonas* showed 30% on both datasets. They concluded that this could be due to the prevalence of *Pseudomonas* in the aerobic matter (Labana et al., 2007; Silby et al., 2011). They reported that the naphthalene or anthracene pathway was incomplete conferring to KEGG observations, but findings of Whynot (2009) indicated that chemical evidence from bio-slurry experiments of soil shows that anthracene was completely easily degraded by the same microbial population. *Pseudomonas* spp. was the most prevalent microorganism in accordance with MG-RAST and SIGEX clones. *Mycobacterium* and *Sphingomonas* spp. were found to possess reduced biodegradable genes of 0.2% and 0.3%, respectively. When MG-RAST datasets of gene annotation were utilised as aromatic metabolism via a named subsystem analysis, about 2.5% of the entire features recognised by raw reads discovered that there was increase in the quantity to 41%. They concluded that enriched aromatic metabolism was present when compared with pristine oil (Martineau et al., 2010).

6.5 CONCLUSION AND PROSPECTS

This chapter provided detailed information on the microbial community analysis of soils contaminated with nickel, arsenic, chromium, and petroleum hydrocarbon. Relevant information was also provided on the application of novel techniques for evaluation of taxonomic and functional properties that could lead to remediation of heavy metal-polluted soil as well as the process involved in the microbial community analysis of soils contaminated. Detailed information was provided on bioremediation techniques such as phytoextraction, phytostabilisation, phytostimulation, phytovolatilisation, immobilisation technique, and soil washing. Relevant information was also provided on numerous techniques for taxonomic profiling of the soil microorganisms The application of nanotechnology could also be a sustainable biotechnological technique that could be applied as an effective bioremediation technique for remediation of soils contaminated with nickel, arsenic, chromium, and petroleum hydrocarbon.

REFERENCES

Ali H, Khan E, Sajad MA (2013) Phytoremediation of heavy metals—Concepts and applications. *Chemosphere* 91 (7), 869–881. doi: 10.1016/j. chemosphere.2013.01.075.

Ashraf A, Bibi I, Niazi NK, Ok YS, Murtaza G, Shahid M, Kunhikrishnan A, Mahmood T (2016) Chromium(VI) immobilisation efficiency of acid-1 activated banana peel over organo-montmorillonite in aquatic environments. *Int J Phytoremed.* doi: 10.1080/15226514.2016.1256372.

Azarbad H, Van Gestel CAM, Niklińska M, Laskowski R, Röling WFM, Van Straalen NM (2016) Resilience of soil microbial communities to metals and additional stressors: DNA based approaches for assessing "stress-on-stress" response. *Int J Mol Sci*, 17 (6), 933. https://doi.org/10.3390/ijms17060933.

Bolan N, Kunhikrishnan A, Thangarajan R, Kumpiene J, Park J, Makino T, Kirkham MB, Scheckel K (2014) Remediation of heavy metal(loid)s contaminated soils – To mobilize or to immobilize? *J Hazard Mater* 266, 141–166.

Bradl H, editor (2002) *Heavy Metals in the Environment: Origin, Interaction and Remediation*, vol. 6. London: Academic Press.

Brevik EC, Slaughter L, Singh BR, Steffan JJ, Collier DD, Pereira P (2020) Soil and human health: Current status and future needs. *Air, Soil Water Res* 13, 1–23. doi: 10.1177%2F1178622120934441.

Cele EN, Maboeta M (2016) A greenhouse trial to investigate the ameliorative properties of biosolids and plants on physicochemical conditions of iron ore tailings: Implications for an iron ore mine site remediation. *J Environ Manag* 165, 167–174.

Cheema S, Lavania M, Lal, B (2015) Impact of petroleum hydrocarbon contamination on the indigenous soil microbial community. *Ann Microbiol* 65, 359–369. https://doi.org/10.1007/s13213-014-0868-1.

Chien C, Yang Z, Cao W, Tu Y, Kao C (2015) Application of an aquatic plant ecosystem for swine wastewater polishment: A full-scale study. *Desalination Water Treat*, 1–10.

Chintakovid W, Visoottiviseth P, Khokiattiwong S, Lauengsuchonkul S (2008) Potential of the hybrid marigolds for arsenic phytoremediation and income generation of remediators in Ron Phibun District, Thailand. *Chemosphere* 70, 1532–1537.

Clarridge JE 3rd (2004) Impact of 16S rRNA gene sequence analysis for identification of bacteria on clinical microbiology and infectious diseases. *Clin Microbiol Rev*, 17 (4), 840–862. https://doi.org/10.1128/CMR.17.4.840-862.2004.

Das PK (April 2018) Phytoremediation and nanoremediation: Emerging techniques for treatment of acid mine drainage water. *Defence Life Sci J* 3 (2), 190–196. doi: 10.14429/dlsj.3.11346.

Delgado-Baquerizo M, Reich PB, Trivedi C et al. (2020) Multiple elements of soil biodiversity drive ecosystem functions across biomes. *Nat Ecol Evol* 4, 210–220. doi: 10.1038/s41559-019-1084–y.

Doni S, Macci C, Peruzzi E (2015) Heavy metal distribution in a sediment phytoremediation system at pilot scale. *Ecol Eng* 81, 146–157.

Duffus JH (2002) Heavy metals-a meaningless term? *Pure Appl Chem*, 74 (5), 793–807.

Fang H, Cai L, Yu Y, Zhang T (2013) Metagenomic analysis reveals the prevalence of biodegradation genes for organic pollutants in activated sludge. *Bioresour Technol* 129, 209–218.

Farraji H, Zaman NQ, Tajuddin RM, Faraji H (2016) Advantages and disadvantages of phytoremediation: A concise review. *Int J Env Tech Sci* 2, 69–75.

Feng NX, Yu J, Zhao HM et al. (2017) Efficient phytoremediation of organic contaminants in soils using plant–endophyte partnerships. *Sci Tot Environ*, 583, 352–368.

Fergusson JE, editor (1990) *The Heavy Elements: Chemistry, Environmental Impact and Health Effects*. Oxford: Pergamon Press.

He ZL, Yang XE, Stoffella PJ (2005) Trace elements in agroecosystems and impacts on the environment. *J Trace Elem Med Biol*, 19 (2–3), 125–140.

Hugerth LW, Andersson AF (2017). Analysing microbial community composition through amplicon sequencing: From sampling to hypothesis testing. *Front Microbiol*, 8, 1561.

Igiri B, Okoduwa S, Idoko GO, Akabuogu EP, Adeyi AO, Ejiogu IK (2018) Toxicity and bioremediation of heavy metals contaminated ecosystem from tannery wastewater: A review. https://doi.org/10.1155/2018/2568038.

Joynt J, Gray M, Turco R, Nakatsu CH (2006) Microbial ecology 51 (2), 209–219. doi: 10.1007/s00248-005-0205-0.

Kapahi M, Sachdeva S (2019) Bioremediation options for heavy metal pollution. *J Health Pollut*, 9 (24), 191203. https://doi.org/10.5696/2156-9614-9.24.191203.

Khalid S, Shahid M, Niazi NK, Murtaza B, Bibi I, Dumat C (2017) A comparison of technologies for remediation of heavy metal contaminated soils. *J Geochem Explor* 182 (part B), 247–268. ISSN 0375-6742.

Kunin V, Copeland A, Lapidus A, Mavromatis K, Hugenholtz P (2008) A bioinformatician's guide to metagenomics. *Microbiol Mol Biol Rev: MMBR*, 72 (4), 557–578. https://doi.org/10.1128/MMBR.00009-08.

Labana S, Kapur M, Malik DK, Prakash D, Jain R (2007) Diversity, biodegradation and bioremediation of polycyclic aromatic hydrocarbons. In *Environmental Bioremediation Technologies*, pp. 409–443. Springer.

Liao X, Li Y, Yan X (2015) Removal of heavy metals and arsenic from a co-contaminated soil by sieving combined with washing process. *J Environ Sci*, 1–9.

Limmer M, Burken J (2016) Phytovolatilization of organic contaminants. *Environ Sci Technol* 50 (13), 6632–6643. doi: 10.1021/acs.est.5b04113. ISSN 0013-936X.

Lone MI, He Z-l, Stoffella PJ, Yang X-e (2008) Phytoremediation of heavy metal polluted soils and water: Progresses and perspectives. *J Zhejiang Univ Sci B* 9 (3), 210–220. doi: 10.1631/jzus.B0710633. ISSN 1673-1581.

Martin TA, Ruby MV (2004) Review of in situ remediation technologies for lead, zinc and cadmium in soil. *Remediation* 14 (3), 35–53.

Martineau C, Whyte LG, Greer CW (2010) Stable isotope probing analysis of the diversity and activity of methanotrophic bacteria in soils from the Canadian high Arctic. *Appl Environ Microbiol* 76, 5773 5784.

Mejeha OK (2016) Biodegradation of petroleum hydrocarbons in soils cocontaminated with petroleum hydrocarbons and heavy metals derived from petroleum, PhD thesis, New Castle University.

Melo ÉECd, Nascimento CWAd, Accioly AMdA, Santos ACQ (2008) Phytoextraction and fractionation of heavy metals in soil after multiple applications of natural chelants. *Sci Agr* 65, 61–68.

Meyer F, Paarmann D, D'Souza M, et al. (2008) The metagenomics RAST server - a public resource for the automatic phylogenetic and functional analysis of metagenomes. *BMC BioInform* 9, 386.

Nam Y-S, Park Y-J, Lee I-S, Bae B-H (2008) A comparative study on enhanced phytoremediation of Pb contaminated soil with phosphate solubilizing microorganism (PSM) and EDTA in column reactor. *J Korean Soc Environ Eng* 30, 500–506.

Ojuederie O, Babalola O (2017) Microbial and plant-assisted bioremediation of heavy metal polluted environments: A review. *Int J Environ Res Public Health*, 14 (12), 1504. doi: 10.3390/ijerph14121504.

Othman YA, Leskovar D (2018) Organic soil amendments influence soil health, yield, and phytochemicals of globe artichoke heads. *Biol Agric Hortic*, 1–10. doi: 10.1080/01448765.2018.1463292. S2CID 91041080.

Pacyna JM (1996) Monitoring and assessment of metal contaminants in the air. In *Toxicology of Metals*, LW Chang, L Magos, T Suzuli, Eds., pp. 9–28. Boca Raton, FL: CRC Press.

Park B, Son Y (2016) Ultrasonic and mechanical soil washing processes for the removal of heavy metals from soils. *Ultrason Sonochem.*

Peng Y, Leung HCM, Yiu SM, Chin FYL (2012) IDBA-UD: A de novo assembler for single-cell and metagenomic sequencing data with highly uneven depth. *Bioinformatics* 28, 1420–1428.

Pilipović A, Orlović, Saša G, Vladislava P-P, LeopoldGalić Z, Vasić V (2008) Environmental application of forest tree species in phytoremediation and reclamation. Needs and priorities for research and education in biotechnology applied to emerging environmental challenges in SEE countries 39.

Pilon-Smits E (2005) Phytoremediation. *Ann Rev Plant Biol* 56 (1), 15–39. doi: 10.1146/annurev.arplant.56.032604.144214. ISSN 1543-5008

Quinn J, Negri M, Hinchman R, Moos L, Wozniak J, Gatliff E (2001) Predicting the effect of deep-rooted hybrid poplars on the groundwater flow system at a large-scale phytoremediation site. *Int J Phytoremediat* 3, 41–60.

Rodrigo-Comino J, López-Vicente M, Kumar V, Rodríguez-Seijo A (2020) Soil science challenges in a new era: A transdisciplinary overview of relevant topics. *Air, Soil Water Res* 13, 1–17. https://doi.org/10.1177%2F1178622120977491.

Römkens P (2002) Potentials and drawbacks of chelate-enhanced phytoremediation of soils. *Environ Pollut* 116, 109–121.

Rupassara SI, Larson RA, Sims GK, Marley KA (2002) Degradation of atrazine by hornwort in aquatic systems. *Bioremediation J*, 6 (3), 217–224. doi: 10.1080/10889860290777576.

Sabir M, Waraich EA, Hakeem KR, Öztürk M, Ahmad HR, Shahid M (2015) *Phytoremediation, Soil Remediation and Plants*. Elsevier Inc. http://dx.doi.org/10.1016/B978-0-12-799937-1.00004-8.

Salam L, Obayori S, Nwaokorie F, Suleiman I, Mustapha R (2017) Metagenomic insights into effects of spent engine oil perturbation on the microbial community composition and function in a tropical agricultural soil. *Environ Sci Pollut Res Int*, 24 (8), 7139–7159. doi: 10.1007/s11356-017-8364-3.

Sarkar J, Kazy S, Gupta A, Dutta A, Mohapatra B, Roy A, Bera P, Mitra A, Sar P (2016) Biostimulation of indigenous microbial community for bioremediation of petroleum refinery sludge. *Front Microbiol,* 21. https://doi.org/10.3389/fmicb.2016.01407.

Schnoor JL, Light LA, McCutcheon SC, Wolfe NL, Carreia LH (1995) Phytoremediation of organic and nutrient contaminants. *Environ Sci Technol*, 29, 318A–323A.

Sebastian B, Chude V (2018) Regional soil partnerships reports on soil pollution regional status of soil pollution: Nigeria, West Africa. *Res J Environ Sci*, 3, 316–320.

Shah, MP (2020) *Advanced Oxidation Processes for Effluent Treatment Plants*. Elsevier.

Shah, MP (2021) *Removal of Emerging Contaminants through Microbial Processes*. Springer.

Shahid M, Xiong T, Masood N, Leveque T, Quenea K, Austruy A, Foucault Y, Dumat C (2014) Influence of plant species and phosphorus amendments on metal speciation and bioavailability in a smelter impacted soil: A case study of food-chain contamination. *J Soil Sediment*, 14, 655–665.

Shakoor M, Niazi N, Bibi I, Rahman M, Naidu R, Dong Z, Shahid M, Arshad M (2015) Unraveling health risk and speciation of arsenic from groundwater in rural areas of Punjab, Pakistan. *Int J Environ Res Public Health*, 12, 12371–12390.

Shakya M, Lo CC, Chain P (2019) Advances and challenges in metatranscriptomic analysis. *Front Genet*, 10, 904. https://doi.org/10.3389/fgene.2019.00904.

Shanker A, Cervantes C, Lozatavera H, Avudainayagam S (2005) Chromium toxicity in plants. *Environ Int*, 31 (5), 739–753. doi: 10.1016/j.envint.2005.02.003.

Silby MW, Winstanley C, Godfrey SAC, Levy SB, Jackson RW (2011) *Pseudomonas* genomes: Diverse and adaptable. *FEMS Microbiol Rev*, 35, 652–680.

Udovic M, Lestan D (2010) Fractionation and bioavailability of Cu in soil remediated by EDTA leaching and processed by earthworms (Lumbricus terrestris L.). *Environ Sci Pollut Res*, 17, 561–570.

USEPA (1997) Recent developments for in situ treatment of metal contaminated soils, Tech. Rep. EPA-542-R-97-004, USEPA.

Van Den Bos A (2002) Phytoremediation of volatile organic compounds in groundwater: Case studies in plume control. Draft report prepared for the US EPA Technology Innovation Office under a National Network for Environmental Management Studies Fellowship.

Venegas A, Rigol A, Vidal M (2015) Viability of organic wastes and biochars as amendments for the remediation of heavy metal-contaminated soils. *Chemosphere*, 119, 190–198.

Ventorino V, Pascale A, Adamo P et al. (2018) Comparative assessment of autochthonous bacterial and fungal communities and microbial biomarkers of polluted agricultural soils of the Terra dei Fuochi. *Sci Rep*, 8, 14281. https://doi.org/10.1038/s41598-018-32688–5

WHO/FAO/IAEA (1996) *Trace Elements in Human Nutrition and Health*. Geneva, Switzerland: World Health Organization.

Whynot C (2009) The efficacy of different bioremediation strategies in removing mutagenic hazard from contaminated soil. (Master's thesis). Retrieved from ProQuest Dissertations and Theses. (Accession Order No. MR47539).

Wood P (1997) Remediation methods for contaminated sites, In *Contaminated Land and Its Reclamation*, R Hester, R Harrison, Eds. Cambridge, UK: Royal Society of Chemistry.

Wuana RA, Okieimen FE (2011) Heavy metals in contaminated soils: A review of sources, chemistry, risks and best available strategies for remediation. *Int Scholar Res Network ISRN Ecol*, 2011, 20 pages.

Yan A, Wang Y, Tan SN, Mohd Yusof ML, Ghosh S, Chen Z (2020) Phytoremediation: A promising approach for revegetation of heavy metal-polluted land. *Front Plant Sci*, 11. doi: 10.3389/fpls.2020.00359.

Zainab S, Ali Jawaid SM, Vishen S, Verma S (2015) Remediation of heavy metals contaminated soil and soil washing. *i-manager's J Civil Eng*, 5 (3).

7 Microbe Performance and Dynamics in Activated Sludge Digestion

Charles Oluwaseun Adetunji
Edo State University Uzairue

Ogundolie Frank Abimbola
Baze University

Kshitij RB Singh
Banaras Hindu University

Olugbemi T. Olaniyan
Rhema University

Ruth Ebunoluwa Bodunrinde
Federal University of Technology Akure

Abel Inobeme
Edo State University Uzairue

John Tsado Mathew
Ibrahim Badamasi Babangida University Lapai

Jay Singh
Banaras Hindu University

Ravindra Pratap Singh
Indira Gandhi National Tribal University

DOI: 10.1201/9781003354147-7

CONTENTS

7.1 INTRODUCTION

The effective performance of the microbiota of activated sludge obtained in the treatment of wastewater can be attributed to the structure, function and diversity of the microorganisms present in the sludge. Activated sludge is often referred to as biosolids, and they contain rich medium with inorganic and organic pollutants, organic chemicals, and nutrients which can be easily degraded by microorganisms. The microbes found in sludge range from fungi, protozoa, algae to bacteria which normally have the highest population in the microbiome (Adetunji et al., 2019; Adetunji and Ugbenyen 2019; Adetunji et al., 2021; Adetunji and Anani 2021; Sangeetha et al., 2021; Dauda et al., 2022a,b). An activated sludge hence is a sludge processed by this consortium of microbes.

During sludge treatment, several microorganisms are involved in the spontaneous utilisation of the starting nutrients and metabolites in order to break down the solid components and reclaim water present in them, thereby improving the quality of the effluent. Also, during this stage bioaccumulation of inorganic materials in the sludge by the microbes occurs. For effective and efficient activation of the sludge, microbial diversity and structure are important. Changes in the sludge such as pH or rate of the dissolved oxygen can aid the overgrowth of *Nocardia* and *Actinomyces* and other filamentous bacteria and *Proteobacteri*a as a result of change adapted by microbes.

Anaerobic digestion observed during sludge action is one of the most promising technologies in wastewater/waste sludge management owing to the importance of its product (Nielsen et al., 2017). These processes utilise mesophilic sludge-based digesters where mesophilic microbes mediated process results in the activation of sludge, resulting in the production of biogas which normally contains methane gas used for electricity generation, cooking gas and fuel (Mata-Alvarez et al., 2014; Shah, 2021; Weiland, 2010; Nielsen et al., 2017).

Activated sludge can be treated in bioreactors when they involve the use of living cells. Several types of bioreactors have been in use over the years such as stirred-tank bioreactors (Hyde et al., 2019; Narayanan and Narayan, 2019), packed-bed biofilms (Willen et al., 2018), membrane bioreactors (Jo et al., 2016; Cheng et al., 2018, Zhang et al., 2019; Zheng et al., 2019), moving bed reactors with biofilm (Qureshi et al., 2005), fluidised-bed reactors (Escudié et al., 2006; Polin et al., 2019; Wang et al., 2021), and semifluidised bed biofilms (Wilen et al., 2018; Shah, 2020).

Therefore, this chapter intends to provide comprehensive information on the performance and microbial community dynamics in the anaerobic digestion of waste-activated sludge.

7.2 PROCESSES INVOLVED IN CORRELATING THE POPULATION DYNAMICS OF PATHOGENS SUCH AS MESOPHILIC SLUDGE DIGESTERS WITH SEVERAL PROCESS PARAMETERS

Ghasimi et al. (2015) utilised the pyrosequencing method to examine the population dynamics of archaeal and bacteria during their work. They closely monitored the variation in the microbial groups of the mesophilic reactor. The microbial taxonomic group was examined using a total of 18,000 orders which showed a disparity in the population between thermophile and mesophile digesters with respect to time. *Bacteroides* was the most predominant genus in the mesophilic digester subsequent to acclimation, with the difference in its comparative large quantity since a maximum of 90% to a minimum of 46% in some time. *Anaerolinea* was the second main genus as a mesophilic digester that accounts for 5%–24%, while *Parabacteroides* had 34% of large quantity initially and later reduced to 1%–12% over time when comparing its population with the entire bacterial community. Figure 7.1 represents the dynamics and relative abundance of bacterial and methanogenic community of the prevailing genus with the function of time.

For archaeal families in the mesophilic digesters, *Methanosaeta* genus had been discovered to have a proportion which ranges from 81% to 94% than *Methanobacterium* that thrives in various anaerobic biogas systems (Vanwonterghem et al., 2014). During their work, they found that the increase in the abundance of *Bacteroides* in large quantities was associated with the experiment in the accumulation through day-to-day research (Feng et al., 2009). Moreover, the community of archaea was reduced by 60% within days of the experimental setup, and syntrophic acetate oxidising (SAO) bacteria are specialised in oxidising acetate to CO_2 and H_2. They reported that despite these organisms' struggle to thrive, they cannot survive and could not hinder VFA accumulation; they said that this could be the reason to the high productivity of VFAs from fermenting *Bacteroides* (Feng et al., 2009). The mesophilic reactor bacteria were more abundant compared to archaea and SAO bacteria. They added that the changing trend of archaeal was found to be similar to bacteria per gram sludge, while SAO bacteria had quality far lower than the two microorganisms. The β-diversity result proved the presence of a bacterial community and stable archaea in mesophilic digesters. Vanwonterghem et al. (2014) suggested that determinants of the microbial community are very important when operational circumstances impose selective pressure. The role of abiotic and biotic factors in the industrial management of methanogenesis in the digestion system under anaerobic conditions cannot be overemphasised.

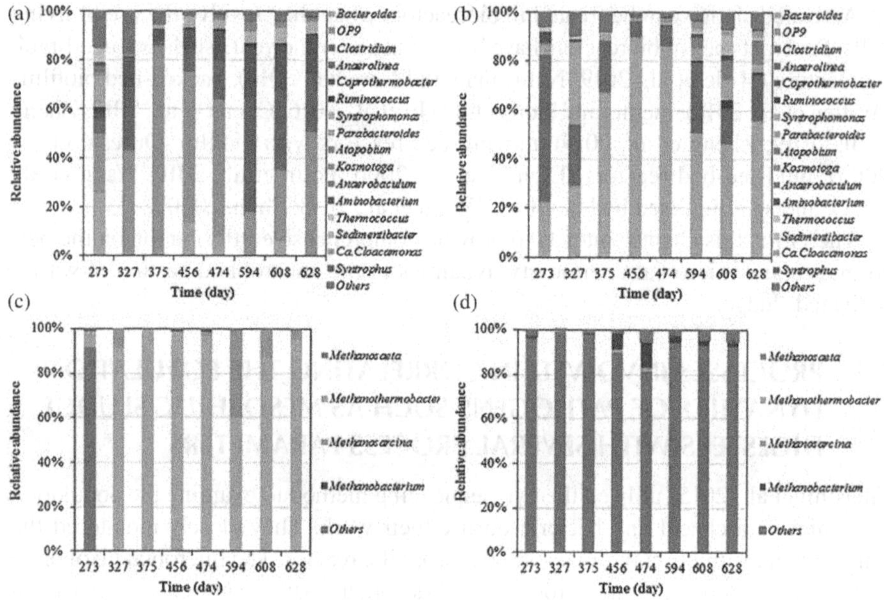

FIGURE 7.1 The relative and dynamics abundance of the dominant genera in the function of time. (a) Bacterial community in 55°C digesters; (b) bacterial community in 35°C digesters; (c) methanogenic community in 55°C digesters; (d) methanogenic community in 35°C digesters. (Reprinted with permission from Ghasimi, D. S., Tao, Y., de Kreuk, M., Zandvoort, M. H., & van Lier, J. B. (2015). *Microbial population dynamics during long-term sludge adaptation of thermophilic and mesophilic sequencing batch digesters treating sewage fine sieved fraction at varying organic loading rates. Biotechnology for biofuels*, *8*(1), 1–15).

Some abiotic factors such as pH and ammonia have been examined to advance the system's stability for the creation of biogas (Ju et al., 2017). Some researchers have recently applied phylogenetic analyses that target 16S rRNA genes to determine the constituent of the microbial system in a bioreactor (De Vrieze et al., 2015). The deterministic parameters guide the succession of the microbial community, thus predicting the trend in a shift under disturbance within the environment (Peces et al., 2018). Most related investigations have been done under pilot-scale bioreactors. The production of biogas such as carbon dioxide and methane through anaerobic digestion using various kinds of feedstocks occurs in a complex community of microbes and the microbiome present in the bioreactor. The anaerobic digestion process can be classified into steps: acidogenesis, acetogenesis, hydrolysis, and syntrophic methanogenesis (Lv et al., 2010). Every one of the phases is carried out by a specific group of microorganisms.

During hydrolysis, polysaccharides including starch, hemicellulose, lipids, as well as proteins are hydrolysed in the presence of hydrolases including xylanase, protease, amylase and cellulase produced through hydrolytic bacteria, which results in the release of oligomers or monomers such as xylose from hemicelluloses, maltose and glucose from starch, amino acids from proteins and glycerol from lipids. The bacteria involved in the process of hydrolysis are diverse phylogenetically, while *Bacteriodes* and *Firmicutes* are the two phyla that contain mainly the hydrolytic microbes that are established in the bioreactors. Generally, hydrolytic bacteria could produce hydrolases fast with lesser sensitivity to ecological modify like pH along with temperature. With the exception of the recalcitrant substrates like lignocelluloses, the hydrolytic step is not considered the rate-determining stage. The hydrolytic bacteria present in the bioreactor are capable of using the products of hydrolysis as the substrate for growth majorly through fermentation (Peces et al., 2018).

The fermentation of the hydrolytic products occurs in the presence of acidogenic microorganisms, giving rise to products such as butyrate, acetate, isobutyrates and propionate. Other products are formed during acidogenesis, such as sulphide, hydrogen and carbon dioxide. Acidogens consist of fermentative and hydrolytic bacteria, which lack hydrolytic potential (Lu et al., 2013). The microbiomes that produce biogas through the use of metagenomics have been investigated in several studies. The first mutagenic research was done in 2008, which used 454 pyrosequencing techniques and assessed the microbiome in terms of functional diversity in a full-scale bio-based reactor (Schluter et al., 2008).

7.3 SPECIFIC EXAMPLES OF MESOPHILIC SLUDGE DIGESTERS AND ARCHAEAL METHANOGENS

The operation of anaerobic digestion depends on microbial groups and communities of high complexity, which usually interact at a close range in the degradation of the waste biomass as well as organic material, resulting in carbon dioxide and methane production. The use of high-throughput methods with molecular techniques has largely expanded the knowledge of networks at trophic level and metabolic diversities involved in the operation of bioreactors (Kleinsteuber, 2019). The microbial breakdown of the complex biomass and organic materials into simple and relatively stabilised compounds, with methane and carbon dioxide being the major groups, is aided by different groups of microorganisms. Some of the communities include acetogenic, fermentative, methanogenic and syntrophic bacteria. These microbial groups adopt varying pathways in evading various harsh environmental conditions found in the digesters where they carry out their biodegradation processes. There are multiple forms of competition within the digesters between the bacteria that produce methane and the other groups that reduce sulphate (Qu et al., 2009).

Stolze et al. (2016) made a remarkable attempt to characterise the metagenomes for some full-scale biogas (out of which one was thermophilic while two were mesophilic) for the digestion of maize silage with manure. Current insights focusing on the compositions of the community of microorganisms in sludge are usually on syntrophic consortia, which are made of microorganisms that are metabolically distinct and strongly connected with methanogens (Narihiro et al., 2015). These microorganisms typically display resilience and resistance to harsh conditions, which is vital in maintaining robust anaerobic performance with time. Furthermore, each step is usually performed by a class of population that are unique taxonomically and redundant functionally but capable of coexisting and replacing each other in the sustenance of the performance of the process (De Vrieze et al., 2016).

The investigations in full-scale anaerobic digesters tend to only genetic information from the taxonomic perspective with limited details on functional potentials, which are capable of being complemented by remarkable throughput technologies that involve genomics for proper exploring of the community of different microorganisms and their structural and physiological status through ecology (Liu et al., 2015). The operation of an anaerobic digester with high stability is controlled by four different groups of bacteria existing in a dynamic equilibrium and takes part through the sequential method of digestion of the complex organic matter into simpler components that are then utilised by methanogens for the generation of methane. The structural and functional compositions of the microbial community are paramount and need to be optimised and intended for higher effectiveness. The transformation of complex polymeric materials into simple, reducing sugars and amino acids is brought about by the activities of hydrolytic microbes. Usually, bacteria are then accompanied by acetogenesis and acidogenesis for the generation of the various organic acids later converted into hydrogen gas, acetates, and carbon dioxide for methanogenesis (Kleinsteuber, 2019). Figure 7.2 shows the overall method of transferring complex

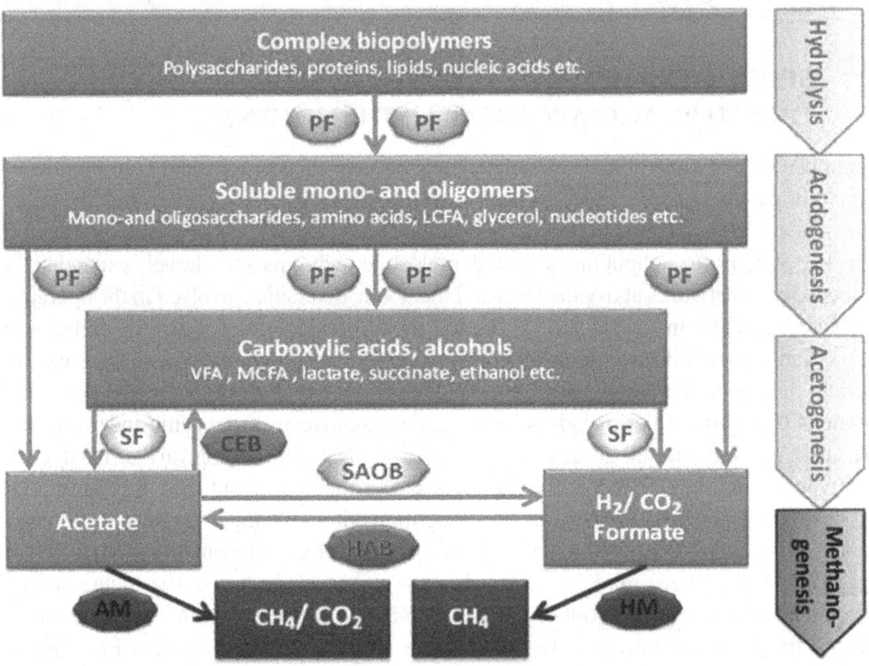

FIGURE 7.2 Schematic illustration of various steps involved in AD and shows different microbes' functional groups. The light grey colour represents all the processing stages carried out through bacteria, as well as their products are also specified in light grey. The dark grey colour represents the processing steps and the products performed by archaea. PF, primary fermenters (hydrolytic and acidogenic bacteria); SF, secondary fermenters (syntrophic proton-reducing bacteria); SAOB, syntrophic acetate oxidising bacteria; HAB, homoacetogenic bacteria (bacteria were performing reductive acetogenesis via the Wood–Ljungdahl pathway); CEB, chain-elongating bacteria (e.g., *Clostridium kluyveri*); AM, acetoclastic methanogens; HM, hydrogenotrophic methanogens. (Reprinted with permission from Kleinsteuber, S. (2018). *Metagenomics of methanogenic communities in anaerobic digesters. Biogenesis of hydrocarbons. Biogenesis of hydrocarbons, handbook of hydrocarbon and lipid microbiology*, 1–23).

biomolecules into methane under anaerobic circumstances in various metabolic stages such as hydrolysis, acidogenesis, acetogenesis and methanogenesis.

More recent studies have also revealed that fungi might have a very important responsibility in hydrolysis, thereby aiding the activities of pathogens in the breakdown of plant biomass (Bengelsdorf et al., 2012). The study of the different groups of the microbial communities beyond the prokaryotic components alone can be done using various techniques that are independent of culture but relying more on the analysis along with the study of 16S rRNA genes like sequence assessment and analysis of library clones, denaturing gradients, electrophoresis, sequencing of 16S amplicon and length polymorphism (Ariesyady et al., 2007). Kirkegaard et al. (2017) in their study utilised 32 full-scale anaerobic digesters found in wastewater treatment plants (WWTPs) which were examined with the aid of 16S rRNA amplicon sequencing, to be able to ascertain those microorganisms that were inherent, profuse and active in this process. Their application of principal component analysis showed the unique feature of the present mesophilic bacterial presence. Vrieze et al. (2015) reported that their findings were reliable with the bacterial groups influenced by the variation in temperature and ammonia present. They also added that despite that their survey took a longer period of 6 years, the digester communities inherent in individual wastewater were not altered with respect to time. Moreover, they also found that the predominant microbial population was the same throughout the reaction following the same technique (Rivière et al., 2009; Sundberg et al., 2013).

Parabacteroides are obligated anaerobic short rods that are constantly found in anaerobic ecology, generating many acids like formic acid, lactic acid, acetic acid as well as propionic acid through carbohydrates (Tan et al., 2012). Shin et al. (2019) accounted that one major prerequisite in the central improvement of anaerobic digesters is understanding the makeup and growth of the population of the microorganism due to the harmonious action and biological conversion of the feedstock of the microorganism (Karakashev et al., 2005). The investigation of De Vrieze et al. (2016) and Muller et al. (2016) showed that when culture-independent molecular techniques like high-throughput sequencing are used, there are disparities/variations in the communities of microorganisms in the anaerobic digesters. (Zhou et al., 2010). The operational conditions of the digester (Meng et al., 2018) and the type of substrate (Cerny et al., 2018) are crucial factors that regulate the population pattern of the microorganism in the anaerobic digester. Sometimes there is a need for the pre-treatment of the substance sludge of anaerobic digesters (Waclawek et al., 2018).

It has been validated that the co-digestion of diverse organic wastes process could improves the production of biogas by synergistic activities Al-Addous et al. (2018), but there must be the availability of a secondary substratum. Zhou et al. reported that waste-activated sludge contains mostly inactive aerobic cells originating from the activated sludge method. Shin et al. (2010) and Bolzonella et al. (2005) reported the utilisation of pre-treatment Carrère et al. (2016) or co-substrate Hidaka et al. (2015), which increased the production of CH_4. *Chloroflexi* which was found to be the primary bacterium dominant in anaerobic digesters on large scale was produced during a global research survey by Rivière et al. (2009). Stams et al. (2012) found Euryarchaeota (strict anaerobes) taking part in the methanogenesis process final stage of the anaerobic digester food chain during the methanogenesis process. Li et al. (2015) revealed anaerobic bacteria and amino-acid-degrading bacteria to be important members of anaerobic digesters and mesophilic organisms; *Mesotaga* was found to predominate in the sludge (Nesbo et al., 2012). The finding of Lee et al. (2018) mentioned that the lack of maximum effectiveness of the microorganisms used in their research could be due to the exposure of digesters and co-digesters to an unfavourable environment, which hampered their growth and ability.

Wang et al. (2018) recorded the colonisation of new microbes in sludge digestion and then investigated it using various theories of microbial ecology (Woodcock et al., 2017; Liu et al., 2019).

7.4 DIFFERENT TYPES OF BIOREACTORS USED IN WASTE-ACTIVATED SLUDGE FOR ANAEROBIC DIGESTION

There is a large amount of waste generated in the modern municipal plants used for wastewater treatment. Hence, the need for effective management of sludge made up of the primary and secondary becomes paramount, and this is one of the most pressing issues in the recent practices of waste treatment. Basically, the treatment plants utilise a great amount of energy for aeration and pumping, which are vital in the management system. However, it is possible to extract energy from the sludge and wastewater (McCarty et al., 2011). Different technologies for conversion of sludge to energy include gasification, pyrolysis, combustion, and anaerobic digestion (AD).

AD is a reliable approach proper for creating bioenergy such as biogas rich in methane using sewage sludge (Kobayashi et al., 2008). The various reactions involved are mediated through the activities of complex microbial groups that cooperate and, at some other times, competition for the utilisation of the organic substrate. Therefore, the understanding of the growth and composition of the different microorganisms is paramount for the enhancement of AD sludge since the conversion of the feedstock through biological means depends on the harmonious activities of the microorganisms present.

Out of the two major categories of sewage sludge in AD, the secondary sludge comprises unprocessed biomass, which remains when the biological treatment stage is over, like the activated sludge stage. The secondary sludge, also known as the waste-activated sludge, is mostly cellular biomass. The secondary sludge can be considered an organic substrate and an immigrant source that is the inactive group in the digester (Saunders et al., 2016). The organic components in the activated sludge have enormous content of cell walls and polymeric materials of extracellular origin hence capable of resisting biochemical attack from the environment; therefore, sludge digesters need a longer retention time. Some cells thrive inside the digester-producing enzymes during the anaerobic process, while others only function as substrate. Studies using culture-independent techniques at the molecular level have shown various microbial communities in the anaerobic digesters (Connelly et al., 2017).

7.5 TYPES OF BIOREACTORS FOR TREATMENT OF ANAEROBIC WASTES

Various types of bioreactors are engaged in wastewater management, in which contaminated water is recycled in an anaerobic or aerobic tank wherever immobilised cells or free poised microbes on a matrix are utilised to metabolise organic materials forming sludge that is discharged or recycled. The bioreactor is based on the pattern of solid particles through tiny diameters (1–4 mm) with the impulsive immobilisation of the anaerobic grouping, which is a vital condition for the efficient process of bioreactor.

7.5.1 Stirred-Tank Bioreactors

This approach is one of the ancient applications of biotechnology in industries. It involves the treatment of industrial effluents using a stirred-tank bioreactor. Aside from its inherent limitations, this method is still popular and has been subject to many diversifications and modifications (Hyde et al., 2019). The traditional process of activated sludge makes use of an aerobic tank which is a vessel usually agitated and seeded using microbial sludge as inoculums, and it is usually the portion of the active sludge that is recycled. The suspended growth of the microorganism takes place here. Air is kept at high pressure to make provision for adequate oxygen inside the medium. However, the extent of dissolution of atmospheric oxygen is low due to the large tank used for this purpose. Moreover, it is expensive to maintain air compressors in this approach, which is a major financial limitation of this type (Narayanan and Narayan, 2019).

7.5.2 Packed-Bed Biofilm

The approach is based on the growth of microorganisms in the medium. The reactors here are multiphased and are concerned with heterogeneous systems. They use support particles (downflow fixed film is an exception to this) like polymers beads, granules of silica, and activated carbon particles. The growth and multiplication of the microbial cells take place inside the biofilm. However, the film tends to detach away from the particles' surface when there is an increase in the thickness of the biofilm beyond a particular range of values ($\delta = 0.3–0.5$ mm). This detachment is known as sloughing and there is a replacement with new cells so that the thickness of the biofilm would remain constant during the bioreactor operation (Wilen et al., 2018).

7.5.3 Moving Bed Reactor with Biofilm

This is commonly used during the treatment of wastewater sludge. However, this reactor is not very common, and no particle beds are present, as suggested by the name. It is a bioreactor made of the stirred tank and usually fed with the particle aggregates of biofilm that remained suspended in the solution of the substrate within the stirred tank. The air compressed and sparged at high pressure helps in keeping the suspension of aggregates. As a result of the agitation produced by the stream of air, there is a movement of these aggregates around the bulk of the liquid, which is why the word moving bed biofilm is commonly used in the description of this system (Qureshi et al., 2005).

7.5.4 Fluidised-Bed Reactor

It is very suitable for a large-capacity installation because it can operate this reactor at a comparably high velocity of the fluid. The admittance of the industrial effluence occurs at the base of the column at a higher speed than the lowest fluidisation velocity but is lower when compared to the terminal velocity of free settling for each particle in the aggregate of the aggregate biofilm. As a result of this, all the aggregates become suspended, which is the basis of the fluidisation in the substrate solution. Also, channelling is absent because each aggregate is surrounded by the substrate. In a similar vein, there is an increase in the overall volume of the bioreactor as expansion occurs in the bed (Escudié et al., 2006).

7.5.5 Semifluidised Bed Biofilm

They are comparably newer when compared to the previous. This technology has outstanding advantages compared to traditional fluidised beds; however, they are also expensive to maintain in comparison. The liquid velocity required for the efficient operation is higher; hence, such reactors are operated at a higher capacity. As a result, there is high fluidisation of the bed (Wilen et al., 2018).

7.6 CONCLUSION AND PROSPECTS

This chapter provided detailed information on the performance and microbial community dynamics in the AD of waste-activated sludge. Relevant information was also provided on the several processes involved in correlating the population dynamics of pathogens such as mesophilic sludge digesters with several process parameters which includes archaeal methanogens. Different types of bioreactors used in waste-activated sludge for AD and information on different types of bioreactors for the treatment of anaerobic wastes were provided.

REFERENCES

Adetunji, C. O., Kumar, D., Raina, M., Arogundade, O., Sarin, N. B. (2019). Endophytic microorganisms as biological control agents for plant pathogens: a panacea for sustainable agriculture. In Varma A, Tripathi S, Prasad R (eds) *Plant Biotic Interactions*. Springer, Cham. https://doi.org/10.1007/978-3-030-26657-8_1

Adetunji, C. O., Ugbenyen, M. A. (2019). Mechanism of action of nanopesticide derived from microorganism for the alleviation of abiotic and biotic stress affecting crop productivity. In: Panpatte D., Jhala Y. (eds) *Nanotechnology for Agriculture: Crop Production & Protection*. Springer: Singapore. https://doi.org/10.1007/978-981-32-9374-8_7

Adetunji, C. O., Anani, O. A. (2021). Plastic-Eating Microorganisms: Recent Biotechnological Techniques for Recycling of Plastic. In: Panpatte, D. G., Jhala, Y. K. (eds) *Microbial Rejuvenation of Polluted Environment. Microorganisms for Sustainability*, vol 25. Springer, Singapore. https://doi.org/10.1007/978-981-15-7447-4_14

Adetunji, C. O., Olaniyan, O. T., Bodunrinde, R. E., Ahamed, M. I. (2021). Bioconversion of poultry waste into added-value products. In: Inamuddin, Khan A. (eds) *Sustainable Bioconversion of Waste to Value Added Products. Advances in Science, Technology & Innovation*. Springer, Cham. https://doi.org/10.1007/978-3-030-61837-7_21

Al-Addous, M., Saidan, M. N., Bdour, M., Alnaief, M. (2018). Evaluation of biogas production from the co-digestion of municipal food waste and wastewater sludge at refugee camps using an automated methane potential test system. *Energies*, 12, 32.

Ariesyady, H. D., Ito, T., Okabe, S. (2007). Functional bacterial and archaeal community structures of major trophic groups in a full-scale anaerobic sludge digester. *Water Res*, 41(7), 1554–1568. doi: 10.1016/j.watres.2006.12.036.

Bengelsdorf, F. R., Gerischer, U., Langer, S., Zak, M., Kazda, M. (2012). Stability of a biogas-producing bacterial, archaeal and fungal community degrading food residues. *FEMS Microbiol Ecol*, 84(1), 201–212. doi: 10.1111/1574–6941.12055

Bolzonella, D., Pavan, P., Battistoni, P., Cecchi, F. (2005) Mesophilic anaerobic digestion of waste activated sludge: Influence of the solid retention time in the wastewater treatment process. *Process Biochem*, 40, 1453–1460.

Carrère, H., Antonopoulou, G., Affes, R., Passos, F., Battimelli, A., Lyberatos, G., Ferrer, I. (2016). Review of feedstock pretreatment strategies for improved anaerobic digestion: From lab-scale research to full-scale application. *Bioresour Technol*, 199, 386–397.

Cˇerný, M., Víteˇzová, M., Víteˇz, T., Bartoš, M., Kushkevych, I. (2018). Variation in the distribution of hydrogen producers from the Clostridiales order in biogas reactors depending on different input substrates. *Energies*, 11, 3270.

Cheng, C., Zhou, Z., Pang, H., Zheng, Y., Chen, L., Jiang, L. M., Zhao, X. (2018). Correlation of microbial community structure with pollutants removal, sludge reduction and sludge characteristics in micro-aerobic side-stream reactor coupled membrane bioreactors under different hydraulic retention times. *Bioresour Technol*, 260, 177–185.

Connelly, S., Shin, S. G., Dillon, R. J., Ijaz, U. Z., Quince, C., Sloan, W. T., Collins, G. (2017). Bioreactor scalability: Laboratory-scale bioreactor design influences performance, ecology, and community physiology in expanded granular sludge bed bioreactors. *Front Microbiol*, 8, 664.

Dauda, W. P., Abraham, P., Glen, E., Adetunji, C. O. , Ghazanfar, S., Ali, S., Al-Zahrani, M., Azameti, M. K., Alao, S. E. L., Zarafi, A. B., Abraham, M. P., Musa, H. (2022a). Robust profiling of Cytochrome P450s (P450ome) in notable *Aspergillus* spp. *Life*, 12(3), 451. https://doi.org/10.3390/life12030451

Dauda, W. P., Morumda, D., Abraham, P., Adetunji, C. O., Ghazanfar, S., Glen, E., Abraham, S. E., Peter, G. W., Ogra, I. O., Ifeanyi, U. J., Musa, H., Azameti, M. K., Paray, B. A., Gulnaz, A. (2022b). Genome-wide analysis of Cytochrome P450s of *Alternaria* Species: evolutionary origin, family expansion and putative functions. *J Fungi*, 8(4), 324. https://doi.org/10.3390/jof8040324

De Vrieze, J., Raport, L., Roume, H., Vilchez-Vargas, R., Jáuregui, R., Pieper, D. H., Boon, N. (2016). The full-scale anaerobic digestion microbiome is represented by specific marker populations. *Water Res*, 104, 101–110.

De Vrieze, J., Saunders, A. M., He, Y., Fang, J., Nielsen, P. H., Verstraete, W., Boon, N. (2015). Ammonia and temperature determine potential clustering in the anaerobic digestion microbiome. *Water Res*, 75, 312–323.

Escudie, R., Epstein, N., John, G., Bi, H. (2006). Layer inversion phenomenon in binary-solid liquid-fluidized beds: Prediction of the inversion velocity. *Chem Eng Sci*, 61(20), 6667–6690. doi: 10.1016/j.ces.2006.06.008

Feng, L., Chen, Y, Zheng, X. (2009). Enhancement of waste activated sludge protein conversion and volatile fatty acids accumulation during waste activated sludge anaerobic fermentation by carbohydrate substrate addition: The effect of pH. *Environ Sci Technol*, 43, 4373–4380.

Ghasimi, D. S. M, Yu Tao, Y., Kreuk, M., Zandvoort, M. H., van Licr, J. B. (2015). Microbial population dynamics during long-term sludge adaptation of thermophilic and mesophilic sequencing batch digesters treating sewage fine sieved fraction at varying organic loading rates. *Biotechnol Biofuels*, 8, 171.

Hidaka, T., Wang, F., Tsumori, J. (2015). Comparative evaluation of anaerobic digestion for sewage sludge and various organic wastes with simple modeling. *Waste Manag*, 43, 144–151.

Hyde, K. D., Xu, J., Rapior, S., Jeewon, R., Lumyong, S., Niego, A. G. T., Stadler, M. (2019). The amazing potential of fungi: 50 ways we can exploit fungi industrially. *Fungal Divers*, 97, 1–136. https://doi.org/10.1007/s13225-019-00430-9.

Jo, S. J., Kwon, H., Jeong, S. Y., Lee, C. H., Kim, T. G. (2016). Comparison of microbial communities of activated sludge and membrane biofilm in 10 full-scale membrane bioreactors. *Water Res*, 101, 214–225.

Ju, F., Lau F., ZhangLinking, T. (2017). Microbial community, environmental variables, and methanogenesis in anaerobic biogas digesters of chemically enhanced primary treatment sludge. *Environ Sci & Technol*, 51, 3982–3992.

Karakashev, D., Batstone, D. J., Angelidaki, I. (2005). Influence of environmental conditions on methanogenic compositions in anaerobic biogas reactors. *Appl Environ Microbiol*, 71, 331–338.

Kirkegaard, R. H., McIlroy, S. J., Kristensen, J. M., Nierychlo, M., Karst, S. M., Morten, S., Dueholm, M. S., Albertsen, M., Nielsen, P. H. (2017). Identifying the abundant and active microorganisms common to full-scale anaerobic digesters. https://doi.org/10.1101/104620.

Kleinsteuber, S. (2019). Metagenomics of methanogenic communities in anaerobic digesters. In: Stams A., Sousa D. (eds) *Biogenesis of Hydrocarbons. Handbook of Hydrocarbon and Lipid Microbiology*. Springer, Cham. https://doi.org/10.1007/978-3-319-78108-2_16.

Kobayashi, T., Li, Y. Y., Harada, H. (2008). Analysis of microbial community structure and diversity in the thermophilic anaerobic digestion of waste activated sludge. *Water Sci Technol*, 57, 1199–1205.

Lee, J., Kim, E., Han, G., Tongco, J. V., Shin, S. G., Hwang, S. (2018). Microbial communities underpinning mesophilic anaerobic digesters treating food wastewater or sewage sludge: A full-scale study. *Bioresour Technol*, 259, 388–397.

Li, J., Rui, J., Yao, M., Zhang, S., Yan, X., Wang, Y., Yan, Z., Li, X. (2015). Substrate type and free ammonia determine bacterial community structure in full-scale mesophilic anaerobic digesters treating cattle or swine manure. *Front Microbiol*, 6, 1337.

Liu, Z., Cichocki, N., Hübschmann, T., Süring, C., Ofi̦teru, I. D., Sloan, W. T., Grimm, V., Müller, S. (2019). Neutral mechanisms and niche differentiation in steady-state insular microbial communities revealed by single cell analysis. *Environ Microbiol*, 21, 164–181.

Liu, S., Wang, F., Xue, K., Sun, B., Zhang, Y., He, Z., Van Nostrand, J. D., Zhou, J., Yang, Y. (2015). The interactive effects of soil transplant into colder regions and cropping on soil microbiology and biogeochemistry. *Environ Microbiol*, 17, 566–576.

Lu, X., Rao, S., Shen, Z., Lee, P. K. H. (2013). Substrate induced emergence of different active bacterial and archaeal assemblages during biomethane production. *Bioresour Technol*, 148, 517–524.

Lv, W., Schanbacher, F. L., Yu, Z. (2010). Putting microbes to work in sequence: Recent advances in temperature-phased anaerobic digestion processes. *Bioresour Technol*, 101, 9409–9414.

Mata-Alvarez, J., Dosta, J., Romero-Güiza, M. S., Fonoll, X., Peces, M., Astals, S. (2014). A critical review on anaerobic co-digestion achievements between 2010 and 2013. *Renew Sustain Energy Rev*, 36, 412–427.

McCarty, P. L., Bae, J., Kim, J. (2011). Domestic wastewater treatment as a net energy producer—Can this be achieved? *Environ Sci Technol*, 45, 7100–7106.

Meng, X., Zhang, Y., Sui, Q., Zhang, J., Wang, R., Yu, D., Wang, Y., Wei, Y. (2018). Biochemical conversion and microbial community in response to ternary pH buffer system during anaerobic digestion of swine manure. *Energies*, 11, 2991.

Müller, B., Sun, L., Westerholm, M., Schnürer, A. (2016). Bacterial community composition and fhs profiles of lowand high-ammonia biogas digesters reveal novel syntrophic acetate-oxidising bacteria. *Biotechnol Biofuels*, 9, 1–18.

Narayanan, C. M., Narayan, V. (2019). Biological wastewater treatment and bioreactor design: A review. *Sustain Environ Res*, 29, 33. https://doi.org/10.1186/s42834-019-0036–1.

Narihiro, T., Nobu, M. K., Kim, N.-K., Kamagata, Y., Liu W.-T. (2015). The nexus of syntrophy-associated microbiota in anaerobic digestion revealed by long-term enrichment and community survey. *Environ Microbiol*, 17, 1707–1720.

Nesbø, C., Bradnan, D., Adebusuyi, A., Dlutek, M., Petrus, A., Foght, J., Doolittle, W. F., Noll, K. (2012). Mesotoga prima gen. nov., sp. nov., the first described mesophilic species of the Thermotogales. *Extremophiles*, 16, 387–393.

Nielsen, M., Holst-Fischer, C., Malmgren-Hansen, B., Bjerg-Nielsen, M., Kragelund, C., Møller, H. B., Ottosen, L. D. M. (2017). Small temperature differences can improve the performance of mesophilic sludge-based digesters. *Biotechnol Lett*, 39(11), 1689–1698.

Peces, M., Astals, S., Jensen, S., Clarke, W (2018). Deterministic mechanisms define the long-term anaerobic digestion microbiome and its functionality regardless of the initial microbial community. *Water Res*, 141, 366–376.

Polin, J. P., Peterson, C. A., Whitmer, L. E., Smith, R. G., Brown, R. C. (2019). Process intensification of biomass fast pyrolysis through autothermal operation of a fluidized bed reactor. *Appl Energy*, 249, 276–285.

Qu, X. et al. (2009). Anaerobic biodegradation of cellulosic material: Batch experiments and modelling based on isotopic data and focusing on aceticlastic and non-aceticlastic methanogenesis. *Waste Manag*, 29(6), 1828–1837.

Qureshi, N., Annous, B. A., Ezeji, T. C., Karcher, P., Maddox, I. S. (2005). Biofilm reactors for industrial bioconversion processes: Employing potential of enhanced reaction rates. *Microb Cell Fact*, 4(1), 24. doi: 10.1186/1475–2859–4–24.

Rivière, D., Desvignes, V., Pelletier, E., Chaussonnerie, S., Guermazi, S.;Weissenbach, J., Li, T., Camacho, P., Sghir, A. (2009b). Towards the definition of a core of microorganisms involved in anaerobic digestion of sludge. *ISME J*, 3, 700–714.

Sangeetha, J., Hospet, R., Thangadurai, D., Adetunji, C. O., Islam, S., Pujari, N., Al-Tawaha, A. R. M. S. (2021). Nanopesticides, nanoherbicides, and nanofertilizers: the greener aspects of agrochemical synthesis using nanotools and nanoprocesses toward sustainable agriculture. In: Kharissova OV, Torres-Martínez LM, Kharisov BI (eds) *Handbook of Nanomaterials and Nanocomposites for Energy and Environmental Applications*. Springer, Cham. https://doi.org/10.1007/978-3-030-36268-3_44

Saunders, A. M., Albertsen, M., Vollertsen, J., Nielsen, P. H. (2016). The activated sludge ecosystem contains a core community of abundant organisms. *ISME J*, 10, 11–20.

Schlüter, A., Bekel, T., Diaz, N. N., Dondrup, M., Eichenlaub, R., Gartemann, K. H., Krahn, I., Krause, L., Krömeke, H., Kruse, O., Mussgnug, J. H., Neuweger, H., Niehaus, K., Pühler, A., Runte, K. J., Szczepanowski, R., Tauch, A., Tilker, A., Viehöver, P., Goesmann, A. (2008). The metagenome of a biogas-producing microbial community of a production-scale biogas plant fermenter analysed by the 454-pyrosequencing technology. *J Biotechnol*, 136, 77–90.

Shah, M. P. (2020). *Advanced Oxidation Processes for Effluent Treatment Plants*. Elsevier.

Shah, M. P. (2021). *Removal of Emerging Contaminants through Microbial Processes*. Springer.

Shin, J., Cho, S. K., Lee, J., Hwang, K., Jae Woo Chung, J. W., Jang, H. N., Shin, S. G. (2019). Performance and microbial community dynamics in anaerobic digestion of waste activated sludge: Impact of immigration. *Energies*, 12, 573. doi: 10.3390/en12030573.

Shin, S. G., Han, G., Lim, J., Lee, C., Hwang, S. (2010). A comprehensive microbial insight into two-stage anaerobic digestion of food waste-recycling wastewater. *Water Res*, 44, 4838–4849.

Stams, A. J., Sousa, D. Z., Kleerebezem, R., Plugge, C. M. (2012). Role of syntrophic microbial communities in high-rate methanogenic bioreactors. *Water Sci Technol*, 66, 352–362.

Stolze, Y., Bremges, A., Rumming, M., Henke, C., Maus, I., Puhler, A., Sczyrba, A., Schluter, A. (2016). Identification and genome reconstruction of abundant distinct taxa in microbiomes from one thermophilic and three mesophilic production-scale biogas plants. *Biotechnol Biofuels*, 9, 156.

Sundberg, C. et al. (2013). 454 Pyrosequencing analyses of bacterial and archaeal richness in 21 full-scale biogas digesters. *FEMS Microbiol Ecol*, 85, 612–626.

Tan, H-Q., Li, T-T., Zhu, C., Zhang, X-Q., Wu, M., Zhu, X-F. (2012). *Parabacteroides chartae* sp. nov., an obligately anaerobic species from wastewater of a paper mill. *Int J Syst Evol Microbiol*, 62(Pt 11), 2613–2617.

Vanwonterghem, I., Jensen, P. D., Dennis, P. G., Hugenholtz, P., Rabaey, K., Tyson, G. W. (2014). Deterministic processes guide long-term synchronised population dynamics in replicate anaerobic digesters. *ISME J*, 8, 2015–2028.

Vrieze, J. De et al. (2015). Ammonia and temperature determine potential clustering in the anaerobic digestion microbiome. *Water Res*, 75, 312–323.

Wacławek, S., Grübel, K., Silvestri, D., Padil, V. V. T., Wacławek, M., Cˇerník, M., Varma, R. S. (2018). Disintegration of wastewater activated sludge (WAS) for improved biogas production. *Energies*, 12, 21.

Wang, P., Yu, Z., Zhao, J., Zhang, H. (2018). Do microbial communities in an anaerobic bioreactor change with continuous feeding sludge into a full-scale anaerobic digestion system? *Bioresour Technol*, 249, 89–98.

Wang, C., Zhu, C., Huang, J., Li, L., Jin, H. (2021). Enhancement of depolymerization slag gasification in supercritical water and its gasification performance in fluidized bed reactor. *Renew Energy*, 168, 829–837.

Weiland, P. (2010). Biogas production: Current state and perspectives. *Appl Microbiol Biotechnol*, 85, 849–860.

Wilén, B.-M., Liébana, R., Persson, F., Modin, O., Hermansson, M. (2018). The mechanisms of granulation of activated sludge in wastewater treatment, its optimization, and impact on effluent quality. *Appl Microbiol Biotechnol*, 102(12), 5005–5020. doi: 10.1007/s00253-018-8990–9.

Woodcock, S., Sloan, W. T. (2017). Biofilm community succession: A neutral perspective. *Microbiology*, 163, 664–668.

Zhang, H., Sun, M., Song, L., Guo, J., Zhang, L. (2019). Fate of NaClO and membrane foulants during in-situ cleaning of membrane bioreactors: Combined effect on thermodynamic properties of sludge. *Biochem Eng J*, 147, 146–152.

Zheng, Y., Cheng, C., Zhou, Z., Pang, H., Chen, L., Jiang, L. M. (2019). Insight into the roles of packing carriers and ultrasonication in anaerobic side-stream reactor coupled membrane bioreactors: Sludge reduction performance and mechanism. *Water Res*, 155, 310–319.

8 Genomic Analysis of Heavy Metal-Resistant Genes in Wastewater Treatment Plants

Charles Oluwaseun Adetunji and Abel Inobeme
Edo University Iyamho

Kshitij RB Singh
Banaras Hindu University

Ruth Ebunoluwa Bodunrinde
Federal University of Technology Akure

John Tsado Mathew
Ibrahim Badamasi Babangida University Lapai

Olugbemi T. Olaniyan
Edo University Iyamho

Ogundolie Frank Abimbola
Federal University of Technology Akure

Jay Singh
Banaras Hindu University

Vanya Nayak
Banaras Hindu University

Ravindra Pratap Singh
Indira Gandhi National Tribal University

DOI: 10.1201/9781003354147-8

CONTENTS

8.1 INTRODUCTION

Constant increase in agricultural and industrial processes aimed at meeting the global demand has resulted in an increase in ecological and environmental problems. Industrial processes such as metal automotive repair and manufacturing, forging, rubber and plastic production, combustion of fossil fuel, mining, storage batteries, smelting and manufacturing of alkaline lead to release of pollutants such as heavy metals and other contaminants into the soils and water bodies (Rajaganapathy et al., 2011; Ahemad and Malik, 2012; Adetunji et al., 2019; Adetunji and Ugbenyen 2019; Adetunji et al., 2021; Adetunji and Anani 2021; Sangeetha et al., 2021; Dauda et al., 2022a,b).

Sludge water release and other agricultural processes, largely due to leaching of pesticides, herbicides and even fertilizers into the aquatic environment, have resulted in problems globally affecting the aquatic ecosystem. This has resulted in increase in the release of various heavy metals into the water bodies. Though some heavy metals such as iron are essential to living organisms, others such as arsenic, lead, nickel, cadmium, zinc, mercury and copper are naturally dense metals and metalloids and are generally referred to as toxic heavy metals even at little concentrations. Irrespective of whether beneficial or not a heavy metal is, they are all concentration-based, and at a certain concentration, they become toxic and their toxicity is influenced by their accumulation in what is called bioaccumulation. When they surpass the threshold, these metals become harmful and result in metabolic, physiological disorders and even death.

In wastewater treatment plants (WWTPs), a mixture of wastewater generated by household wastewater, industrial wastewater and rainwater are treated in order to be safe for reuse. This treatment can be achieved either through physical/chemical methods, biological methods or the both. The use of biological methods (microbes) or exposure of microbes to chemical stress (such as antibiotics and toxic metals) (Di Cesare et al., 2016) during the process of purification has over time resulted in heavy metal-resistant microorganisms. Analyzing the microbiome to understand the genes responsible for their heavy-metal resistance will further assist in understanding these heavy metal-resistant microbes. Hence, this chapter provides a detailed overview of genomic analysis of heavy-metal resistance genes (HMRGs) present in wastewater treatment plants.

8.2 HMRGS IN THE ENVIRONMENT

Yang et al. (2020) carried out a study on the resistance to heavy metals in *Salmonella* and *Escherichia coli* and how it correlates with genes that are resistant to disinfectants and genes resistant to antibiotics. The polymerase chain reaction was used in detecting the HMRGs. The findings from the study showed the wide presence of genes resistant to heavy metals in *Salmonella* and *E. coli* obtained from retail meats and chicken farms, and their association with disinfectant resistance genes that could bring about co-resistance to other antimicrobials and heavy metals.

Sultan et al. (2020) assessed some bacteria in Wular and Dal lakes in India. The microbial isolates were screened for resistance to heavy metals and antimicrobials and their correlation with genetic elements that are mobile in nature. From the result obtained, there was multidrug resistance in most of the isolates. Moreover, molecular characterization confirmed the presence of drug-resistant determinants as well as genes resistant to heavy metals.

Di Cesare et al. (2016) worked on the fate of different genes resistant to antibiotics and heavy metals. In their work, they correlated the abundance of various genetic markers with each other and their connections with abiotic (such as total nitrogen and total organic carbon) and biotic factors that influence the community of microorganisms. Analysis was done on water samples collected for the presence and abundance of antimicrobial-resistant genes. The findings from the study showed a vital role of heavy-metal resistance genes in the dispersion and aided by the presence of genes resistant to antimicrobials.

Chen et al. (2019) studied the relationship between the population of soil bacteria and pollution due to heavy metals in a tailings dam for copper processing in China. They reported the abundance of genes that show resistance to macrolides and arsenic in the area investigated. The high abundance of genes resistant to heavy metals showed a strong correlation with cadmium content, which showed the relevance of cadmium in the choice of metal resistance genes. The study enhanced the understanding of the interrelationship between multimetal pollution in the environment the microbial resistance.

Furthermore, Ramos et al. did a characterization of the genotypes of *Pseudomonas* spp. that were obtained from different water sources, including streams, rivers, lakes, and plants for sewage treatment. Sequencing of 16s rDNA was used to identify bacteria while resistance profile to heavy metals and antimicrobials was determined using minimum inhibitory concentration. A total number of 23 *Pseudomonas* spp. were isolated and characterized, which majorly included *Pseudomonas aeruginosa*, *Pseudomonas hunanensis* and *Pseudomonas asiatica*. In a related study, Long et al. (2021) investigated the concentration of heavy metals and the constituents of the microbial community of sediment collected from eight different rivers. The result showed that the microbial community was made of 13 different phyla, which majorly included *Proteobacteria*, *Firmicutes*, *Actinobacteria* and some other unclassified bacteria. Furthermore, it was observed that there was a rise in the population of *Firmicutes* along with the levels of the pollutant. They concluded that the findings from the study can enhance the comprehension of the variation of the communities of microorganisms in the e-waste recycling area.

8.3 APPLICATION OF HIGH-THROUGHPUT SEQUENCING-BASED METAGENOMIC APPROACH FOR THE ASSESSMENT OF DIVERSITY, OCCURRENCE, AND THE LEVEL OF ASSESSMENT OF MOBILE GENETIC ELEMENTS AND ANTIBIOTIC RESISTANCE GENES IN AEROBIC AND ANAEROBIC SLUDGE

Metagenomics captures the diversity of microorganisms and the disparity in their ecological feature from strains, thereby producing microbial populations of varying properties. Coleman et al. (2006) studied the comparison of the genome of *Prochlorococcus* from a sea shotgun clone with the genome from a pure culture which produces a sequence of homogeneous variants and showed that some parts of the varied population can go into extinction (Dumont and Murrell, 2005). Researchers reported farming of different species in their homogeneous form, the microbial population that seems uncultivable is usually done through metagenomics tools.

Similarly, Venter et al. (2004) researched an ocean environment and discovered low diversity of microbes inhabiting that location using a shotgun sequencing survey method. They collected cells from microorganisms and viruses with varying fraction sizes and extracted their DNA, and their study heralds a change in the environment due to microbial influence, thereby calling attention to the cogent present challenges related to storing information, incorporating and evaluating data sets of numerous metagenomics. The assemblage of the sea microbial plankton of over 1,400 individual rRNA genes was an effective metric for calculating taxonomy. The proteins they encountered were consistent with those usually found in the ocean. They made a stunning report that sequencing large-inserted DNA along with single-cell genome would be needed to characterize all the communities of microorganisms without modification fully. The limitation they encountered during their work was the contamination caused by microorganisms conceded a larger percentage of the sample restraining utilization for interpretation of ecological data (Delong and Karl, 2005); Mahenthiralingam et al. (2006) indicated that metagenomics study needs much attention in cautious sampling, verifying measures, independent sampling when sequencing larger-scale metagenomic. Nevertheless, they came about the Sargasso sea dataset, which served as a useful resource in combating indigenous genomic complexities; however, certain problems were encountered during sampling.

Tringe et al. (2005), in their studies, conducted gene-by-gene analysis, considering the variations and similarities existing within microbial communities from the carcass of a whale on the seafloor, Sargasso sea and the drainage of a community acid, which showed that a more number of sampling systems can enormously facilitate the comparison of genomic evolutionary variation of the microbial population from community to community (Delon et al., 2006).

Verma et al. (2001) and Alam et al. (2011) researched identifying heavy metals and bacteria that are antibiotics resistant from tannery wastewater. Mobile genetic elements are known to convey antibiotic resistance genes which are integrons, plasmids and transposons (Zhang et al., 2011; Tennstedt et al., 2005; Ma et al., 2011).

The application of DNA cloning and microarray techniques have indicated Proteobacteria to be most prevalent in activated sludge, where carbohydrates and proteins were found in higher concentrations (Xia et al., 2010; Murugananthan et al., 2004; Snaidr et al., 1997). It was also found that the *Bacteroidetes* efficiently degraded several organic materials in the presence of several carbohydrates and proteins encoding genes, which can degrade enzymes (Thomas et al., 2011). The concentration of oxygen was found as a determining factor in branding the microorganism community in wastewater treatment and contributes to the disparities of microorganism communities, which could be aerobic and anaerobic sludge (Cheong and Hanson, 2006).

Moreover, Verma and Chand Sharma (2020) stated that tanneries act as a major source of environmental contamination by generating large amounts of tannery waste. In their studies, they sequenced two metagenomes that represent the tannery waste dumpsites using the illumine pattern. The microbial diversity from the study showed the presence of a high amount of *Firmicutes*, *Proteobacteria*, *Actinobacteria* and *Bacteroidetes*. Furthermore, the availability of contaminants degrading microorganisms also showed the potencies of the bioremediation. Han and Yoo (2020) studied the occurrence and distribution of antibiotic resistance genes (ARGs) and the structural composition of the community of bacteria present in the activated sludge. Through the use of broad scanning of the metagenomic profiling, they were able to identify about 153 ARG subcategories which belong to 19 ARG categories. In addition, there was also a relatively higher amount of some particular ARG associated with tetracycline, sulfonamide and macrolide.

Lafebve et al. reported the occurrence of 4% *Synergistetes* sp. present while treating aerobic sludge. An amino acid from the degradation of proteins and some peptides was utilized by *Synergistetes*, providing sulfate and fatty acids with short chains that perform terminal degradation with bacteria that produce sulfate and methanogens. *Bacteroides*, *Desulfovibrio* and *Clostridium* were discovered to predominate the anaerobic sludges, while *Pseudomonas* and *Burkholderia* predominated the aerobic sludges (Wang et al., 2013). They found *Clostridia* to be most abundant due to its ability to produce endospores to enable survival under hostile environmental conditions (Cheong and Hanson, 2006), while bacteria that reduce sulfate were abundant in anaerobic sludge, which is a common pollutant of tannery effluent (Murugananthan et al., 2004).

Godon et al. (1997) and Santegoeds et al. (1999) found out that the population of *Archaea* was greater than that of other bacteria in anaerobic sludge and recorded that this could be a result of lack of oxygen and reduced physical disorder since bacteria that resisted the efficacy of antibiotics have the potential to transfer mobile genetic element in ground and surface water into tannery wastewater in the environment, thereby posing serious public health concern (Chee-Sanford et al., 2001). They concluded that the high-throughput sequence technique provided an in-depth knowledge of the community of the microorganisms, structures, functions and interactions within the aerobic and anaerobic tannery wastewater treatment sludge.

Martinez (2009), Ju et al. (2016), and Jia et al. (2015) reported that integrons are found to facilitate the exchange and incorporation of antibiotics resistance genes, leading to the proliferation of antibiotics resistance in bacteria produced from wastewater. Therefore, when there is no proper treatment of wastewater by-products and bioaerosols, it could constitute great environmental and public health challenges (Han and Yoo, 2020; Shah, 2020). Moreover, the use, abuse, and prolonged usage of antibiotics resulted in the selection of antibiotics by bacteria and some genes from the unset of production of antibiotics (Walsh, 2003). The rate of microbial resistance to antibiotics is also fast alarming, leading to death and lack of specificity in action. This has called for the use of plants in effectively treating wastewater (O'Neill, 2016; Bouki et al., 2013). Freely moving genetic materials are abundant in the plants used in treating wastewater (Guo et al., 2017) and were reported to be responsible for the horizontal transfer of genes. Currently, metagenomic assay together with the high-throughput approach in sequencing was found reliable in the determination of abundance and microbial diversity of ARGs in drinking water, soil and activated sludge. It was observed that when the ARG and antibiotic-treated bacteria were inserted as the influent in the wastewater treatment plant, they survived all the treatment units and were released along with the effluents as illustrated in Figure 8.1, which proved that the release of antibiotics and heavy metals in the sewage systems should be reduced (Mao et al., 2015; Burch et al., 2014; Ross and Topp, 2015). This technique helped in the proper profiling of the microorganism community (Yoo et al., 2017; Zhou et al., 2018).

More recently, metagenomic studies in combination with high-throughput sequencing have been viewed as a potential method that is independent of culture in determining the abundance and diversity of antimicrobial resistance genes in different kinds of environments, including activated sludge, soil, drinking water, and

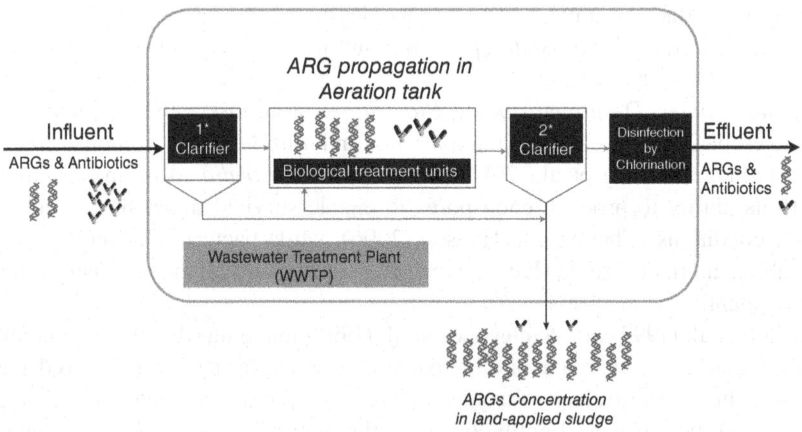

FIGURE 8.1 Insertion of ARGs and antibiotics as influents in the wastewater treatment plant. The influents are treated in four chambers of the wastewater treatment plant and are then released as effluents that consist of ARGs and antibiotics. (Reprinted with permission from Mao, D. et al. (2015), *Prevalence and proliferation of antibiotic resistance genes in two municipal wastewater treatment plants. Water research*, *85*, 458–466.)

sediments. In addition, the approach has been shown to possess various advantages in the profiling of microbial community function and structural composition (Hvistendahl, 2012).

Various investigations have reported the resistance of some bacteria to antibiotics in several kinds of environments. The compilation of drug-resistant genes is a representation of the wastewater resistome. This is a promising technique for resisting the various kinds of antibiotics found in their environment (Guo et al., 2017; Shah, 2021). It is possible to transmit antimicrobial resistance from environmental to clinical bacteria. Both ways around, the wastewater environment therefore acts as a suitable ground for antibiotics resistance microorganisms that cause serious diseases in humans and other organisms. The major stimulating parameter is the sub lethal dose of the drugs.

However, environmental and physiological stress from various factors within the environment could also be driver agents in the accumulation of antibiotics and other drugs (Chen et al., 2015). Methods that depend on culture have also been employed to study the position of dynamic genetic elements during the transmission of genes that are resistant to antibiotics. Studies have revealed that such genes can influence the adaptation and evolution of bacteria and are vital in the emergence, transmission and recombination of antibiotic resistance (Jackson et al., 2011). In order to ensure a thorough understanding of the wide distribution of the problems related to resistance to antibiotics, various scientific and modern techniques have been put forward to aid the understanding of the comprehensive antibiotic resistance in various ecosystems which are known as the resistome (Jałowieki et al., 2017). Majeed et al. (2021), in their work, focused on the detailed evaluation of the patterns involved in the metagenomic obtained indicators of resistance to antibiotics through different steps of the treatment. They observed a decline in the abundance of the ARGs by approximately 50% from the influent and each of the sampling points described by a specific resistome. They also reported the presence of about 90% of the ARGs that were found in the effluents also in the influents. The analysis of the discriminatory and fundamental resistomes and the general ARGs cut across the whole samples.

8.4 RELEVANT INFORMATION ON THE GENES CODING FOR ANTIBIOTIC RESISTANCE IN NUMEROUS COMMUNITIES AS WELL AS THE APPLICATION OF BLAST ANALYSIS AGAINST ANTIBIOTIC RESISTANCE

One factor inherent in the successful utilization of a given therapeutic agent is the possible development of resistance and tolerance to the compound under consideration from when it was first utilized for the purpose. The assertion is a fact for chemical agents used for the treatment of various illnesses caused by parasites, fungi, bacteria and viruses and other forms of degenerative diseases such as diabetes and cancer. This also includes other diseases faced by other types of organisms such as fish, insects, and plants. Rapid and cost-effective sequencing technologies have emerged recently together with platforms of bioinformatics, which have the tendency to bring about a revolution in microbial surveillance.

Through investigations of the characterization of antimicrobial resistance genes in the genomes of bacteria and metagenomes with high complexities, it is possible to reveal the scope of antimicrobial resistance present in a particular bacteria or complex communities of an organism. Currently, there are suites for bioinformatics pipelines and ARG databases useful for metagenomic and genomic details (Cytryn, 2013). Previously, the characterization of ARGs was on clinically-based bacteria. However, there is growing awareness that bacteria present in natural ecosystems could also develop resistance to various groups of antibiotics, which is an indication that a larger component of the clinically connected ARGs emerged from bacteria found within the environment (Forsberg et al., 2014).

The discovery and emergence of various antimicrobial agents formed the basis of the transformation of modern medicine in its therapeutic paradigm. Thus, antibiotics have formed the development of significant medical interventions in the therapeutic sector. However, there is a threatening issue emerging from the development of resistance among microorganisms. More recently, the WHO has identified resistance to antibiotics as one of the most significant public health issues (WHO, 2014). Infectious disease due to multidrug resistance is connected with the recent rise in death rate compared to those that are due to susceptible bacteria having a serious burden economically, which is estimated to be above 20 billion dollars each year in the United States alone. Also, antibiotic resistance, based on the present report, is estimated to be responsible for over 300 million premature deaths by 2050.

Bacteria possess significant plasticity genetically that permits their response to varieties of environmental threats such as the presence of antibiotic compounds capable of jeopardizing their occurrence. Furthermore, bacteria that share similar ecological niches with other organisms that produce antimicrobials have developed different mechanisms for withstanding the impact of harmful antibiotic compounds. Hence, their inherent resistance makes it possible for them to thrive in their presence. Thus, from the perspective of evolution, bacteria developed two genetic strategies for adapting to antibiotic attacks. The first involves a mutation in their genetic makeup, while the other consists of the acquisition of DNA with foreign coding for resistance development through a horizontal gene transfer process (Tong et al., 2017).

Wang et al. (2013) aimed at exploring the resistance of microbial genes in wastewater from tanneries and antibiotics comprehensively. They made a comparison of numerous techniques of high throughput. As a result, they observed from the BLAST technique that about 0.0081% of the anaerobic sludge and 0.0101% of aerobic sludge were assigned to different genes that show resistance to antibiotics. Zhang et al. (2009) documented that treatment plants used for sewage are a vital reservoir of resistance to environmental antibiotics. They used the BLAST technique for ARDB-based assessment in their investigation but observed that MG-RAST depended on BLAT but of lesser sensitivity.

Ma et al. (2015) in their work reported that the potential of the WWTP is capable of affecting the community of bacteria through various operation pathways such as the characteristics of influent, the retention time of the sludge, reactor volumes, hydraulic retention time and physical constrain. More current investigations have shown that the community of bacteria plays a vital role in the occurrence and enrichment of ARGs. They also observed that the co-election process of the community of bacteria and ARGs are enhanced during the process of treating wastewater as a result of a large number of nutrient materials and certain conditions of operation.

Wang et al. (2013) adopted the Illumina sequencing pattern to assess the abundance, occurrence, and distribution of genes resistant to antibiotics and mobile elements in anaerobic and aerobic sludge from tannery waste. Tong et al. (2017), in their study, investigated the diversity, abundance and occurrence of ARGs as well as mobile elements of genetic origin and the distribution of plasmids within broiler and layer feces through the use of the illumining sequencing technique of high throughput. Yoo and Lee (2021) aimed to assess the impact of the persistence and prevalence of ARGs on WWTPs. They employed shotgun metagenomics since it is a vital approach in gaining an intense understanding of the entire genomic pools from the combined microbiome from the environment.

8.5 MECHANISMS OF ANTIMICROBIAL RESISTANCE

Several physiological and biochemical mechanisms have been put forward in accounting for antimicrobial resistance. The lack of basic knowledge in this regard is responsible for the poor achievement in the prevention and management of resistance development and has been recognized as a major problem in the medical sector (D'Costa et al., 2007). It is not surprising that bacteria have developed complex mechanisms for resistance to drugs to avoid the effect of the antimicrobial agent. The achievement of resistance is developed through several biochemical routes. Also, a single bacterial cell can use several mechanisms in surviving particular antibiotics.

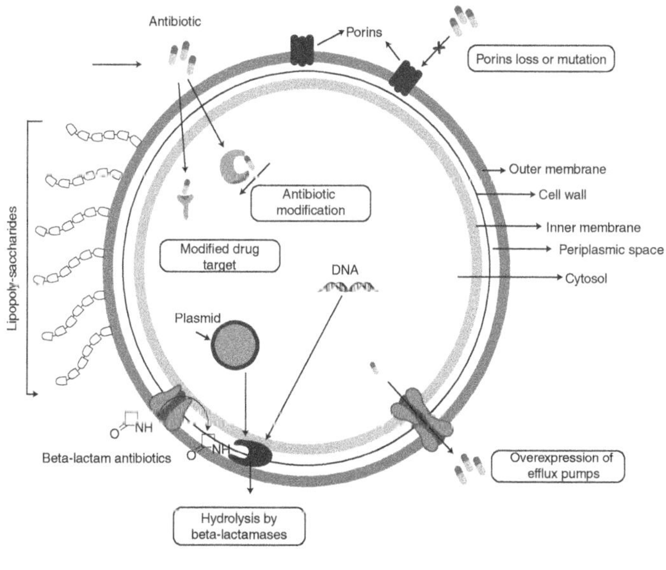

FIGURE 8.2 Structure of Gram-negative bacteria and their mechanism of resistance of beta-lactamases. (Reprinted with permission from Breijyeh, Z., Jubeh, B., & Karaman, R. (2020). *Resistance of Gram-negative bacteria to current antibacterial agents and approaches to resolve it. Molecules*, *25*(6), 1340 [CC BY 4.0].)

For example, fluoroquinolone resistance may occur through three varying biochemical pathways, all of which are capable of coexisting in the same bacteria simultaneously, thus creating additive input and increasing the extent of the resistance. In fluoroquinolone (FQ) resistance, there may be a mutation of the genes encoding the site of target and the overexpression of the efflux pumps (Thomas and Nielsen, 2005).

Another example is the mechanism of resistance to beta-lactams in bacteria (Gram-negative) which is the production of beta-lactamases, while the resistance in the case of Gram-positive bacteria is mostly acquired through modifications of their target area which is the protein that binds penicillin. It has been explained that this phenomenon results from the significant differences in the envelope covering the cells of the Gram-positive and -negative. In Gram-negative bacteria, the outer membrane helps control molecule entrance into the periplasmic space (Manson et al., 2010). Moreover, the beta-lactamase resistance mechanism in Gram-negative bacteria arises from the expression of antibiotic-inactivating enzymes and nonenzymatic path, as shown in Figure 8.2.

8.6 CONCLUSION AND PROSPECTS

Therefore, this chapter has provided detailed information on the effectiveness of genomic analysis of HMRGs present in the wastewater treatment plant. Also, detailed information on the modes of action on how these genes could regulate how the level of antibiotic resistance in these wastewaters was provided. The application of next-generation sequencing will also go a long way in the characterization of HMRGs that are present in wastewater treatment plants.

REFERENCES

Adetunji CO, Anani OA (2021) Plastic-Eating Microorganisms: Recent Biotechnological Techniques for Recycling of Plastic. In: Panpatte DG, Jhala YK (eds) *Microbial Rejuvenation of Polluted Environment. Microorganisms for Sustainability*, vol 25. Springer: Singapore. https://doi.org/10.1007/978-981-15-7447-4_14

Adetunji CO, Ugbenyen MA (2019) Mechanism of action of nanopesticide derived from microorganism for the alleviation of abiotic and biotic stress affecting crop productivity. In Panpatte D, Jhala Y (eds) *Nanotechnology for Agriculture: Crop Production & Protection*. Springer: Singapore. https://doi.org/10.1007/978-981-32-9374-8_7

Adetunji CO, Kumar D, Raina M, Arogundade O, Sarin NB (2019) Endophytic microorganisms as biological control agents for plant pathogens: a panacea for sustainable agriculture. In Varma A, Tripathi S, Prasad R (eds) *Plant Biotic Interactions*. Springer: Cham. https://doi.org/10.1007/978-3-030-26657-8_1

Adetunji CO, Olaniyan OT, Bodunrinde RE, Ahamed MI (2021) Bioconversion of poultry waste into added-value products. In Inamuddin, Khan A (eds) *Sustainable Bioconversion of Waste to Value Added Products. Advances in Science, Technology & Innovation*. Springer: Cham. https://doi.org/10.1007/978-3-030-61837-7_21

Ahemad M, Malik A (2012) Bioaccumulation of heavy metals by zinc resistant bacteria isolated from agricultural soils irrigated with wastewater. *Bacteriol J*, 2: 12–21.

Alam MZ, Ahmad S, Malik A (2011) Prevalence of heavy metal resistance in bacteria isolated from tannery effluents and affected soil. *Environ Monit Assess*, 178: 281–291. https://doi.org/10.1007/s10661-010-1689–8.

Bouki C, Venieri D, Diamadopoulos E (2013) Detection and fate of antibiotic resistant bacteria in wastewater treatment plants: A review. *Ecotoxicol Environ Saf*, 91: 1–9.

Breijyeh Z, Jubeh B, Karaman R (2020) Resistance of Gram-negative bacteria to current antibacterial agents and approaches to resolve it. *Molecules*, 25(6): 1340.

Burch TR, Sadowsky MJ, LaPara TM (2014) Fate of antibiotic resistance genes and class 1 integrons in soil microcosms following the application of treated residual municipal wastewater solids. *Environ Sci Technol*, 46: 5620–5627.

Chee-Sanford JC, Aminov RI, Krapac IJ, Garrigues-Jeanjean N, Mackie RI (2001) Occurrence and diversity of tetracycline resistance genes in lagoons and groundwater underlying two swine production facilities. *Appl Environ Microbiol*, 67: 1494–1502. https://doi.org/10.1128/AEM.67.4.1494–1502.2001.

Chen B, Hao L, Guo X, Wang N, Ye B (2015) Prevalence of antibiotic resistance genes of wastewater and surface water in livestock farms of Jiangsu Province, China. *Environ Sci Pollut Res Int*, 22(18): 3950–3959.

Chen J, Li J, Zhang H, Shi W, Liu Y (2019) Bacterial heavy-metal and antibiotic resistance genes in a copper tailing dam area in Northern China. *Front Microbiol*, 10: 1916. doi: 10.3389/fmicb.2019.01916.

Cheong DY, Hansen CL (2006) Bacterial stress enrichment enhances anaerobic hydrogen production in cattle manure sludge. *Appl Microbiol Biotechnol*, 72: 635–643. https://doi.org/10.1007/s00253-006-0313–x.

Coleman ML, Sullivan MB, Martiny AC, Steglich C, Barry K, Delong EF, Chisholm SW (2006) Genomic islands and the ecology and evolution of Prochlorococcus. *Science*, 311(5768): 1768–1770.

Cytryn E (2013) The soil resistome: The anthropogenic, the native, and the unknown. *Soil Biol Biochem*. doi: 10.1016/j.soilbio.2013.03.017.

Dauda WP, Abraham P, Glen E, Adetunji CO, Ghazanfar S, Ali S, Al-Zahrani M, Azameti MK, Alao SEL, Zarafi AB, Abraham MP, Musa H (2022a) Robust profiling of Cytochrome P450s (P450ome) in notable *Aspergillus* spp. *Life*, 12(3): 451. https://doi.org/10.3390/life12030451

Dauda WP, Morumda D, Abraham P, Adetunji CO, Ghazanfar S, Glen E, Abraham SE, Peter GW, Ogra IO, Ifeanyi UJ, Musa H, Azameti MK, Paray BA, Gulnaz A (2022b) Genome-wide analysis of Cytochrome P450s of *Alternaria* Species: evolutionary origin, family expansion and putative functions. *J Fungi*, 8(4): 324. https://doi.org/10.3390/jof8040324

D'Costa VM, Griffiths E, Wright GD (2007) Expanding the soil antibiotic resistome: Exploring environmental diversity. *Curr Opin Microbiol*, 10: 481–489.

DeLong EF, Karl DM. (2005) Genomic perspectives in microbial oceanography. *Nature*, 437(7057): 336–442.

DeLong EF, Preston CM, Mincer T, Rich V, Hallam SJ, Frigaard NU, Martinez A, Sullivan MB, Edwards R, Brito BR, Chisholm SW, Karl DM (2006) Community genomics among stratified microbial assemblages in the ocean's interior. *Science*, 311(5760): 496–503.

Di Cesare A, Eckert EM, D'Urso S, Bertoni R, Gillan DC, Wattiez R, Corno G (2016) Co-occurrence of integrase 1, antibiotic and heavy metal resistance genes in municipal wastewater treatment plants. *Water Res*, 94, 208–214. doi: 10.1016/j.watres.2016.02.049.

Dumont MG, Murrell JC (2005) Stable isotope probing—linking microbial identity to function. *Nat Rev Microbiol*, 3(6): 499–504.

Forsberg KJ, Patel S, Gibson MK, Lauber CL, Knight R, Fierer N, Dantas G (2014) Bacterial phylogeny structures soil resistomes across habitats. *Nature*. doi: 10.1038/nature13377.

Godon JJ, Zumstein E, Dabert P, Habouzit F, Moletta R (1997) Molecular microbial diversity of an anaerobic digester as determined by small-subunit rDNA sequence analysis. *Appl Environ Microbiol*, 63: 2802–2813.

Guo J, Li J, Chen H, Bond PL, Yuan Z (2017) Metagenomic analysis reveals wastewater treatment plants as hotspots of antibiotic resistance genes and mobile genetic elements. *Water Res*, 123: 468–478.

Han I and Yoo K (2020) Metagenomic profiles of antibiotic resistance genes in activated sludge, dewatered sludge and bioaerosols. *Water*, 12(6): 1516. https://doi.org/10.3390/w12061516.

Hvistendahl M (2012) Public health. China takes aim at rampant antibiotic resistance. *Science*, 336: 795. doi: 10.1126/science.336.6083.795.

Jackson RW, Vinatzer B, Arnold DL, Dorus S, Murillo J (2011) The influence of the accessory genome on bacterial pathogen evolution. *Mob Genet Element*, 1(1): 55–65.

Jałowiecki Ł, Chojniak J, Dorgeloh E, Hegedusova B, Ejhed H, Magnér J, Płaza G (2017) Using phenotype microarrays in the assessment of the antibiotic susceptibility profile of bacteria isolated from wastewater in on-site treatment facilities. *Folia Microbiologica*, 62(6): 453–461.

Jia S, Shi P, Hu Q, Li B, Zhang T, Zhang XX (2015) Bacterial community shift drives antibiotic resistance promotion during drinking water chlorination. *Environ Sci Technol*, 49: 12271–12279.

Ju F, Li B, Ma L, Wang Y, Huang D, Zhang T (2016) Antibiotic resistance genes and human bacterial pathogens: Co-occurrence, removal, and enrichment in municipal sewage sludge digesters. *Water Res*, 91: 1–10.

Long S, Tong H, Zhang X, Jia S, Chen M, Liu C (2021) Heavy metal tolerance genes associated with contaminated sediments from an e-waste recycling river in Southern China. *Front Microbiol*, 12: 665090. doi: 10.3389/fmicb.2021.665090

Ma L, Zhang XX, Cheng S, Zhang Z, Shi P et al. (2011) Occurrence, abundance and elimination of class 1 integrons in one municipal sewage treatment plant. *Ecotoxicology* 20: 968–973. https://doi.org/10.1007/s10646-011-0652-y.

Ma X, Zhang Q, Zhu Q, Liu W, Chen Y et al. (2015) A robust CRISPR/Cas9 system for convenient, high-efficiency multiplex genome editing in monocot and dicot plants. *Mol Plant*, 8: 1274–1284.

Mahenthiralingam E, Baldwin A, Drevinek P, Vanlaere E, Vandamme P, Lipuma JJ, Dowson CG (2006) Multilocus sequence typing breathes life into a microbial metagenome. *PLoS One*, 1: e17.

Majeed HJ, Riquelme MV, Davis BC, Gupta S, Angeles L, Aga DS, Garner E, Pruden A, Vikesland PJ (2021) Evaluation of metagenomic-enabled antibiotic resistance surveillance at a conventional wastewater treatment plant. *Front Microbiol*. https://doi.org/10.3389/fmicb.2021.657954.

Manson JM, Hancock LE, Gilmore MS (2010) Mechanism of chromosomal transfer of Enterococcus faecalis pathogenicity island, capsule, antimicrobial resistance, and other traits. *Proc Natl Acad Sci U S A*., 107(27): 12269–12274.

Mao D, Yu S, Rysz M, Lou Y, Yang F, Li F, Hou J, Mu Q, Alvarez PJJ (2015) Prevalence and proliferation of antibiotic resistance genes in two municipal wastewater treatment plants. *Water Res*, 85: 458–466.

Martinez JL (2009) Environmental pollution by antibiotics and by antibiotic resistance determinants. *Environ Pollut*, 157: 2893–2902.

Murugananthan M, Bhaskar Raju G, Prabhakar S (2004) Separation of pollutants from tannery effluents by elector flotation. *Sep Purif Technol*, 40: 69–75. https://doi.org/10.1016/j.seppur.2004.01.005.

O'Neill J (2016) Tackling drug-resistant infections globally: Final report and recommendations. https://amr-review.org/sites/default/files/160525_Final%20paper_with%20cover.pdf (accessed on 10 December, 2018).

Rajaganapathy V, Xavier F, Sreekumar D, Mandal PK (2011) Heavy metal contamination in soil, water and fodder and their presence in livestock and products: A review. *J Environ Sci Technol*, 4: 234–249.

Ross J, Topp E (2015) Abundance of antibiotic resistance genes in bacteriophage following soil fertilization with dairy manure or municipal biosolids, and evidence for potential transduction. *Appl Environ Microbiol*, 81: 7905–7913.

Sangeetha J, Hospet R, Thangadurai D, Adetunji CO, Islam S, Pujari N, Al-Tawaha ARMS (2021) Nanopesticides, nanoherbicides, and nanofertilizers: the greener aspects of agrochemical synthesis using nanotools and nanoprocesses toward sustainable agriculture. In: Kharissova OV, Torres-Martínez LM, Kharisov BI (eds) *Handbook of Nanomaterials and Nanocomposites for Energy and Environmental Applications*. Springer: Cham. https://doi.org/10.1007/978-3-030-36268-3_44

Santegoeds CM, Damgaard LR, Hesselink G, Zopfi J, Lens P et al. (1999) Distribution of sulfate-reducing and methanogenic bacteria in anaerobic aggregates determined by microsensor and molecular analyses. *Appl Environ Microbiol*, 65: 4618–4629.

Shah MP (2020) *Advanced Oxidation Processes for Effluent Treatment Plants*. Elsevier.

Shah MP (2021) *Removal of Emerging Contaminants through Microbial Processes*. Springer.

Snaidr J, Amann R, Huber I, Ludwig W, Schleifer KH (1997) Phylogenetic analysis and in situ identification of bacteria in activated sludge. *Appl Environ Microbiol*, 63: 2884–2896.

Sultan I, Ali A, Gogry FA, Rather IA, Sabir JSM, Haq QMR. (2020) Bacterial isolates harboring antibiotics and heavy-metal resistance genes co-existing with mobile genetic elements in natural aquatic water bodies. *Saudi J Biol Sci*, 27(10): 2660–2668. doi: 10.1016/j.sjbs.2020.06.002.

Tennstedt T, Szczepanowski R, Krahn I, Pühler A, Schlüter A (2005) Sequence of the 68,869 bp IncP-1α plasmid pTB11 from a waste-water treatment plant reveals a highly conserved backbone, a Tn*402*-like integron and other transposable elements. *Plasmid*, 53: 218–238. https://doi.org/10.1016/j.plasmid.2004.09.004.

Thomas CM, Nielsen KM (2005) Mechanisms of, and barriers to, horizontal gene transfer between bacteria. *Nat Rev Microbiol*, 3(9): 711–721.

Thomas F, Hehemann JH, Rebuffet E, Czjzek M, Michel G (2011) Environmental and gut *bacteroidetes*: The food connection. *Front Microbiol*, 2: 93.

Tong, H., Cai, L., Zhou, G., Yuan, T., Zhang, W., Tian, R., et al. (2017) Temperature shapes coral-algal symbiosis in the South China Sea. *Sci. Rep*, 7:40118. doi: 10.1038/srep40118

Tringe SG, von Mering C, Kobayashi A, Salamov AA, Chen K, Chang HW, Podar M, Short JM, Mathur EJ, Detter JC, Bork P, Hugenholtz P, Rubin EM (2005) Comparative metagenomics of microbial communities. *Science*, 308(5721): 554–557.

Venter JC, Remington K, Heidelberg JF, Halpern AL, Rusch D, Eisen JA, Wu D, Paulsen I, Nelson KE, Nelson W, Fouts DE, Levy S, Knap AH, Lomas MW, Nealson K, White O, Peterson J, Hoffman J, Parsons R, Baden-Tillson H, Pfannkoch C, Rogers YH, Smith HO (2004) Environmental genome shotgun sequencing of the Sargasso Sea. *Science*, 304(5667): 66–74. doi: 10.1126/science.1093857

Verma SK, ChandSharma P (2020) NGS-based characterization of microbial diversity and functional profiling of solid tannery waste metagesnomes. *Genomics*, 112(4): 2903–2913.

Verma T, Srinath T, Gadpayle RU, Ramteke PW, Hans RK et al. (2001) Chromate tolerant bacteria isolated from tannery effluent. *Biores Technol*, 78: 31–35. https://doi.org/10.1016/S0960-8524(00)00168-1.

Walsh, C (2003) *Antibiotics: Actions, Origins, Resistance*. ASM Press: Washington, DC.

Wang Z, Zhang X-X, Huang K, Miao Y, Shi P et al. (2013) Metagenomic profiling of antibiotic resistance genes and mobile genetic elements in a tannery wastewater treatment plant. *PLoS One*, 8(10): e76079. doi: 10.1371/journal.pone.0076079.

WHO (2014) *Antimicrobial Resistance: Global Report on Surveillance 2014*. World Health Organization. http://www.who.int/drugresistance/documents/surveillancereport/en/, last accessed on March 4, 2015.

Xia S, Duan L, Song Y, Li J, Piceno YM et al. (2010) Bacterial community structure in geographically distributed biological wastewater treatment reactors. *Environ Sci Technol*, 44: 7391–7396. https://doi.org/10.1021/es101554m.

Yang S, Deng W, Liu S, Yu X, Mustafa GR, Chen S, He L, Ao X, Yang Y, Zhou K, Li B, Han X, Xu X, Zou L (2020) Presence of heavy metal resistance genes in Escherichia coli and Salmonella isolates and analysis of resistance gene structure in E. coli E308. *J Glob Antimicrob Resist*. 21: 420–426. doi: 10.1016/j.jgar.2020.01.009.

Yoo K, Lee G (2021) Investigation of the prevalence of antibiotic resistance genes according to the wastewater treatment scale using metagenomic analysis. *Antibiotics*, 10: 188. https://doi.org/10.3390/ antibiotics10020188.

Yoo K, Lee TK, Choi EJ, Yang J, Shukla SK, Hwang SI, Park J (2017) Molecular approaches for the detection and monitoring of microbial communities in bioaerosols: A review. *J Environ Sci*, 51: 234–247.

Zhang T, Zhang XX, Ye L (2011) Plasmid metagenome reveals high levels of antibiotic resistance genes and mobile genetic elements in activated sludge. *PLoS One*, 6: e26041. https://doi.org/10.1371/journal.pone.0026041.

Zhang XX, Zhang T, Fang HH (2009) Antibiotic resistance genes in water environment. *Appl Microbiol Biotechnol*, 82: 397–414. https://doi.org/10.1007/s00253-008-1829-z.

Zhou H, Wang X, Li Z, Kuang Y, Mao D, Lou Y (2018) Occurrence and distribution of urban dust-associated bacterial antibiotic resistance in Northern China. *Environ Sci Technol Lett*, 5: 50–55.

9 Molecular Characterization of Multidrug-Resistant Genes in Wastewater Treatment Plants

Charles Oluwaseun Adetunji
Edo University Iyamho

John Tsado Mathew
Ibrahim Badamasi Babangida University Lapai

Kshitij RB Singh
Banaras Hindu University

Ruth Ebunoluwa Bodunrinde
Federal University of Technology Akure

Abel Inobeme and Olugbemi T. Olaniyan
Edo University Iyamho

Ogundolie Frank Abimbola
Federal University of Technology Akure

Jay Singh
Banaras Hindu University

Vanya Nayak and Ravindra Pratap Singh
Indira Gandhi National Tribal University

DOI: 10.1201/9781003354147-9

CONTENTS

9.1 INTRODUCTION

The rate of resistance to multidrug antibiotics in both pathogenic and non-pathogenic microbes is on the constant rise worldwide due to the multidrug antibiotic-resistant genes, which is alarmingly a major concern because soon common microbial infections will be hard to treat with the set of antibiotics available presently. The multidrug-resistant genes observed to be the cause of multidrug resistance in these microbes. It has been reported that horizontal genetic materials transfer such as transposons, DNA, genes, RNA, bacteriophages, plasmids can give rise to antibiotic resistance in pathogens as a result of the treatment conditions of the wastewater treatment plants (WWTPs) which actually results to biofilms formed during purification, the response to effluent discharged from hospitals which largely contains drugs and antibiotics and heavy metal-induced stress (Rizzo *et al.*, 2013a; Karkman *et al.*, 2018).

Molecular characterization using different methods is essential in understanding the origin of the resistance developed in these wastewater. To better understand the diversity of these multi-drug-resistant microbes in a microbiota, the gene acquisition systems operate as a genetic mechanism that allows the incorporation, expression and exchange of free genes called integrons (Stalder *et al.*, 2012; Marathe *et al.*, 2013; Domingues *et al.*, 2012; Gillings, 2014; Deng *et al.*, 2015). Metagenomics could be carried out using some of these novel techniques such as next-generation sequencing, quantitative polymerase chain reaction (qPCR) (Karkman *et al.*, 2016, 2018) PCR analysis of integrons and INTEGRALL regions (Park *et al.*, 2018) using PCR, application of high-throughput qPCR arrays (Stedtfeld *et al.*, 2008; Muziasari *et al.*, 2016; Muurinen *et al.*, 2017), whole genome sequencing (Mahfouz *et al.*, 2018) of microbial community DNA, and traditional cultivation on antibiotics (Hultman *et al.*, 2018; Park *et al.*, 2018). Hence, this chapter elaborates on the application of molecular techniques for characterization of multi-resistance genes in wastewater.

9.2 APPLICATION OF PCR AND OTHER RELEVANT MOLECULAR TECHNIQUES SUCH AS 16S RDNA SEQUENCING AND PCR GENOTYPING

In parasitology, normal laboratory diagnosis entails using traditional methods for morphological identification of parasites, such as optical microscopy. To improve parasite identification and characterization, molecular biology approaches are increasingly being applied to diagnose parasite structures. The current potential diagnostic approaches for parasitic disease confirmation are PCR, real-time PCR (RT-PCR), loop-mediated isothermal amplification (LAMP), LuminexxMAP, random amplified polymorphic DNA (RAPD), amplified fragment length polymorphism (AFLP), microsatellites, and restriction fragment length polymorphism (RFLP). The classification of the variety of classes residing inside microsatellites is crucial for understanding how these environments function. It has also become a public concern because it is important to perform biodiversity restoration or conservation. Traditionally, species have been identified and classified based on morphological criteria that are strongly related to climate factors or reach their limitations, particularly in groups in which many of them tend to be difficult to obtain, as in the case with numerous microbe classes. The PCR has become a crucial device designed for studying the performance of biological schemes due to the necessity to comprehend molecular pathways in organisms. A variety of symbols for detecting nuclear DNA polymorphisms are presently available. Microsatellites are part of the commonly utilized markers in genetic diversity investigations. The detailed study opened an innovative limit that necessitates high-throughput molecular techniques, high computer memory, novel data analysis techniques, and the incorporation of interdisciplinary expertise (Kadri, 2020).

However, the organisms in pleural effusion have been identified mostly by using single-species PCR and standard pathogens culture. The effectiveness of a commercially available multiplex pathogenic PCR assay established in pneumonia in identifying bacteria implicated in pleural disease, especially empyema, was studied. In total, 51/197 issues had transudate, whereas 146/197 cases had exudate. The study showed that there was an experimental doubt of para-pneumonic effusion in 42% of cases ($n=90/214$), and the definitive clinical diagnosis of empyema was prepared in 29% of cases ($n=61/214$). It was found that anaerobes showed most prevalence in cases of empyema, followed by means of Gram-negative rods and Gram-positive cocci. The multiplex PCR technique acknowledged more pathogens on the panel (23.3%) (7/30) than traditional approaches 6.7% (2/30) ($p=0.008$). Furthermore, when only the microorganisms in the pneumonia panel were considered, the multiplex PCR-based assay demonstrated higher specificity than traditional microbiology. However, it has been suggested that to cover the majority of pathogens involved in pleural infection, a specialized pleural empyema multiplex PCR panel incorporating anaerobes might be required (Franchetti *et al.*, 2020).

In order to detect and differentiate *Vibrio cholera* groups, researchers employed an innovative, effective gene touchdown-multiplex PCR (TMPCR), which would be composed of two amplifications related to multiplex PCR as well as conditional touchdown method. *V. cholera*-exact amplicons (588 bp) were obtained through the ompW gene, which encodes the serogroup-specific amplicons and external membrane protein. These amplicons were generated using a panel of biochemical marker-based TMPCR methods. According to reciprocal analysis, the TMPCR has a limit of detection as low as 100 pg of the O139, non-O1/non-O139 or O1 in reactions comprising evenly or unequally mixed gDNAs two-template combinations employing *V. cholerae* O139, O1, or evenly combined O139 and O1. Furthermore, when tested for serotyped *V. cholerae* strains isolated along with those retrieved from experimental samples, the O serogroup-specific TMPCR method demonstrated 100% (Bhumiratana *et al.*, 2014).

Vibrio is a large genus of freshwater bacteria with about 72 species, 12 of which have been found in human experimental samples. The bulk of *Vibrio* infections in humans is caused by 3 of these 12 species: *V. parahaemolyticus, V. cholerae* and *V. vulnificus.* Since phenotypic traits vary between biochemical and species, identification takes 2 or more days, and reliable and rapid identification of *Vibrio* species has proven difficult. A multiplex PCR was designed that used species-specific primers to amplify gene areas in four species to facilitate the identification of *V. parahaemolyticus, V. cholerae, V. mimicus* and *V. vulnificus*. The experiment was run on 309 *Vibrio* isolated from 26 different species (including 12 human infections). Among the other 119 isolates, the assay discovered four more *V. parahaemolyticus* isolates. The multiplex results for these four isolates were validated using sequencing based on rpoB, and they all grouped with the other *V. parahaemolyticus* sequences. Moreover, in a phylogenetic analysis, the rpoB genes of 12–15 formerly unknown isolates grouped by means of other *Vibrio* species, and three isolates looked to be of nameless *Vibrio* species. Therefore, the PCR assay was considered as a straightforward, fast, and reliable method for identifying the main *Vibrio* infections in experimental specimens, as well as rpoB genotyping was recommended as an alternative method for identifying other *Vibrio* species (Tarr *et al.*, 2007).

9.3 UTILIZATION OF PHENOTYPIC SUSCEPTIBILITY PATTERNS USING THE KIRBY–BAUER DISK DIFFUSION TECHNIQUE AND SOME OTHER RELEVANT TECHNIQUES

Wearable self-powered sensors are a popular topic in the literature because of their versatility. The Internet of Things (IoT) and implanted devices are making progress. The blending of various materials is a frontier to be investigated in harvesting energy through the physical environment or movement to act as an active component or power sensor of analyte revealing. This review summarizes the most recent research on integrating nanogenerators in wearable devices depending on the contact of triboelectric as well as piezoelectric schemes into more easy along with high harvesting arrangement for charging batteries or identifying multiple analytes in self-powered biosensors and sensors (Oliveira, 2021).

Antibiotic susceptibility tests (ASTs) are widely utilized in clinical diagnosis of the antibiotic resistance of bacterial isolates, but it is a slow, tedious and time-consuming process. The key factor which is used to determine whether bacteria are resistant or susceptible to an antibiotic is the minimum inhibition concentration (MIC). MIC is considered to predict the lowest concentration of the antibiotic which is responsible to avoid the growth of bacteria. If the breakpoint is recorded equal to or less than the determined MIC, the bacterial isolate is considered as resistant to the antibiotic. These experimental breakpoints are regularly updated and organized by various national organizations like Laboratory Clinical Standards Institute (LCSI) in the USA. Therefore, the MIC value plays a significant role in analyzing bacteria's phenotypic resistance, determining the efficiency of innovative antibiotics and monitoring the resistance of the global drug status. Although the process is slow, this technique can efficiently detect whether the antibiotic would inhibit pathogenic growth or not. On the other hand, this impediment increases the time to make effective antibiotic therapy and decisions for appropriate, which results in poor clinical outcomes as well as increased patient mortality. Therefore, it has now become important to develop a rapid technology which can recognize antibiotic susceptibility at the earliest possible treatment stage. Various AST technologies are been used such as microfluidic technologies which are utilized for rapid ASTs at the single-cell level, and they include both gene-based antimicrobial resistance (AMR) detection (droplet digital analysis) and phenotypic analysis (microfluidic-based single bacterial culture) as shown in Figure 9.1 (Zhang *et al.*, 2020).

The majority of disinfectants show antibacterial activity of varying degrees. The isolates were completely susceptible to hydrogen peroxide and formalin but do not exhibit susceptibility against sodium hypochlorite at the suggested user dilution. Because several of the bacteria isolated in the study were resistant to standard antibiotics and disinfectants, it is suggested that all the aspects should be included when

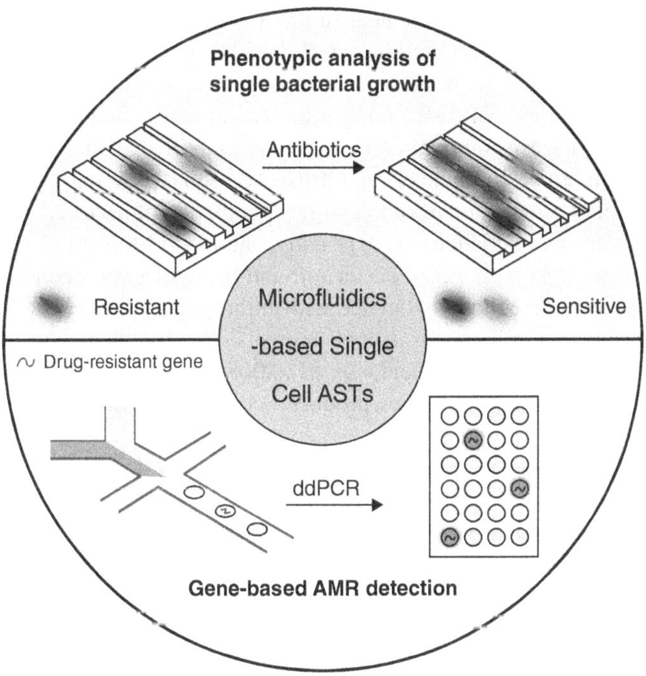

FIGURE 9.1 Rapid microfluidic technology including both phenotypic analysis and gene-based antimicrobial resistance (AMR) detection at the single-cell level. The phenotypic analysis is based on the single bacterial culture, and the AMR test uses droplet digital analysis. (Reprinted with permission from Zhang, K., Qin, S., Wu, S., Liang, Y., & Li, J. (2020). *Microfluidic systems for rapid antibiotic susceptibility tests (ASTs) at the single-cell level. Chemical Science, 11*(25), 6352–6361 [CC BY-NC 3.0].)

considering certain therapeutic decisions. The antibiotic resistance could spread resistant strains to other fish microorganisms as well as human pathogens, making treatment of the resulting disease(s) difficult; hence, this antibiotic resistance may well be spread from person to person who eats or handles the carrier fish. As a result, it is best to boil fish thoroughly before eating it to destroy any bacteria that might be present (Zhu *et al.*, 2018). Timely detection of bacterium and their susceptibility to antimicrobial drugs and successive effective antibiotic diagnosis are extremely important for appropriate clinical satisfaction. Traditional detection techniques of bacterial resistance, including broth microdilution, automated instruments, and disc diffusion, are continuously and extensively used. However, findings cannot be received until 48 hours after collecting a sample, resulting in the administration of broad-spectrum antibiotics for an extended period or overuse. As a result, there is a push to design and implement newer, quicker, sensitive, standardized and particular procedures with repeatable results into everyday microbiological laboratory practice.

MALDI-TOF MS (matrix-assisted laser ionization/desorption time-of-flight mass spectrometry) has recently been brought into laboratory practice, and microfluidics, as well as microdroplet-based technologies, may be adopted in the near future. This study focuses on the techniques and equipment that are now being used and those that may be utilized in the near future to measure antibiotic efficacy in clinical microbiology laboratories (Benkova *et al.*, 2020).

Water bodies were reported to be the most crucial habitat for microorganisms. This is a primary way of distributing microbes in nature and as a substantial reservoir of antibiotic-resistant organisms (Rizzo *et al.*, 2013; Shah, 2021; Baquero *et al.*, 2008). In their study, Tamames *et al.* (2010) used 16S rRNA and concluded that fresh water and soil wastewater harbor the largest population of bacterial groups. Martinez (2009), Andersson and Hughes (2011), and Cantón and Morosin (2011) reported that some opportunistic pathogenic bacteria found to be of great economic importance during the study were those associated with anthropogenic activities. Bacteria's properties to resist antibiotics occur in environments with little to no anthropogenic influences (D'Costa *et al.*, 2011; Dantas *et al.*, 2008; Segawa *et al.*, 2013; Shah, 2020). They revealed that the antibiotics produced could have also lost their effectiveness due to long years of production.

Wastewater treatment was the target point for antibiotic resistance genes and bacterial spread in the community (Vaz-Moreira *et al.*, 2014; Rizzo *et al.*, 2013). The presence of these bacteria with the resistant gene increases the possibility of non-antibiotic-resistant bacteria (Bennett, 2008). *Acinetobacter* spp. and *Pseudomonas* spp, among others, were found by Bonomo and Szabo (2006), Kaskhedikar and Chhabra (2010) to possess the potential to develop multidrug resistance in the environment having the ability to recuperate in any environment (Trautmann *et al.*, 2005; Silge *et al.*, 2014). These bacteria were found in effluent wastewater (Sautour *et al.*, 2003). Zhang *et al.* (2013) also reported the reason for increased antibiotic-resistant potential in *Acinetobacter* spp.

Numberger *et al.* (2019) utilized a full-length 16S ribosomal RNA gene with some boundaries and were able to accomplish distinguished and consistent taxonomic results; they were able to identify some enteric bacteria in effluent samples such as *Campylobacter*, *Salmonella*, *Yersinia* and some environmental bacteria such as

Pseudomonas, *Acinetobacter* and *Aeromonas* that could act as an antibiotic-resistant bacterium (Figueira *et al.*, 2011; Khan *et al.*, 2007). Huse *et al.* (2008) compared the full-length 16S rRNA with other molecular methods of identification of antibiotic-resistant bacteria in water effluent and found out that this method was able to demonstrate improved species-level resolution along with recovering phylogenetic tree results compared to others, providing an enhanced classification of uncommon bacterial clusters that are cost high-priced on some long-read platforms for sequencing. Ye and Zhang (2013), Marti *et al.* (2013), and Cai *et al.* (2014) also studied the use of this technique in WWTPs founded on data sequencing, and they were able to distinguish compositional disparities among the microbiomes of wastewater treatment plants, water inflow and effluent with the use of full-length 16S ribosomal RNA sequencing method.

Bacteria could be very injurious to humans and even animals when they carried antibiotic resistance genes and were found in environments like water bodies and effluents (McLellan *et al.*, 2010). 16S rRNA amplicon data was discovered to imitate a total large quantity of microorganisms owing to the PCR amplification procedures and bearing in mind the inconsistency in the copies of the gene in the 16S rRNA bacterial taxonomic group (Větrovský and Baldrian, 2013; Piwosz *et al.*, 2018). Researchers reported a significant reduction in the abundance of antibiotic resistance genes with detection by 16S rRNA (Huang *et al.*, 2015). Another study reported eliminating 16S rRNA, tet gene, and intl from different waste treatment systems ranging from 33% to 99% (Huang *et al.*, 2017). The difference they encountered in 16S rRNA then indicated that the use could eliminate bacteria from wastewater. An increased proportion of the antibiotic resistance gene could be risky, leading to proliferated antibiotic-resistant bacterial strains in the surroundings (Liu *et al.*, 2013).

9.4 ROLE OF BETA-LACTAMASE RESISTANCE TO BETA-LACTAM ANTIBIOTICS AS WELL AS A NEW TYPE OF BETA-LACTAMASE ENZYME, WHICH ENTAILS AMPC BETA-LACTAMASE AND EXTENDED-SPECTRUM BETA-LACTAMASE

Beta-lactamase formed through Gram-negative microbes is formed continuously at a constant rate at high concentrations or by direct stimulation of the beta-lactam antibiotics by means of the regulatory scheme (Zhu *et al.*, 1992). Safo *et al.* (2005) reported that the production level in Gram-negative organisms is in a reduced state. Harris and Ferguson (2012) extensively studied the molecular mode of beta-lactam resistance action. Gram-negative bacteria have advanced various approach, like the invention of new PBPs, which has lessened the affinity to beta-lactam antibiotics to elude the bacteria efficacy of beta-lactam antibiotics (Jacoby, 2009), helping to eradicate beta-lactam antibiotics alterations in the bacterial cell. In the *Enterobacteriaceae* family, AmpC expression was stimulated from antibiotics by beta-lactam. AmpC was known in beta-lactam-sensitive *E. coli* strains via the stepwise selection on a medium containing beta-lactam antibiotics (Eriksson-Grennbereg, 1968). Juan *et al.* (2005) reported the health implications of prolonging the use of beta-lactam antibiotics, which might lead to resistance strain of *P. aeruginosa*, making the purpose of treatment abortive and lethal. Ximin and Jun (2013), in their research, elucidated the molecular basis of the induction of beta-lactam, bringing about the development of an efficient mixture treatment approach by inhibiting the stimulation of beta-lactamase. Furthermore, they reported that bacteria from the *Enterobacteriaceae* family express AmpC via stimulation by beta-lactam antibiotics, thereby linking the expression resulting from treatment with beta-lactam antibiotics in cell wall recycling.

Harris and Ferguson (2012) reported that the detection of beta-lactamase inhibitors is a capable strategy to fight the predominant issue of resistance in beta-lactam antibiotics. They added that the inhibitors that target the stimulation of the beta-lactamase way could avert the advent of resistance to beta-lactam antibiotics, thereby enhancing the effectiveness of the antibiotics as they detected the inhibitors from the effluent pump (Lomovskaya and Bostain, 2006). In order to support their hypothesis, the rate of rising of a resistant strain of *P. aeruginosa* to ceftazidime was far lower than the discovery limit when compared with the wild-type parental isolate (Moya *et al.*, 2009).

Moreover, there are various other inhibitors which inhibit beta-lactamase. Figure 9.2 shows the interaction of the beta-lactamase with different beta-lactamase inhibitors. Figure 9.2a and c shows the interaction of vaborbactam, which inhibits the β-lactamases of classes A, but at the same time, it does not enlarge the activity of the metallo-beta-lactamases (MBLs) or OXA enzymes. Figure 9.2b shows the interaction of beta-lactamase with the avibactam interacting with representative enzymes of classes A, C and D. However, overall, the most potent inhibitors are the bicyclic boronates, which efficiently inhibit both MBLs and a serine beta-lactamase through mimicking the tetrahedral oxyanion produced through β-lactam hydrolysis as shown in Figure 9.2d.

Furthermore, researchers have also established that gene expression of chromosomal AmpC by *Citrobacter freundii*, *Morganella morganii*, *Serratia macescens* and *Enterobacter cloacae* is induced by beta-lactam antibiotics like cefoxitin, etc., but there is less stimulation by both the third or/and fourth generation of cephalosporins (Hanson and Sanders, 1999). For stimulation to take place, there is a need for the DNA-binding protein AmpR, and this process is known to be reversible; previously, the stimulating agent was detached. During their work (Lindberg and Normark, 1987; Lindquist *et al.*, 1989), they utilized *Enterobacter cloacae* and *Citrobacterfreundii*. Beta-lactam antibiotics affect the cell wall of the bacterial cell, and the antibiotics then bind together to be able to carry out the inhibition process of the carboxypeptidase and transpeptidases; these are enzymes that are catalyzed by Ala-D-Ala cross-linkages of peptidoglycan which surrounds the bacteria. Deepti and Deepthi (2010) showed that the beta-lactam-resistant potential could be inherent in certain microbial species, for example, Enterococci, which are integrally non-sensitive to PBP. Instead, this property could also be gotten via impulsive alteration of the genetic makeup or transfer of DNA. Functionally, resistance properties in beta-lactam may be due to the production of enzyme beta-lactamases, impenetrability, modification of target and efflux.

Livermoore and Paterson (2006) reported that the usual cause for resistance in Gram-positive bacteria in methicillin-resistant *S. aureus* and *Pneumococci* are the charges existing in the usual PBPs or the acquisition of extra beta-lactam non-sensitive PBPs, while in Gram-negative bacteria, resistance is usually as a result of

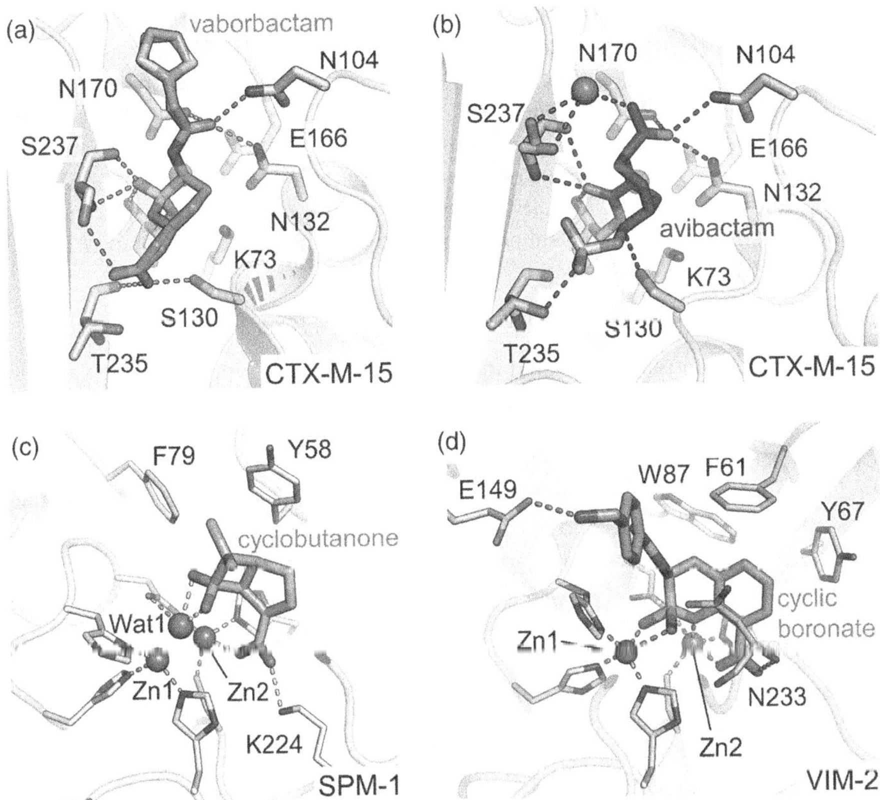

FIGURE 9.2 Different interactions of beta-lactamase inhibitors with beta-lactamase. (a) Vaborbactam binds to class A CTX-M-15. (b) Avibactam binds with the CTX-M-15 (PDB 4hbu). (c) A cyclobutanone shows its binding with the subclass B2 metallo-β-lactamase SPM-1 (PDB 5ndb). (d) A bicyclic boronate is found to be bonded with the subclass B1 metallo-beta-lactamase VIM-2 (PDB 5fqc). The sticks show the inhibitor molecules and the protein residues during their interactions. Moreover, the interaction between ligands is represented as colored dashes. (Reprinted with permission from Tooke, C. L., Hinchliffe, P., Bragginton, E. C., Colenso, C. K., Hirvonen, V. H., Takebayashi, Y., & Spencer, J. (2019). *β-Lactamases and β-Lactamase Inhibitors in the 21st Century. Journal of molecular biology, 431*(18), 3472–3500 [CC BY 4.0].)

the combination of endogenic acquired beta-lactamases, coupled with natural up-controlled impenetrability and efflux. Bradford (2001) reported that beta lactamases could be chromosomally programmed or plasmid-mediated. Chromosomal beta-lactamases were supposed to have advanced from PBPs and they both exhibited similar sequence distinction. This finding could also be due to the selective pressure used by beta-lactam-producing soil organisms inherent in the environment. TEM-1 enzymes, which spread universally, are now discovered in several species of Enterobacteriaceae such as *Hemophilus influenza*, *Pseudomonas aeruginosa* and *Neisseria gonorrhoeae*. SHV1, also known as sulphydra variable type 1, is also a beta-lactamase enzyme mainly found in *E. coli* and *Klebsiella*. Researchers found plasmid-mediated AmpC beta-lactamases to be universal but less shared compared with extended-spectrum beta-lactamases (ESBLs); these enzymes are less cause of cefoxitin resistance in *Escherichia coli* than the increased population of chromosomal AmpC β -lactamases. The CMY-2 beta-lactamase was reported to have the broadest geographic diversity and common cause for the beta-lactam antibiotics resistance in non-typhoid strains of *Salmonella* in many countries (Miriagou *et al.*, 2004). Stains with AmpC usually possess a resistant ability to multiple antibiotic agents, making a particular agent problematic (Pai *et al.*, 2004; Vimont *et al.*, 2007).

9.5 CONCLUSION AND PROSPECTS

This chapter has provided detailed information on diverse types of molecular techniques that are applied in the characterization of multi-resistance genes that are available in wastewater. Detailed information on the usage of PCR and other relevant molecular techniques such as 16S rDNA sequencing and PCR genotyping, and the application of phenotypic susceptibility patterns using the Kirby–Bauer disk diffusion technique and some other relevant techniques were highlighted. Special attention was paid to the role of beta-lactamase resistance to beta-lactam antibiotics as well as a new type of beta-lactamase enzymes, which entails AmpC beta-lactamase and extended-spectrum beta-lactamase. The application of next-generation sequencing will also go a long way toward the detection of numerous genes that might be responsible for a different level of antibiotic resistance in diverse wastewater.

REFERENCES

Adetunji CO, Kumar D, Raina M, Arogundade O, Sarin NB (2019). Endophytic microorganisms as biological control agents for plant pathogens: a panacea for sustainable agriculture. In Varma A, Tripathi S, Prasad R (eds) *Plant Biotic Interactions*. Springer: Cham. https://doi.org/10.1007/978-3-030-26657-8_1

Adetunji CO, Ugbenyen MA. (2019). Mechanism of action of nanopesticide derived from microorganism for the alleviation of abiotic and biotic stress affecting crop productivity. In Panpatte D, Jhala Y (eds) *Nanotechnology for Agriculture: Crop Production & Protection*. Springer: Singapore. https://doi.org/10.1007/978-981-32-9374-8_7

Adetunji CO, Anani OA (2021). Plastic-Eating Microorganisms: Recent Biotechnological Techniques for Recycling of Plastic. In: Panpatte DG, Jhala YK (eds) *Microbial Rejuvenation of Polluted Environment. Microorganisms for Sustainability*, vol 25. Springer: Singapore. https://doi.org/10.1007/978-981-15-7447-4_14

Adetunji CO, Olaniyan OT, Bodunrinde RE, Ahamed MI (2021). Bioconversion of poultry waste into added-value products. In Inamuddin, Khan A (eds) *Sustainable Bioconversion of Waste to Value Added Products. Advances in Science, Technology & Innovation*. Springer: Cham. https://doi.org/10.1007/978-3-030-61837-7_21

Andersson DI, Hughes D (2011). Persistence of antibiotic resistance in bacterial populations. *FEMS Microbiol Rev* 35: 901–911.

Baquero F, Martínez JL, Cantón R (2008). Antibiotics and antibiotic resistance in water environments. *Curr Opin Biotechnol* 19: 260–265.

Barbosa TM, Levy SB (2000). The impact of antibiotic use on resistance development and persistence. *Drug Resist Updat* 3: 303–311.

Benkova M, Soukup O, Marek J (2020). Antimicrobial susceptibility testing: Currently used methods and devices and the near future in clinical practice. *J Appl Microbiol*. doi: 10.1111/jam.14704.

Bennett PM (2008). Plasmid encoded antibiotic resistance: Acquisition and transfer of antibiotic resistance genes in bacteria. *Br J Pharmacol* 153: S347–S357.

Bhumiratana A, Siriphap A, Khamsuwan N, Borthong J, Chonsin K, Sutheinkul O (2014). O serogroup-specific touchdown-multiplex polymerase chain reaction for detection and identification ofvibrio choleraeO1, O139, and non-O1/Non-O139. *Biochem Res Int*: 1–10. doi: 10.1155/2014/295421

Bonomo RA, Szabo D (2006). Mechanisms of multidrug resistance in *Acinetobacter* species and *Pseudomonas aeruginosa*. *Clin Infect Dis* 43: S49–S56.

Bradford PA (2001). Extended-spectrum-lactamases in the 21st century: Characterization, epidemiology, and detection of this important resistance threat. *Clin Microbiol Rev* 14: 933–951.

Cai L, Ju F, Zhang T (2014). Tracking human sewage microbiome in a municipal wastewater treatment plant. *Appl Microbiol Biotechnol* 98: 3317–3326.

Cantón R, Morosini MI (2011). Emergence and spread of antibiotic resistance following exposure to antibiotics. *FEMS Microbiol Rev* 35: 977–991.

Cateau E, Verdon J, Fernandez B, Hechard Y, Rodier M-H (2011). *Acanthamoeba* sp. promotes the survival and growth of *Acinetobacter baumanii*. *FEMS Microbiol Lett* 319: 19–25.

Dantas G, Sommer MO, Oluwasegun RD, Church GM (2008). Bacteria subsisting on antibiotics. *Science* 320: 100–103.

Dauda WP, Abraham P, Glen E, Adetunji CO, Ghazanfar S, Ali S, Al-Zahrani M, Azameti MK, Alao SEL, Zarafi AB, Abraham MP, Musa H (2022a) Robust profiling of Cytochrome P450s (P450ome) in notable *Aspergillus* spp. *Life*, 12(3): 451. https://doi.org/10.3390/life12030451

Dauda WP, Morumda D, Abraham P, Adetunji CO, Ghazanfar S, Glen E, Abraham SE, Peter GW, Ogra IO, Ifeanyi UJ, Musa H, Azameti MK, Paray BA, Gulnaz A (2022b). Genome-wide analysis of Cytochrome P450s of *Alternaria* Species: evolutionary origin, family expansion and putative functions. *J Fungi*, 8(4): 324. https://doi.org/10.3390/jof8040324

D'Costa VM et al. (2011). Antibiotic resistance is ancient. *Nature* 477: 457–461.

Deepti R, Deepthi N (2010). Extended-spectrum β-lactamases in Gram negative bacteria 2(3): 263–274. doi: 10.4103/0974-777X.68531.

Deng Y et al. (2015). Resistance integrons: Class 1, 2 and 3 integrons. *Ann Clin Microbiol Antimicrob* 14(1): 1–11.

Domingues S, da Silva GJ, Nielsen KM (2012). Integrons: Vehicles and pathways for horizontal dissemination in bacteria. *Mob Genet Element* 2(5): 211–223.

Eriksson-Grennberg KG (1968). Resistance of *Escherichia coli* to penicillins. II. An improved mapping of the ampA gene. *Genet Res* 12: 147–156.

Figueira V, Vaz-Moreira I, Silva M, Manaia CM (2011). Diversity and antibiotic resistance of *Aeromonas* spp. in drinking and waste water treatment plants. *Water Res* 45: 5599–5611.

Franchetti L, Desiree MS, Tamm M, Kathleen J, Daiana S (2020). Multiplex bacterial polymerase chain reaction in a cohort of patients with pleural effusion. *BMC Infect Dis* 20(99): 1–10. https://doi.org/10.1186/s12879-020-4793-6.

Gillings MR (2014). Integrons: Past, present, and future. *Microbiol Mol Biol Rev* 78(2): 257–277.

Hanson ND, Sanders CC (1999). Regulation of inducible AmpC β-lactamase expression among Enterobacteriaceae. *Curr Pharmaceut Des* 5: 881–9411.

Harris PN, Ferguson JK (2012). Antibiotic therapy for inducible AmpC beta-lactamase-producing Gram-negative bacilli: What are the alternatives to carbapenems, quinolones and aminoglycosides? *Int J Antimicrob Agents* 40: 297–305.

Huang X, Liu C, Li K, Su J, Zhu G, Liu L (2015). Performance of vertical up-flow constructed wetlands on swine wastewater containing tetracyclines and tert genes. *Water Res* 70: 109–117.

Huang X, Zheng J, Liu C, Liu L, Liu Y, Fan H (2017). Removal of antibiotics and resistance genes from swine wastewater using vertical flow constructed wetlands: Elect of hydraulic flow direction and substrate type. *Chem Eng J* 308: 692–699.

Hultman J, Tamminen M, Pärnänen K, Cairns J, Karkman A, Virta M (2018). Host range of antibiotic resistance genes in wastewater treatment plant influent and effluent. *FEMS Microbiol Ecol* 94(4): fiy038. doi: 10.1093/femsec/fiy038.

Huse SM et al. (2008). Exploring microbial diversity and taxonomy using SSU rRNA hypervariable tag sequencing. *PLoS Genet* 4: e1000255.

Jacoby GA (2009). AmpC beta-lactamases. *Clin Microbiol Rev* 22: 161–182.

Jernberg C, Lofmark S, Edlund C, Jansson JK (2010). Long-term impacts of antibiotic exposure on the human intestinal microbiota. *Microbiology* 156: 3216–3223.

Juan C, Gutierrez O, Oliver A, Ayestaran JI, Borrell N, Perez JL (2005). Contribution of clonal dissemination and selection of mutants during therapy to Pseudomonas aeruginosa antimicrobial resistance in an intensive care unit setting. *Clin Microbiol Infect* 11: 887–892.

Kadri, K (2020). Polymerase chain reaction (PCR): Principle and applications. *Synth Biol – New Interdis Sci.* doi: 10.5772/intechopen.86491.

Karkman A, Do TT, Walsh F, Virta MP (2018). Antibiotic-resistance genes in waste water. *Trends Microbiol* 26(3): 220–228.

Karkman A, Johnson TA, Lyra C, Stedtfeld RD, Tamminen M, Tiedje JM, Virta M (2016). High-throughput quantification of antibiotic resistance genes from an urban wastewater treatment plant. *FEMS Microbiol Ecol* 92(3).

Karumathil DP, Yin H-B, Kollanoor-Johny A, Venkitanarayanan K (2014). Effect of chlorine exposure on the survival and antibiotic gene expression of multidrug resistant *Acinetobacter baumannii* in. *Water Int J Environ Res Public Health* 11: 1844–1854.

Kaskhedikar M, Chhabra D (2010). Multiple drug resistance in *Aeromonas hydrophila* isolates of fish. *Food Microbiol* 28: 157–168.

Khan NH et al. (2007). Isolation of *Pseudomonas aeruginosa* from open ocean and comparison with freshwater, clinical, and animal isolates. *Microb Ecol* 53: 173–186.

Kristiansson E, Fick J, Janzon A, Grabic R, Rutgersson C, Weijdegård B, Larsson DJ (2011). Pyrosequencing of antibiotic-contaminated river sediments reveals high levels of resistance and gene transfer elements. *PLoS One* 6(2): e17038.

Kumar M, Ram B, Sewwandi H, Honda R, Chaminda T (2020). Treatment enhances the prevalence of antibiotic-resistant bacteria and antibiotic resistance genes in the wastewater of Sri Lanka, and India. *Environ Res* 183: 109179.

Lindberg F, Normark S (1987). Common mechanism of AmpC β-lactamase induction in Enterobacteria: Regulation of the cloned *Enterobacter cloacae* P99 β-lactamase gene. *J Bacteriol* 169: 758–763.

Lindquist S, Lindberg F, Normark S (1989). Binding of the *Citrobacter freundii* AmpR regulator to a single DNA site provides both autoregulation and activation of the inducible *ampC* β-lactamase gene. *J Bacteriol* 171: 3746–3753.

Liu L, Liu C, Zheng J, Huang X, Wang Z, Liu Y, Zhu G (2013). Elimination of veterinary antibiotics and antibiotic resistance genes from swine wastewater in the vertical flow constructed wetlands. *Chemosphere* 91: 1088–1093.

Livermoore DM, Paterson DL (2006). *Pocket Guide to Extended Spectrum β-Lactamases in Resistance*. New Delhi: Springer.

Lomovskaya O, Bostian KA (2006). Practical applications and feasibility of efflux pump inhibitors in the clinic – a vision for applied use. *Biochem Pharmacol* 71: 910–918.

Mahfouz N, Caucci S, Achatz E, Semmler T, Guenther S, Berendonk TU, Schroeder M (2018). High genomic diversity of multi-drug resistant wastewater Escherichia coli. *Sci Rep* 8(1): 1–12.

Marathe NP, Regina VR, Walujkar SA, Charan SS, Moore ER, Larsson DJ, Shouche YS (2013). A treatment plant receiving waste water from multiple bulk drug manufacturers is a reservoir for highly multi-drug resistant integron-bearing bacteria. *PLoS One* 8(10): e77310. doi: 10.1371/journal.pone.0077310.

Marti E, Jofre J, Balcazar JL (2013). Prevalence of antibiotic resistance genes and bacterial community composition in a river influenced by a wastewater treatment plant. *PLoS One* 8: e78906.

Martinez JL (2009). Environmental pollution by antibiotics and by antibiotic resistance determinants. *Environ Pollut* 157: 2893–2902.

McLellan SL, Huse SM, Mueller-Spitz SR, Andreishcheva EN, Sogin, ML (2010). Diversity and population structure of sewagederived microorganisms in wastewater treatment plant influent. *Environ Microbiol* 12: 378–392.

Miriagou V, Tassios PT, Legakis NJ, Tzouvelekis LS (2004). Expanded-spectrum cephalosporin resistance in non-typhoid *Salmonella. Int J Antimicrob Agents* 23: 547–555.

Moya B, Dotsch A, Juan C, Blazquez J, Zamorano L, Haussler S (2009). β-Lactam resistance response triggered by inactivation of a nonessential penicillin-binding protein. *PLoS Pathog* 5: e1000353. doi: 10.1371/journal.ppat.1000353.

Muziasari WI, Pärnänen K, Johnson TA, Lyra C, Karkman A, Stedtfeld RD, Virta M (2016). Aquaculture changes the profile of antibiotic resistance and mobile genetic element associated genes in Baltic Sea sediments. *FEMS Microbiol Ecol* 92(4): fiw052.

Numberger D, Ganzert L, Zoccarato L, Mühldorfer K, Sauer S, Hans-Peter Grossart HP, Greenwood AD (2019). Characterization of bacterial communities in wastewater with enhanced taxonomic resolution by full-length 16S rRNA sequencing 9: 9673. https://doi.org/10.1038/s41598-019-46015-z.

Oliveira HPd (2021). Wearable nanogenerators: Working principle and self-powered biosensors applications. *Electrochem* 2: 118–134. https://doi.org/10.3390/electrochem 2010010.

Pai H, Kang CI, Byeon JH, Lee KD, Park WB, Kim HB, Kim EC, Oh MD, Choe KW (2004). Epidemiology and clinical features of bloodstream infections caused by AmpC-type-β-lactamase-producing *Klebsiella pneumoniae. Antimicrob Agents Chemother* 48: 3720–3728

Park JH, Kim YJ, Seo KH (2018). Spread of multidrug-resistant Escherichia coli harboring integron via swine farm waste water treatment plant. *Ecotoxicol Environ Safety* 149: 36–42.

Piwosz K et al. (2018). Determining lineage-specific bacterial growth curves with a novel approach based on amplicon reads normalization using internal standard (ARNIS). *ISME J* 12: 2640–2654.

Pruesse E, Peplies J, Glöckner, FO (2012). SINA: Accurate high-throughput multiple sequence alignment of ribosomal RNA genes. *Bioinformatics* 28: 1823–1829.

Rizzo L, Manaia C, Merlin C, Schwartz T, Dagot C, Ploy MC, Fatta Kassinos D (2013). Urban wastewater treatment plants as hotspots for antibiotic resistant bacteria and genes spread into the environment: A review. *Sci Total Environ* 447: 345–360.

Safo MK, Zhao Q, Ko TP, Musayev FN, Robinson H, Scarsdale N (2005). Crystal structures of the BlaI repressor from *Staphylococcus aureus* and its complex with DNA: Insights into transcriptional regulation of the bla and mec operons. *J Bacteriol* 187: 1833–1844.

Sangeetha J, Hospet R, Thangadurai D, Adetunji CO, Islam S, Pujari N, Al-Tawaha ARMS (2021). Nanopesticides, nanoherbicides, and nanofertilizers: the greener aspects of agrochemical synthesis using nanotools and nanoprocesses toward sustainable agriculture. In: Kharissova OV, Torres-Martínez LM, Kharisov BI (eds) *Handbook of Nanomaterials and Nanocomposites for Energy and Environmental Applications*. Springer: Cham. https://doi.org/10.1007/978-3-030-36268-3_44

Sautour M, Mary P, Chihib NE, Hornez JP (2003). The effects of temperature, water activity and pH on the growth of *Aeromonas hydrophila* and on its subsequent survival in microcosm water. *J Appl Microbiol* 95: 807–813.

Segawa T, Takeuchi N, Rivera A, Yamada A, Yoshimura Y, Barcaza G, Shinbori K, Motoyama H, Kohshim S, Ushida K (2013). Distribution of antibiotic resistance genes inglacier environments. *Environ Microbiol Rep* 5: 127–134.

Shah MP (2020). *Advanced Oxidation Processes for Effluent Treatment Plants*. Elsevier.

Shah MP (2021). *Removal of Emerging Contaminants through Microbial Processes*. Springer.

Silge A et al. (2014). Identification of water-conditioned *Pseudomonas aeruginosa* by Raman microspectroscopy on a single cell level. *Syst Appl Microbiol* 37: 360–367.

Stalder T, Barraud O, Casellas M, Dagot C, Ploy MC (2012). Integron involvement in environmental spread of antibiotic resistance. *Front Microbiol* 3: 119.

Stedtfeld RD et al. (2008). Development and experimental validation of a predictive threshold cycle equation for quantifiation of virulence and marker genes by high-throughput nanoliter-volume PCR on the OpenArray platform. *Appl Environ Microbiol* 74: 3831–3838.

Tamames J, Abellan JJ, Pignatelli M, Camacho A, Moya A (2010). Environmental distribution of prokaryotic taxa. *BMC Microbiol* 10: 85.

Tarr CL, Patel JS, Puhr ND, Sowers EG, Bopp CA, Strockbine NA (2007). Identification of vibrio isolates by a multiplex PCR assay and rpoB sequence determination. *J Clin Microbiol* 45(1): 134–140. doi: 10.1128/jcm.01544–06.

Tooke CL, Hinchliffe P, Bragginton EC, Colenso CK, Hirvonen VH, Takebayashi Y, Spencer J (2019). β-Lactamases and β-Lactamase Inhibitors in the 21st Century. *J Mol Biol* 431(18): 3472–3500.

Trautmann M, Lepper PM, Haller M (2005). Ecology of *Pseudomonas aeruginosa* in the intensive care unit and the evolving role of water outlets as a reservoir of the organism. *Am J Infect Control* 33: S41–S49.

Vaz-Moreira I, Nunes OC, Manaia CM (2014). Bacterial diversity and antibiotic resistance in water habitats: Searching the links with the human microbiome. *FEMS Microbiol Rev* 38: 761–778. https://doi.org/10.1111/1574-6976.12062.

Větrovský T, Baldrian P (2013). The Variability of the he 16S rRNA gene in bacterial genomes and its consequences community analyses. *PLoS One* 8: e57923.

Vimont S, Aubert D, Mazoit JX, Poirel L, Nordmann P (2007). Broad-spectrum β-lactams for treating experimental peritonitis in mice due to *Escherichia coli* producing plasmid-encoded cephalosporinases. *J Antimicrob Chemother* 60: 1045–1050.

Ximin Z, Jun L (2013). Beta-lactamase induction and cell wall metabolism in Gram-negative bacteria. *Front Microbiol* 4: 128. doi: 10.3389/fmicb. 2013.00128.

Ye L, Zhang T (2013). Bacterial communities in different sections of a municipal wastewater treatment plant revealed by 16S rDNA 454 pyrosequencing. *Appl Microbiol Biotechnol* 97: 2681–2690.

Zhang Y, Marrs CF, Simon C, Xi C (2009). Wastewater treatment contributes to selective increase of antibiotic resistance among *Acinetobacter* spp. *Sci Total Environ* 407: 3702–3706.

Zhang K, Qin S, Wu S, Liang Y, Li J (2020). Microfluidic systems for rapid antibiotic susceptibility tests (ASTs) at the single-cell level. *Chem Sci* 11(25): 6352–6361.

Zhang CL, Wang J, Dodsworth JA, Williams AJ, Zhu C, Hinrichs KU, Zheng F, Hedlund BP (2013). In situ production of branched glycerol dialkyl glycerol tetraethers in a great basin hot spring (USA). *Front Microbiol* 4: 1–12.

Zhu X-D, Chu J, Wang Y-H (2018). Advances in microfluidics applied to single cell operation. *Biotechnol J* 13(2): 1700416. doi: 10.1002/biot.201700416.

Zhu Y, Englebert S, Joris B, Ghuysen JM, Kobayashi T, Lampen JO (1992). Structure, function, and fate of the BlaR signal transducer involved in induction of beta-lactamase in *Bacillus licheniformis*. *J Bacteriol* 174: 6171–6178.

10 Microbes and Events in Contaminant Biotransformation

Sreejita Ghosh
Maulana Abul Kalam Azad University of Technology

Dibyajit Lahiri and Moupriya Nag
University of Engineering & Management

Sougata Ghosh
RK University

Rina Rani Ray
Maulana Abul Kalam Azad University of Technology

CONTENTS

DOI: 10.1201/9781003354147-10

10.1 INTRODUCTION

Increased environmental pollution has led to the increased search for eco-friendly, low-cost and more effective environmental cleaning approaches. Exploitation of microbes for remediation of waste-contaminated sites is a beneficial choice because microorganisms exist widely in our biosphere and are metabolically active and capable of growing in various atmospheric conditions. Versatility of nutrients of microbes makes them a wise choice to biologically degrade organic contaminants. Bioremediation is a bioprocess used for the purpose of recycling wastes in some other forms that can be used and reused by another group of microbes. Thus, microbes are important in finding the major alternative strategy to overcome various barriers. Various bioremediation strategies include eradication, degradation, detoxification or immobilization of a wide variety of physical as well as chemical hazardous wastes by microbial actions. The underlying principle of bioremediation is the degradation and transformation of organic pollutants such as oils, hydrocarbons, dyes, pesticides, heavy metals and chemicals. Indigenous microbes called autochthonous microbes are present in contaminated sites and they can provide a promising solution for biodegradation and proper bioremediation of inorganic as well as organic contaminants under appropriate environmental conditions for their metabolism and growth. Hence, bioremediation is a stream of environmental biotechnology which can serve as an alternative.

Certain microbes have the capacity of converting, modifying and utilizing harmful pollutants to obtain energy and produce biomass through this process (Abatenh et al., 2017). Instead of just collecting and storing the pollutants, microbial bioremediation is a well-organized and orderly procedure for degradation and transformation of contaminants into less harmful or non-harmful compounds or elemental forms. Microorganisms make use of some bioremediation approaches such as biofilm formation, enzyme-based reduction, bioaccumulation and biosorption. Sometimes, there develop conditions, under which optimal bioremediation of contaminants cannot be achieved; in such cases, the microbial strains can be genetically modified and improvised. Through methods such as genetic engineering and recombinant

DNA technology, microbial strains can be modified as necessary (Rao et al., 2010). Scientists have so far developed and fabricated different bioremediation approaches, but because of the varied types of pollutants, there is no single foolproof bioremediation method for decontaminating polluted sites.

Microbes capable of performing bioremediation are called bioremediation aiding in the biological cleaning of waste sites. Among the primary bioremediation are fungi, algae and bacteria. The terms bioremediation and biotransformation can be used interchangeably. Using microorganisms to decontaminate contaminated areas have many advantages over various chemical procedures. The advantages of using microbial approaches include eco-friendliness and low cost. Microbes help in restoration of natural environmental conditions and prevent further pollution. This chapter aims to focus on the various bioremediation strategies employed by different microbes and how omics approaches such as metagenomic analyses help in analyzing the microbial communities capable of bioremediation.

10.2 MECHANISM OF BIOREMEDIATION

Application of microorganisms for environmental pollutant degradation is called bioremediation. It is an ecologically safe and sound modernized method utilizing microbial bioprocesses for complete removal of toxic contaminants. Not only are microbes important in maintaining the biogeochemical cycles for perpetuating the environment, keeping all the plants and animals free of diseases and healthy but also necessary for environmental clean-up (Malla et al., 2018). Microbial bioremediation is a promising alternative due to its easier application, environmentally friendly nature, and cost-effectiveness in comparison to the 'muck, suck and truck' non-biological remedial processes where the pollutants are just dug out or pumped up and then shipped somewhere else.

However, these bioremediation procedures are yet to be totally utilized. The major reason for this is that bioremediation strategies effective at one site may not be effective at another site. Moreover, the microbial bioremediation approaches that perform well under laboratory conditions may not perform well in field applications. The cause of these failures is still not understood, and thus many environmentalists are not interested in using bioremediation as an alternative approach for environmental clean-up. Actually, the mechanism controlling microbial growth and activities in contaminated sites is not clearly known, thereby restricting the application of bioremediation. Ability to adapt to extreme environmental conditions and acquire resistance against contaminants such as heavy metals, nutritional versatility and dynamic behavior makes microbes the most appropriate life forms capable of endurance. This characteristic feature of the microorganisms is beneficial and advantageous for toxic compounds and pollutant elimination from the environment. Microorganisms can biologically degrade pollutants through various enzyme-based processes, thereby removing or mitigating environmental contaminations (Lovley, 2003).

Microbes can perform this wide array of bioremediation processes by oxidation, reduction, binding, immobilization, volatilization, complexation, adsorption, precipitation, ion exchange, electrostatic interactions, biosorption, bioaccumulation and biological pollutant transformation. The most common bioremediation technique is the harmful organic pollutant oxidation to non-harmful products. Oxygen is the general electron acceptor for microbial respiration and aerobic biodegradation of various organic pollutants such as pesticides, xenobiotics and benzene (Wackett and Hershberger, 2001). Although aerobic pollutants can be biodegraded by a wide variety of microbes, *Pseudomonas* spp. is the most widely used microorganism for bioremediation because of its ability to biodegrade a vast array of different contaminants (Wackett and Hershberger, 2001).

Extrapolymeric substances (EPSs), which are present on the cell wall of biomass, can get attached to contaminants via mechanisms such as micro precipitation of pollutants or proton exchange pumps. Surfaces of biomass possess a negative charge due to the presence of amino, carboxyl, sulfhydryl and phosphoryl groups acting as sites for ion exchange and metal sinks.

Microbes initiate metal immobilization/mobilization via redox reactions. Bioremediation is carried out by conversion of an insoluble elemental form to its stationary form in sediments into its soluble as well as mobile phase. Mobilization may have some disadvantages when toxic pollutants may get released or redistributed from the solid phase in sediments to the mobile phase (Fomina and Gadd, 2014). This magnifies their bioavailability, and thus, the pollutants can reach the microbial metabolic system and get reduced. Microbial reduction allows leaching of pollutants from the soil. It has been reported that microbes from various natural aquifers mainly *Aspergillus niger* can reduce arsenic (Pokhrel and Viraraghavan, 2006). Biomethylation of heavy metals is a significant bioremediation process in water and soil for toxicity modification, mobility and volatility. This is because methylated species that are volatile can be eliminated from the cells (Bolan et al., 2014). Fraction of organic matter present in the soil acts as methyl donors. There also exists an indirect pollutant mobilization mechanism involving the decomposition of organic matter through microbes. *Schizoplyllum commune* has been found to decompose organic wastes (Wengel et al., 2006). Metabolite excretion like amino acids and carboxylic acids by microorganisms is an important driving force for carrying out pollutant chelation. Thus, it can be concluded that bioremediation strategies should be fabricated on the basis of the knowledge of the microbes present in the sites of pollution, their metabolic activities and the way in which the microbes are reacting to certain changes in the environmental conditions.

10.2.1 Bioremediation by Bacteria

Bacteria are widely present in the environment. Bacterial biosorption is an efficient and cheap technique for pollutant removal both organic and inorganic. Bacterial biomass is composed of dead or living bacterial cells. Several species of bacteria have mastered the survival strategy in presence of pollutants acquiring resistance against them and aiding in the bioremediation of such pollutants (Mustapha and Halimoon, 2015).

Rapid removal of pollutants by bacteria can be achieved through effective biosorption due to the different bacterial cellular structures with respect to the peptidoglycans such as N-acetyl glucosamine and N-acetyl muramic acid. The cell wall of the bacteria is the main physical linkage between the bacterial biomass and the pollutant. The total negative charge because of the anionic functional groups such as hydroxyl, amine, sulfate, carboxyl and phosphate (van der Waals et al., 1997) in Gram-positive bacteria having teichuronic acids, teichoic acids and peptidoglycan in their cell walls and Gram-negative bacteria having phospholipids, lipopolysaccharide (LPS) and peptidoglycan in their cell walls are responsible for binding the pollutants within or on the cells. Pollutant removal through dead cells of biomass is completely an extracellular process.

The carboxyl groups are known to bind the pollutants via surface complexation (Yee and Fein, 2001). The amino groups are responsible for electrostatic interactions and chelation. However, the bacterial strains at first need to be exposed to the pollutants for activation of enzymatic processes for bioremediation. For bacterial species like *Desulfovibrio*, *Pseudomonas*, *Geobacter* and *Bacillus*, there is a minimum necessity of pollutant concentration for initiating enzymatic bioremediation processes (Adenipekun and Lawal, 2012).

10.2.2 Phycoremediation

In marine ecosystems, different algal species exist in large quantities. Algae are autotrophs with low nutritional requirements and the ability to produce a large quantity of biomass. Out of the three major algal groups which are Chlorophyta (green algae), Rhodophyta (red algae) and Phaeophyta (brown algae), brown algae is known to exhibit the highest biosorption capacity (Abbas et al., 2014). Biosorption of pollutants may vary due to the structure and type of algal biomass and chemical constitution of the pollutants. Various algae either in dead or live forms are used as a mixture or as a single entity in column or batch for in situ bioremediation. For heavy metal bioremediation, algae make use of the algal proteins containing hydroxyl, amine, phosphate, sulfate and carboxyl functional groups for complexation. Sodium, magnesium and calcium ions, which are present on the algal cell wall get replaced by the ions of heavy metals due to ion exchange.

10.2.3 Mycoremediation

Fungi are prevalently present in the natural surroundings and are widely used for industrial applications. Fungi can adapt (in terms of their metabolism, ecology and morphology) under different atmospheric conditions and are able to perform processes such as nutrient recycling and decomposition under natural conditions. Fungi are able to endure and survive under stressful conditions such as nutrients, moisture and pH. Mycoremediation may be referred to as the utilization of fungi either dead or living for pollutant elimination from various environmental sectors. Mycoremediation is a cheaper process and does not lead to the production of harmful wastes and by-products. Thus, it can bring about total pollutant mineralization naturally. The success rate of mycoremediation actually depends on the identification and utilization of the appropriate species of fungi for the target contaminant or pollutant. Fungi are capable of accumulating pollutants inside their fruit bodies, thereby decreasing the pollutant concentration and making them unavailable in the surrounding media. The future availability of the pollutants or contaminants in the surrounding media is dependent on the fungal life span, the chemical nature of the elements and the absence or presence of fungi following sequestration. *Saccharomyces cerevisiae* has been found to sequester heavy metals like cadmium (Cd) and lead (Pb) in approximately 65%–79% of heavy metal-contaminated soils (Damodaran et al., 2011). Biosorption via interactions such as adsorption, complexation and ion exchange is carried out by cell walls of fungi composed of proteins, chitin, lipids, glucans, polysaccharides and pigments and other functional groups like sulfate, hydroxyl, amino, carboxyl and phosphate. *Aspergillus* spp. was found to eliminate 65% chromium (Cr) from tannery wastewater and 85% Cr from textile industrial wastewater (Srivastava and Thakur, 2006).

Phylum Basidiomycetes involve mushrooms, white and brown rot fungi (wood-decaying species) and some other fungi. Mushrooms have so far proven to be an essential food in the human diet because of their medicinal and nutritional properties. Apart from being used as foods, they are also used for bioremediation purposes because of their high capacity for uptaking heavy metals. The rate of metal uptake in mushrooms is dependent on mycelial age, time of contact and fructification. Few wild edible mushroom varieties can lead to pollutant accumulation and scatter them disproportionately inside the fruiting bodies of the mushrooms. Some species of white rot fungi like *Termitomyces clypeatus* and *Pleurotus ostreatus* have been reported to biodegrade persistent contaminants (Isildak et al., 2004).

10.2.4 Bioaccumulation and Biosorption

Both bioaccumulation and biosorption are effective methods for bioremediation of pollutants over traditional methods. Bioaccumulation can be defined as the uptake of pollutants/contaminants by living biomass cells (active uptake/metabolism-dependent). Using living biomass cells for bioremediation may not be a suitable choice because highly toxic pollutants can get accumulated within the cells disrupting metabolic functions of the cells, ultimately resulting in cell death.

On the other hand, biosorption by dead biomass cells is not affected by the toxicity of the pollutants and also no nutritional/growth medium is required for the dead cells and so they can be used under any sort of environmental conditions. Pollutants get adsorbed on the biomass surfaces passively without requiring any expenditure of energy (metabolism independent) till equilibrium is reached (Velasquez and Dussan, 2009). Thus, it can be said that biosorption is more effective and advantageous over bioaccumulation or direct uptake since biosorption is independent of metabolism but dependent on the type of biosorbent/biomass and also the nature of the contaminants. Due to such advantages, microbial biomasses of yeast, algae or fungi are used for in situ bioremediation procedures. Bioremediation of heavy metals in the form of nanoparticles (NPs) by bacteria or genetically mutated microbes has been used so far (Klaus-Joerger et al., 2001).

Intracellular sequestration is defined as the concentration of pollutants inside microbial cells. It includes pollutant complexation via surface interactions and subsequently transfers the pollutants within the cells. Extracellular sequestration involves concentration of the pollutants within the periplasm or pollutant complexation forming insoluble precipitates. Precipitation of heavy metals such as Cd has been reported in cells of *Klebsiella planticola* and *Pseudomonas aeruginosa* (Wang et al., 2002).

10.3 STRATEGIES OF MICROBIAL BIOREMEDIATION

There are various types of techniques for microbial bioremediation. The most common strategies that are involved in microbial bioremediation are discussed below.

10.3.1 In Situ Bioremediation Strategy

This technology includes removal of contaminants from the sites of pollution itself. There is no need for any excavation and so it causes few or no disturbances at all to the structure of the soils. Moreover, these strategies are cheaper than ex situ bioremediation strategies because there is no need for any extra cost for any excavation but the cost of designing and on-site fitting of the bioremediation equipment for enhancement of microbial effectiveness may be a little higher. In situ bioremediation strategies are successful in the removal of chlorinated compounds, dyes, hydrocarbons and heavy metals from waste sites (Frascari et al., 2015). For the in situ bioremediation strategy to be fully successful few parameters such as moisture content, type of electron acceptor, temperature, pH and availability of nutrients must be appropriately maintained. Porosity of soil also affects the success rates of in situ bioremediation, unlike the ex situ bioremediation techniques.

10.3.1.1 Biostimulation

This strategy of biostimulation involves certain nutrient injection within the polluted sites (groundwater/soil) in order to stimulate or activate the functioning of the indigenous microbes. It mainly focuses on stimulating the naturally occurring or indigenous fungal and bacterial communities. At first, trace minerals, growth supplements and fertilizers are injected followed by maintenance of some environmental factors such as oxygen, temperature and pH in order to speed up the metabolic rates and initiation of the microbial metabolic pathways (Kumar et al., 2011). The existence of trace quantities of pollutants can even turn on the operons responsible for bioremediation in microbes. This strategy is mainly followed when it is possible to supply oxygen and nutrients to the indigenous microorganisms for speeding up their growth and metabolism. Such nutrients act as the main building blocks of life allowing the microorganisms to use up the energy, cellular biomass and enzymes for biodegradation of the pollutants. The main requirements of the microbes for carrying out biodegradation are carbon (C), nitrogen (N) and phosphorous (P).

10.3.1.2 Bioattenuation/Natural Attenuation

Bioattenuation also known as natural attenuation is the elimination of pollutants from the surrounding environment. This process is carried out via various techniques including biological methods of anaerobic and aerobic biodegradation, physical methods such as dispersion, advection, diffusion, dilution, desorption/sorption and volatilization and chemical reactions such as complexation, ion exchange and abiotic transformation. The main types of natural attenuation are biotransformation and intrinsic bioremediation.

When the environment gets polluted with pollutants, it can clean up the pollutants naturally in four different ways which are described in the following (Li et al., 2010):

Microbes or little bugs living in groundwater or soils utilize some chemicals as their foods. On completed digestion of these chemicals, these microbes change the pollutants into non-toxic gases and water.

Pollutants can remain attached to the soils holding them in place. Although this does not remove the pollutants, through this process, the pollutants remain concentrated within the sites and do not reach to pollute the groundwater.

Diluting the pollutants results in the reduction of their concentration in the soils and groundwater.

Some pollutants such as solvents and oils get evaporated that is they are converted to gases from liquids inside the soil. On leaving the ground surfaces, these gases get exposed to sunlight and become non-toxic. If it is found that natural attenuation is not enough to remediate the pollutants, bioremediation will come into action either via bioaugmentation or via biostimulation.

10.3.1.3 Bioaugmentation

This is one of the types of biodegradation. Bioaugmentation is described as the process in which some exotic/natural/engineered microbes capable of degrading pollutants are added further to augment the biodegradation capacity of the indigenous microorganisms already present in the polluted site. This is done for increasing the growth rates of the naturally present microbial population for enhanced degradation within the contaminated sites. These microbes are collected separately from the contaminated sites, and they are further cultured followed by genetic modification and then these genetically modified microbes are again returned to their sites of collection. Most of the microorganisms essential for bioremediation are found in groundwater and soils polluted with chlorinated ethenes like trichloroethylene and tetrachloroethylene (Niu et al., 2009). Thus bioaugmentation confirms that the in situ microbes can fully eliminate and modify such pollutants to chloride and ethylene, which are potentially non-toxic.

Bioaugmentation is, therefore, the addition of genetically engineered microbes within a system, where they can act as bioremediators for quick and total elimination of complex contaminants. Genetically engineered microbes (GEMs) have the ability to increase the biodegradation capacities of a diverse array of environmental contaminants because the GEMs possess a wide variety of metabolic profiles for converting into less complex and more non-toxic end products (Gomez and Sartaj, 2014). Naturally occurring microorganisms are not quick enough in breaking down certain pollutants so, in order to enhance the biodegradation capacity, GEMs are preferred over natural wild-type microbes. GEMs have the potential for cleaning up groundwater, activated sludge and soils polluted with a diverse range of chemical as well as physical pollutants (Thapa et al., 2012).

10.3.1.4 Bioventing

Bioventing is the process involving controlled airflow activity by oxygen supply to vadose/unsaturated zone for enhancing bioremediation by an increase of indigenous microbial activities. In this process, modifications are done by addition of nutrients and moisture for enhancing bioremediation so that biotransformation of the pollutants to non-toxic compounds is finally achieved. This strategy is preferred over other in situ bioremediation strategies mainly in sites of restoration contaminated with lightly spilled products of petroleum (Hohener and Ponsin, 2014). In an experiment conducted by Sui and Li (2011), the rate of air injection on biodegradation, biotransformation and volatilization of sites polluted with toluene by bioventing was modeled. It was found that at two different rates of air injection (81.504 and 407.52 dm^3/d), there was no notable change in toluene elimination at the end of the period of study (200 days). In the beginning of the experiment (around 100th day), it was reported that higher rate of air injection removed much toluene by volatilization in comparison to lower rates of air injection. Moreover, high rate of air injection does not increase biodegradation or biotransformation of pollutants because of the early air saturation (low or high air injection rate) at the subsurface due to demand for oxygen in biodegradation. Low rate of air injection can also notably increase the rate of biodegradation. Thus, this study demonstrated that the rate of air injection is one of the basic parameters for re-distribution, dispersal and surface loss of pollutants. In this way, Frutos et al. (2010) observed the efficacy of bio venting technique for bioremediation of soil contaminated with phenanthrene and found that >93% of the pollutant was removed after a period of 7 months. Intervals and intensities of airflow produced no noteworthy changes in the removal of diesel from clayey soil showing that longer interval and a lower rate of air injection is more economically viable for bioventing of diesel in clayey soils (Thome et al., 2010). Rayner et al. (2007) found that within a sub-Antarctic site contaminated with hydrocarbons, bioventing with single wells was not much effective in hydrocarbon removal because of the thin soil layer and shallow water table, leading to channel development, while micro-bioventing by utilization of nine little injection rods (placed 0.5 m apart) at the same spot under similar conditions resulted in considerable degradation and removal of hydrocarbons due to more uniform oxygen distribution. It is evident that although intervals and rates of air injection are some of the basic factors for bioventing, the effectiveness of the process actually depends on the number of points for air injection helping in uniform air distribution. In spite of designing bioventing processes for aeration inside unsaturated zones, it can even be used for anaerobic processes of bioremediation mainly inside vadose zones contaminated with chlorinated compounds that recalcitrant in an aerobic environment. In anaerobic bioventing processes, instead of using pure oxygen or air, nitrogen is mixed with carbon dioxide at very low concentrations and hydrogen can be injected to reduce chlorinated vapors along with hydrogen serving as an electron donor. In soils having low permeability, injecting pure oxygen may increase the concentration of oxygen in comparison to that of air injection. Besides, ozonation also aids in the partial oxidation of pollutants that are recalcitrant for accelerating biodegradation.

In contrast to bioventing, which depends on enhanced microbial degradation in the vadose zone by moderate injection of air, the process of soil vapor extraction (SVE) increases the volatilization of volatile organic pollutants by the extraction of vapors. Though both these strategies use similar configuration, hardware and fabrication design, there is a huge difference in their modes of operation. The rate of airflow is more in SVE as compared to that in bio venting. Thus, SVE can be called a physical bioremediation method because of its technique employed for removal of pollutants although the techniques for removal of pollutants in both these strategies are not mutually exclusive.

While using bioventing for field trials, results similar to laboratory conditions may not be achieved due to some environmental constraints and various attributes of the unsaturated zones, in which air is being injected thus increasing the time for bio venting. Higher rate of airflow causes organic volatile compounds to get transferred to the vapor soil phase requiring off-gas treatment for the resultant gases before being released into the atmosphere (Burgess et al., 2001). This problem can be overcome by a combination of bio-trickling with bioventing techniques in order to reduce the pollutant and resulting levels of emitted gases, thereby lowering the prolonged time for treatment related only to bioventing.

10.3.1.5 Biosparging

Biosparging is quite similar to bioventing in which air is injected into the subsurface of the soil for stimulation of the activities of microbes for promoting the removal of pollutants from polluted sites. However, contrary to bioventing, the air is injected at the zone of saturation causing the organic volatile compounds to move upwards to the zone of unsaturation for promoting biodegradation. The efficacy of biosparging mainly depends on two important factors which are the biodegradability of pollutants and soil permeability enhancing the bioavailability of pollutants to microbes. Similar to SVE and bio venting, biosparging also has the same operational constraints with a closely associated technique called in situ air sparging (IAS) depending on high rates of airflow for achieving volatilization of pollutants while biosparging increases biodegradation. Both the mechanisms for removal of pollutants are not mutually exclusive for both strategies. Biosparging is increasingly exploited for treatment of aquifers polluted with petroleum products mainly kerosene and diesel. Kao et al. (2008) observed that biosparging in aquifer plumes contaminated with benzene, toluene, ethylbenzene and xylene (BTEX) produced a drift to aerobic from anaerobic conditions and this was demonstrated via an increase in dissolved oxygen, redox potential, sulfate, nitrate and total heterotrophs that are culturable along with a correlating reduction in dissolved sulfide, ferrous iron, methane, methanogens and total anaerobes. The total reduction of >70% in BTEX shows that biosparging is capable of bioremediating groundwater polluted with BTEX. The main disadvantage of biosparging is to predict airflow direction.

10.3.2 Ex Situ Bioremediation Strategies

These strategies actually involve pollutant excavation from polluted sites and consequently transferring them to a different treatment site. Ex situ techniques of bioremediation are dependent on factors such as pollution depth, treatment expenditure, pollution degree, pollutant type, geology and geographic location of the contaminated sites. Criteria of performance also aid in the determination of ex situ techniques of bioremediation.

10.3.2.1 Biopile

Biopile-associated bioremediation includes piling excavated contaminated soil above the ground followed by nutrient modification and aeration for enhancing bioremediation by simply improving activities of microbes. The constituents required for this process include irrigation, aeration, leachate and nutrient collection systems and a bed for treatment. This technique of bioremediation is particularly being used for its constructive characteristics such as cost-effectiveness enabling efficient biodegradation only if aeration, temperature and nutrients are appropriately controlled. Applying the process of biopiling to contaminated sites restricts low-molecular-weight (LMW) pollutant volatilization and causes effective bioremediation of contaminated extreme environments like very cold areas. With reference to this, Gomez and Sartaj (2014) observed the results due to various rates of application (3 and 6 ml/m^3) of microbial consortia along with developed compost (5% and 10%) on reduction of total petroleum hydrocarbon (TPH) in on-site biopiles at low temperatures, utilizing response surface methodology (RSM) on the basis of the factorial design of experiment (DoE) tone. After the study ended (94 days), it was found that 90.7% TPH was reduced in biostimulated and bioaugmented setups in comparison to that of the control setups showing only an average of 48% reduction of TPH. The maximum rate of reduction of TPH was due to the synergistic interplay between biostimulation and bioaugmentation, thereby showing the versatility of biopiles in bioremediation.

Dias et al. (2014) observed a 71% reduction in the concentration of total hydrocarbon and drift in the structure of bacteria over a study period of 50 days, followed by prior treatment of polluted soil samples before the formation of biopile and consequent biostimulation by fishmeal. They also reported on the versatility of biopile for bioremediation of various soil samples sandy and clayey soil. Biopile versatility shortens the bioremediation time because of the incorporation of the heating system within the design of biopile for increasing the microbial activities and availability of contaminants, thereby increasing the biodegradation rate. Moreover, hot air may be injected inside the biopile design for delivering heat and air in tandem and facilitating bioremediation.

In a study conducted by Sanscartier et al. (2009), it was observed that humid biopile had a very low concentration of TPH in comparison to the passive and hot biopiles because of decreased leaching, optimum content of moisture and minimum volatilization of less biodegradable pollutants. Additionally, biopile can be used for treating huge volumes of contaminated soil inside a limited space. Setup of biopile may be scaled up to a pilot plant, achieving similar results to that attained with laboratory studies. Biopile efficiency depends on the aeration and sieving of polluted soil before processing. For improvement of bioremediation through biopiles, bulking agents like sawdust, straw, wood chips or bark and few other organic components may also be added.

Some of the disadvantages of using biopile systems include operational and maintenance costs, robust engineering, and unavailability of power supply in some remote areas for providing uniform air supply in polluted sites through air pumping machines. Moreover, use of excessively hot air may lead to the drying up of the soil, on which bioremediation is carried out, thereby inactivating the microbes and promoting volatilization rather than biodegradation.

10.3.2.2 Windrows

Being one of the techniques of ex situ bioremediation, windrows are dependent on the periodical turning of contaminated piled soil in order to magnify bioremediation via increased biodegradation activities of transient and/or indigenous hydrocarbonoclastic bacteria existing in contaminated soils. Periodic polluted soil tuning along with water addition increases aeration, uniform pollutant distribution, microbial activities of degradation and nutrients, thereby speeding up the bioremediation rate via biotransformation, assimilation and mineralization. The Windrow technique as compared to biopile exhibited a high carbon removal rate, but the efficiency of Windrow mainly depends upon the type of contaminated soil. Thus, because of the use of periodic turning, it is not considered a good choice for the remediation of soils contaminated with harmful volatiles. The Windrow technique was first used in the removal of methane (greenhouse gas) because of the formation of an anaerobic zone inside polluted piled soil that usually takes place successive to decreased aeration (Hobson et al., 2005).

10.3.2.3 Bioreactor

Bioreactor is actually a vessel, wherein raw materials get converted to a certain product(s) due to an array of biochemical reactions. There are various operating systems of bioreactors including fed-batch, batch, continuous, sequencing batch and multistage. Selection of the mode of operation depends on capital cost and market economy. The conditions within a bioreactor allow the cells to carry on their natural processes by maintaining and mimicking their natural atmosphere for providing them with their optimum growth conditions.

Polluted soil samples may be fed inside a bioreactor either as slurry or dry matter and bioreactor treatment is considered more advantageous than other ex situ bioremediation strategies. There is absolute control of the bioprocess parameters (pH, temperature, aeration rates, agitation, concentrations of inoculums and substrates) in a bioreactor, making it one of the best methods of bioremediation.

The capacity of manipulating and controlling the parameters of the process in bioreactors shows that the biological reactions inside can be improved efficiently for reducing the time taken for bioremediation. The limiting factors of a bioremediation process include the addition of nutrients, bioaugmentation, mass transfer (contact between microbes and pollutants) and bioavailability of pollutants can be efficiently maintained within a bioreactor, thereby making bioremediation associated with bioreactor more effective than other methods. It can further be used for treating water or soil contaminated with volatile organic compounds (VOCs) including BTEX. Different bioreactors employed for various processes of bioremediation removed diverse types of pollutants. The flexibility of the bioreactor designs ensures maximum biodegradation with minimum abiotic loss. Long- or short-term operation of bioreactor holding soil slurry contaminated with crude oil helps in tracking the alterations of population dynamics of the microbes, thereby allowing simple characterization of the main bacterial communities responsible for carrying out the bioremediation processes (Zangi-Kotler et al., 2015). Moreover, bioreactors allow the usage of various substances as bioaugmenting agents or biostimulants, as well as sewage sludge. Additionally, since the bioreactor is an enclosed vessel, GEMs may be utilized for bioaugmentation, following which the GEMs are destroyed prior to the transfer of treated soils to the field for landfilling. This process of containing GEMs within a bioreactor and then subsequent destruction of them will assure that any foreign genetic materials cannot be released into the environment in succession to bioremediation. In bioreactors, the function of biosurfactants was studied to be not so essential because of effective mixing related to the operations of the bioreactor (Mustafa et al., 2015).

Although bioreactor-associated bioremediation has proved effective because of the easy control of all the parameters, the establishment of best operational conditions in relation to every parameter making use of the one-factor-at-a-time (OFAT) approach may need many experiments, which is quite tedious. This obstacle can be mastered by the use of DoE tone providing information about the optimum parameter range via an independent variable set (uncontrollable and controllable factors) within a specified level/region. For optimization of bioremediation procedures, a good understanding of microbial processes is of great significance. However, bioreactor-mediated bioremediation is not a fully practical process because of some limitations. First, as the bioreactor-based bioremediation is an ex situ strategy, the volume of contaminated soil or any other substances to be treated is large, which requires more manpower, more capital costs and increased safety guidelines for transportation of the pollutant to the site of treatment thereby rendering this strategy as cost-ineffective. Second, because of the numerous bioprocess variables or parameters of bioreactors, if some of the parameters are not appropriately controlled and/or maintained at the optimal level may act as a limiting factor thereby reducing activities of the microbes and making the bioremediation by bioreactor as not much an effective process. Third, pollutants may respond differently in presence of different conditions of bioreactors so the most suitable design for bioreactor is of utmost importance. Moreover, the cost of setting up a suitable bioreactor design on both pilot and laboratory scales renders this bioreactor-associated bioremediation technique to be capitally intensive.

10.3.2.4 Landfarming

This is the easiest technique of bioremediation because of its cost-effectiveness and low requirement of machinery for operation. It is regarded as both in situ as well as ex situ bioremediation technique according to its treatment site and depth of the pollutants. Polluted soils are actually tilled or excavated but the treatment site actually dictates the type of bioremediation to be used. When this excavated soil is treated at the same site of its excavation, landfarming is called to be an in situ technique or else it is regarded as an ex situ technique of bioremediation. It has been observed that if a pollutant is present <1 m below the surface of the ground, then bioremediation can be carried out without any excavation but if the pollutant is present at a depth >1.7 m, then it is needed to be transported to the surface of the ground for effective bioremediation (Nikolopoulou et al., 2013). Usually, polluted soils after excavation are delicately applied over a fixed layer of support above the surface of the ground allowing pollutants to be biodegraded aerobically by means of autochthonous microbes. Tillage increases some major operational criteria such as aeration, nutrient (potassium, phosphorous and nitrogen) addition and irrigation stimulating autochthonous microbial activities for improved bioremediation via landfarming. Silva-Castro et al. (2014) reported that irrigation and tillage without adding nutrients in a soil possessing suitable biological activities increased diesel-biodegrading and heterogeneous counts of bacteria thereby increasing bioremediation rate and also identified activity of dehydrogenase to serve as an excellent indicator of treatment by biostimulation and thus it can be used a biological factor to be taken into consideration during landfarming technology. In another field trial conducted by Paudyn et al. (2008), they found that more than 80% of diesel (pollutant) was eliminated by aeration via rototilling strategy in a remote location like Arctic Canada for a study period of 3 years, thereby demonstrating that aeration plays an important role in the removal of pollutants essentially in cold areas. Land farming is mainly used for sites contaminated with hydrocarbons including polyaromatic hydrocarbons (Cerqueira et al., 2014) and so volatilization/weathering and biodegradation are two main mechanisms required for removal of pollutants. The technique of landfarming works in accordance with the regulations prescribed by the government and can be utilized in any climatic conditions. The fabrication of an appropriate design for landfarming having an impermeable liner reduces the leaching of pollutants in neighboring areas during the bioremediation process. Above all, the technique of land farming is very easy to fabricate and establish, requiring less capital cost and having the ability to treat a huge volume of polluted soil having minimum side effects on the environment and requirement of energy.

However, some of the limitations of the land farming technique are decreased activities of microbes because of inappropriate environmental parameters, large operational space, decreased effectiveness in inorganic pollutant removal and extra cost for excavation. In fact, land farming is not suitable for treatment of harmful volatiles because of its pollutant volatilization (removal) mechanism essentially in tropical (hot) climatic areas. These disadvantages and many others make land farming-associated bioremediation tedious and less effective as compared to other techniques of ex situ bioremediation.

One of the main advantages of ex situ bioremediation strategies is that they do not need substantial primary investigation of the contaminated sites before bioremediation, making the primary step cheaper, less tedious and less time-consuming. Because of excavation procedures, related to ex situ techniques of bioremediation, inhomogeneity of pollutants due to depth, intermittent distribution and concentration can be simply reduced by effective optimization of some process parameters (mixing, pH and temperature) for enhancement of any ex situ bioremediation techniques. These strategies allow alterations of physico-chemical, chemical and biological parameters and conditions required for effective and proper bioremediation.

Essentially, the major effect of soil porosity controlling the transport phenomenon during bioremediation is decreased when contaminated soils are being excavated. Ex situ techniques of bioremediation are not suitable to be applied at some of the sites like working sites, inner cities and under buildings. However, the features of excavation in ex situ bioremediation may degrade soil structure thereby polluting and disturbing the neighboring sites. From moderate to intense engineering, there is a requirement for an ex situ bioremediation strategy that involves more capital investment and workforce to construct any such bioremediation setup. In most of the cases, these strategies require a lot of space for their operation. Usually, techniques of ex situ bioremediation are quicker, easier to control and are used for treating a wide variety of pollutants (Philp and Atlas, 2005).

10.4 MICROBES WITH BIOFILM-ASSOCIATED REMEDIATION

The extracellular compounds and the surface components of bacteria such as lipopolysaccharide (LPS), exopolysaccharide (EPS) and flagella in combination with different environmental signals like quorum sensing (QS) aid in biofilm formation, autoaggregation, host colonization and bacterial cell survival in adverse conditions. There are two types of EPS produced by bacteria which are succinoglycan (EPS I) and galactoglucan (EPS II). These EPSs are involved in conferring host specificity, identification and prevention of host plant pathogenesis, maintenance of cell integrity, and signaling of biomolecules in the stages of development, thereby protecting the cells of bacteria from a diverse array of environmental stresses (Nocelli et al., 2016). EPS serves as a mechanism of resistance in bacteria against environmental pollutants, which will otherwise affect the bioprocesses within the bacterial cells. Moreover, EPS is not only involved in biofilm formation and providing resistance against pollutants to the host bacteria but can also protect other non-EPS-producing bacteria from the pollutants thereby exhibiting a 'rescuing' mechanism (Nocelli et al., 2016).

EPS I is an octasccharide composed of one galactose and seven glucose units alongside pyruvyl, acetyl or succinyl subunits while EPS II consists of alternating galactose and glucose residues which are pyruvated and acetylated, respectively. The two forms in which EPS is known to exist are LMW and high-molecular-weight (HMW) fractions. The LMW fraction of EPS is a natural and inactive form that is secreted at normal conditions while the HMW fraction of EPSs is secreted in the presence of stress and the HMW fraction in combination with the LMW fraction forms the EPS. The biofilm so formed on exposure to the pollutants confers resistivity to the bacterial cells against pollutants preventing pollutant diffusion into the cells. Thus, biofilms are actually considered as a potential shielding process that allows the bacterial cells to thrive and survive stress conditions.

10.4.1 Biofilm-Mediated Bioremediation

Bioremediation through biofilms shows higher efficacy in converting harmful wastes due to increased bioavailability of the contaminants to degrading microorganisms and improved adaptability of the degrading microbes to various harmful compounds. This process generally takes place due to microbial metabolism and is dependent on the enzymatic attack by microorganisms for the conversion of environmental contaminants to non-toxic products. There exist several microbes that can biotransform a wide range of environmental contaminants into non-toxic forms.

In comparison to their planktonic counterparts, microbes living inside biofilms show increased tolerance to pollutants, a greater possibility of survival and adaptation and also a strong capacity of decomposing various pollutants via catabolic pathways (Dillewijn et al., 2009). Bacteria forming biofilms are effectively used for the process of remediation since their cells are encased inside an EPS matrix, which confers immunity against numerous environmental pollutants. Besides this, biofilms impart the necessary conditions required for intercellular transfer of genes, cell–cell communication through QS, cohesion and diffusion of metabolites and characteristics for bacterial chemotaxis (Santos et al., 2018).

Biofilm-associated bioremediation can nurture numerous anaerobic and aerobic bacterial species that mostly utilize pollutant degradation as a source of energy. In the case of aerobic degradation, bacteria utilize oxygen as the final acceptor of electrons for breaking down toxic products into non-toxic products, namely, water and carbon dioxide (Azubuike et al., 2016). In the case of anaerobic degradation, sulfate and nitrates act as electron acceptors and thus carry out the function of oxygen for pollutant transformation to harmless or non-toxic products and the formation of by-products is mainly dependent on the electron acceptor.

Nowadays, biofilm-mediated bioremediation is increasingly used to remove various types of environmental pollutants such as persistent organic pollutants (POPs like polychlorinated biphenyls, polychlorinated ethane and polycyclic aromatic hydrocarbons), oil spills, dyes, heavy metals, pesticides, explosives and pharmaceutical compounds (Edwards and Kjellerup, 2013). Thus, bioremediation via biofilms is employed in industries for bioremediation of contaminated groundwater and soil. It was observed that bacteria such as *Dehalococcides, Pseudomonas, Bacillus, Arthrobacter, Cycloclasticus, Alcanivorax, Rhodococcus* and *Burkholderia* are able to effectively remove these pollutants (Dasgupta et al., 2013; Yoshikawa et al., 2016) (Tables 10.1 and 10.2).

TABLE 10.1
Microbes Involved in the Bioremediation of Organic Pollutants

Organic Pollutant	Bioremediation Mechanism	Microbe Involved	References
Fenamiphos (organic pesticide)	Biosorption followed by biodegradation	*Brevibacterium* spp.	Singh et al. (2003)
Polyaromatic hydrocarbons (PAHs)	Biodegradation by biosurfactants	*Sapindus mukurossi*	Shiau et al. (1995)
Toluene	Colonization followed by complexation	Solvent-tolerant bacteria and mycobacteria	Huertas et al. (1998)
Quinalophos and monocrotophos (organophosphorus pesticides)	Bioaccumulation followed by biotransformation	*Scenedesmus bijugatus, Chlorella vulgaris, Nostoc linckia* and *Phormidium tenue*	Megharaj et al. (1987)
Dichlorodiphenyl trichloroethane (DDT)	Biotransformation	*Aulosira fertilissima* and *Anabaena* spp.	Lal et al. (1987)
Tryne (herbicide)	Bioaccumulation and degradation	*Chlamydonmonas reinhardtii, C. vulgaris, Dunaliella tertiolecta* and *Isochrysis galbana*	Weiner et al. (2004)
Pyridaphenthion (organophosphorus insecticide)	Biotransformation	*Scenedesmus* spp. and *Chlorococcum*	Jonsson et al. (2001)
Pyrene	Physicochemical biosorption	*Selenestrum capricornutum*	Lei et al. (2002)
Naphthalene	Biodegradation and biotransformation	*Oscillatoria* spp.	Narro et al. (1992)
Dimethyl phthalate (DMP)	Biodegradation	*Closterium lunula*	Yan and Pan (2004)
Tributyltin (TBT)	Biodegradation	*Chlorella* spp.	Luan et al. (2006)
Trinitrotoluene (TNT)	Biodegradation	*Enterobacter cloacae* strain PB2	Dhankher et al. (2012)

TABLE 10.2
Microbes Involved in the Bioremediation of Inorganic Pollutants

Inorganic Pollutants	Microbes Involved	References
Strontium	*Platymonas subcordiformis*	Mei et al. (2006)
Zinc and lead	*Chlorella* spp.	Kumar and Goyal (2010)
Cadmium and copper	Mixed culture of *Tetrahedron* spp., *Scenedesmus* spp., *Chlorococcus* spp., *Chlorella* spp., *Chroococcus* spp., *Pseudoanabaena* spp. and *Leptolynbya* spp.	Loutseti et al. (2009)
Uranium	*Cytoseria indica*	Edgington et al. (1970)
Thorium	*Isochrysis galbana*	Manikandan et al. (2011)

10.5 ANALYSIS OF REMEDIATING MICROBIAL COMMUNITIES BY METAGENOMIC APPROACHES

Amidst the emergence of next-generation sequencing (NGS) techniques, the area of metagenomics has rapidly evolved. Metagenomic technology has provided an alternative to cloning and allowed the new technique of comparative metagenomics to emerge. Advanced high-throughput techniques have brought a revolution in the field of microbiology. Metagenomics is considered as a rapidly evolving and new research field aiming for the investigation of uncultured organisms for better understanding of the actual microbial diversity, their interactions and cooperation, their metabolic functions and their evolutionary traits under diverse atmospheric conditions. Thus, the newer approach of metagenomics in the field of molecular biology was first introduced by Handelsman et al. (1998) during the study of unknown soil microbial chemistry. Metagenomics is a sequence-based and culture-independent DNA analysis of the microbial samples isolated from the environment, and this sample is also known as the metagenome (Daniel, 2005). Sequence phylogeny can provide information about the diversity like the type of microorganisms that are present and their functions like what these microbes are capable of doing. Thus, metagenomics has aided in bridging the gap left behind by the culturing techniques and has provided us with a new dimension into in situ structures of microbes, their functional dynamics and their capability in increasing the process of bioremediation.

10.5.1 METAGENOMIC APPROACHES

There are mainly two approaches of metagenomics which are sequence-based and function-based. Both these approaches in combination have enhanced our understanding of the microbial diversity that is mostly unculturable and has helped in providing adequate information about the world of the prokaryotes that would otherwise remain obscure.

10.5.1.1 Function-Based Metagenomic Approach

The function-based metagenomics approach is a powerful and potent method to study the functional roles of genes. Functional metagenomics is defined as the extraction of DNA from environmental samples for studying the various functions performed by encoded proteins (Lam et al., 2015). In the functional metagenomic approach, the fragments of DNA are cloned and expressed inside a laboratory host followed by screening for enzyme-based activities. The functional metagenomic approach facilitates novel gene discovery, followed by metagenomic sequencing for exploring some novel conditions of environment that are not studied yet.

In this analysis, desired clones are screened on the basis of detecting heterologous desired trait expression such as the capacity of growing on a selective medium having the toxic waste as the main electron acceptor or energy and carbon source, metabolic functions such as coloration or clear zones near colonies obtained on certain solid media and inhibition of growth of the indicator bacteria (Handelsman et al. 1998). This analysis can be further enhanced by high performance and automatic colony picking, self-regulated pipetting, utilization of microtiter plates, management of data through information assistance and sensitive assays to target a wide array of biomolecules. Usually, screening hits within a metagenomic library are not very high. The extent of obtaining positive hits can be enhanced by using various other strategies for selecting/enriching community genomes having traits of interest before metagenomic library construction, using varied hosts for expression and developing numerous high-yielding screening techniques in relation to the activated fluorescence within the cell shorter (Stenuit et al., 2009).

Over the recent years, Mirete et al. (2007) found new nickel resistive determinants from the rhizospheric metagenomes of plants, which are already adapted to acid mining drainage through the construction of metagenomic libraries and sequencing. Thus, the function-based analyses perform an important function in these experiments by providing crucial knowledge about the functional and metabolic diversity (Prakash and Taylor, 2012) (Figure 10.1).

10.5.1.2 Sequence-Based Metagenomic Approach

The sequence-based metagenomic approach gives information about the microbes independent of cultures. It involves full sequencing of the clones having phylogenetic tie-ups like the 16s rRNA genes of any functional genes encoding the main processes. Probes that are target-specific or appropriate primers are fabricated utilizing the knowledge from databases via common methods of screening such as hybridization and PCR. Nowadays, enormous parallel sequencing may be carried out by the use of microarray. For example, a comprehensive microarray known as the 'Geochip' was used to obtain direct ancestry between functional roles and biochemical processes of communities of microbes in a wide array of environmental conditions (Liu et al., 2010). In most of the cases, sequences of a full community are assessed through shotgun sequencing of every insert inside the metagenomic library. Here, the main aim is to obtain a global view of the microbial communities rather than some specific organisms, pathways or genetic mechanisms. Sequence data are generally assembled as scaffolds and sequence contigs *in silico*.

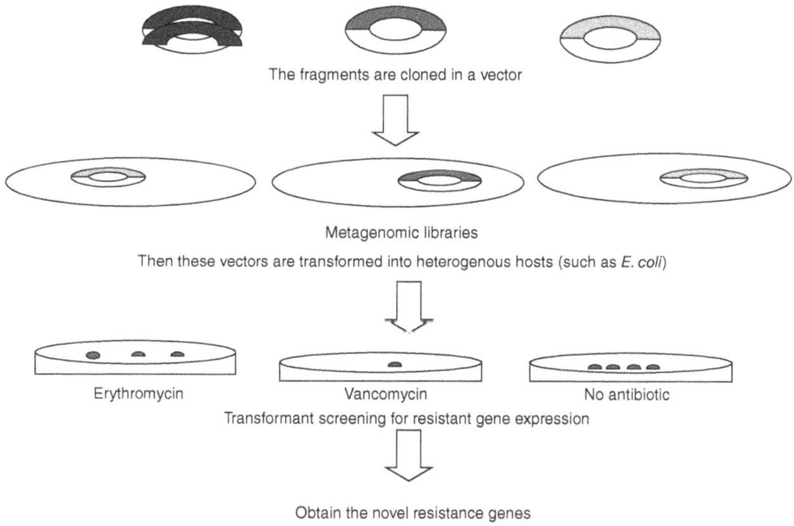

FIGURE 10.1 Function-based metagenomic analyses.

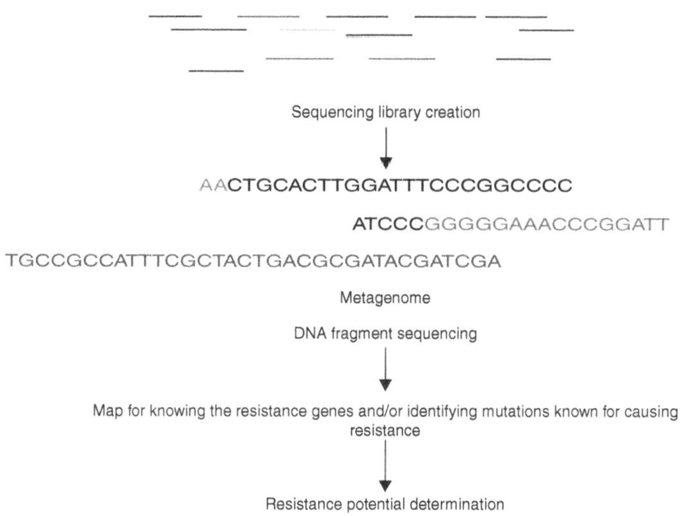

FIGURE 10.2 Sequence-based metagenomic analyses.

In contrast to functional screening, the sequence-based metagenomic analysis is dependent on the analysis of sequence for providing a basis to predict the functions. Considerable databases are indexed in the database of the 'Environmental Genome Sequence' for making the tasks of sequencing more informative and instructive with time since data gets repeatedly compiled from various sources. Sequence-based metagenomic approaches are used to recognize genes, arrange genomes, clarify and finish metabolic pathways and compare organisms from different communities (Figure 10.2).

10.5.2 Major Steps in Metagenomic Approaches

10.5.2.1 Study Site Selection

Metagenomic analyses of the microbial communities are conducted by referring to the scientific queries which are fundamental for a specific site. Prior knowledge regarding the ecological, physical and chemical characteristics of the selected environment is used to deduce more information from the metagenomic database. There are certain hypotheses fabricated on the basis of sites that are fully characterized. For example, in order to fabricate bioremediation techniques for uranium, scientists extensively characterized the areas before starting with metagenomic analyses (He et al., 2011). Multidisciplinary associations with scientists engaged in investigating the non-microbial parameters of the selected sites are utilized. Various habitats need varied sequencing depths on the basis of their extent of completeness and complexity required to refer to the questions being asked. Pilot studies are conducted for the determination of the necessary sequencing depth (Cardenas et al., 2010).

10.5.2.2 Collection of Samples and Extraction of Nucleic Acids

All metagenomic studies begin with nucleic acid extraction from the samples. While selecting a community for metagenomic studies, its composition of species must be evaluated in relation to the number of sequences allotted. A complicated community of microbes generally involves viruses, archaeal, bacterial and microbial eukaryotes. Thus, the nucleic acid pool that is extracted belongs to all the organisms existing inside the sample. A majority of the metagenomic analyses done to date mainly take into account the microbial communities. On the other hand, sequencing through metagenomic studies of the communities including eukaryotes, particularly protists, is most expensive due to their large sizes of genomes and low densities of gene-coding capabilities. Hence, selection of communities excluding eukaryotes or from which eukaryotic DNA is removed is a significant criterion to be considered.

Moreover, only eukaryotic exclusion from a metagenomic study is not appropriate from an ecological perspective because this restricts our capacity for evaluating a microbial community as a whole. There is a substitutive or alternative method that can be used to find molecular data at the protein (metaproteomics) or RNA (metatranscriptomics) level, thereby solving the problem of the enormous eukaryotic noncoding sequences (Cowan et al., 2005). Over recent years, there have been developed certain methods for the extraction of nucleic acids.

Two main techniques for metagenomic data recovery include recovery of cells and their direct lysis. Generally, chemical and physical methods are used for complete genome recovery. Bead beating has proven to be more effective to recover greater diversity in comparison with the chemical method (Siddhapura et al., 2010). Anyhow, the chemical method is gentler and is capable of recovering DNA with HMW, thereby selecting certain taxa based on their specific biochemical attributes. RNA recovery technologies from environmental samples are greatly homogeneous to those employed for the isolation of DNA. Protocols may be optimized to increase mRNA extract yield and most of the time, some co-extraction technologies may also be considered (Popova et al., 2010).

10.5.2.3 Enrichment of Genome and Gene

Active populations of microbes inside contaminated sites can be evaluated via enrichment of the genome and finally metagenomic analyses (Chen and Murrell, 2010). The stable isotope probing (SIP) technique of labeling may be selectively utilized for enriching RNA, DNA or fatty acid-derived phospholipids from active populations of microbes. For analysis through SIP, substrates labeled with stable isotopes such as ^{18}O, ^{15}N and ^{13}C were applied to the environment under analysis for studying active communities of microbes. On consumption of these substrates by active organisms, the labeled atoms get integrated within their RNA or DNA. Then the total nucleic acid is extracted from these samples and non-labeled and labeled RNA or DNA is segregated via density gradient centrifugation and can thus be utilized for constructing metagenomic libraries. Bromodeoxyuridine (BrdU is a structural analog of thymidine) is also used as a complementary substrate for labeling. Microbes that are growing actively within the contaminated environment having BrdU will integrate these labeled molecules within their DNA. This labeled DNA is then isolated from the extracted pool of DNA through immunocapture or density gradient centrifugation (Cowan et al., 2005).

10.5.2.4 Metagenomic Library Construction

Environmental RNA or DNA analysis by metagenomic library construction is one of the potential tools to explore the diversity of microbes. The key steps used for years that are used for so many years to construct metagenomic libraries include synthesizing a suitable size of DNA fragment followed by cloning this DNA fragment inside an appropriate vector. For metagenomic library construction, cloning vectors like bacterial artificial chromosome (BAC), cosmid and fosmid are used since they have a large insert size (Gilbert and Dupont, 2011).

10.5.3 Use of Metagenomics in Bioremediation

Areas, where mainly anthropogenic activities are carried out, are frequently contaminated with diverse types of harmful compounds (Pacwa-Płociniczak et al., 2011). The range of this contamination may differ and influence the most significant components in our ecosystems like soil, water and air (Saharan et al., 2011). The relationship between the functions of the ecosystem and the diversity of species has been an area of interest for a long time (Sutherland et al., 2013). Pollution due to various anthropogenic activities like release of heavy metals and hydrocarbons is capable of strongly disturbing the diversity of microbes along with their structures and composition (Pessoa-Filho et al., 2015). NGS has provided a plethora of opportunities to extensively analyze environmental genomes. Metagenomics in combination with some other techniques of molecular biology has transfigured the microbiological field by shedding light on the adaptation, evolution and diversity of microbes (Riesenfeld et al., 2004). Various studies have been performed on the microbial communities from various environmental conditions like marine water and sediments (Yooseph et al., 2007; DeLong et al., 2006), gut of humans (Turnbaugh et al., 2007), soils (Smets and Barkay, 2005) and mining acid drainage systems (Tyson et al., 2005) to provide some new insights into the structural communities, metabolism, function, evolution and genetic make-up of these microbial communities.

The metagenomic approach provides an excellent opportunity for comprehensive analysis of the response of an ecosystem to certain environmental alterations, however, till now, there have been no such studies reported that have investigated the response and adaptation of communities of microbes toward the environmental contaminants. Metagenomics is a very promising field in bioremediation studies since it can shape the approaches to bioremediation in a wide number of intertwined ways (Tripathi et al., 2018). The bioremediation approach coupled with metagenomics provided better results with increased ratios of degradation in comparison to other bioremediation approaches (Kosaric, 2001).

Metagenomic analysis has significantly enhanced the understanding of the processes as to how the microorganisms form 'bucket-brigades' to degrade the xenobiotic compounds, thereby facilitating the development of polluted sites into sites, in which the native microbial communities can bioremediate the environmental conditions by utilization of intensive ex situ techniques or through bioaugmentation in situ. Moreover, metagenomics aid in the identification of important microbial processes and specification of the way, through which components of the community can be best suited for enabling pollutant mineralization, while the metabolic debate is necessary among diverse species and these metabolic functions can be carried out by a consortium of bacteria rather than by a single bacterial species (Thomas et al., 2012). Thus, metagenomics provides an appropriate database, which can offer a wide array of genes to construct novel strains of microbes for targeted usage in bioremediation purposes. Hence, scientists regard metagenomics as a potent and important technique to eradicate contaminants from the environment (Das and Chandran, 2011).

10.6 CONCLUSION

Microbial bioremediation is an emerging field and can pose as a potential alternative approach to degrade environmental contaminants in a much safer, faster and cheaper way. Thus, the goal of bioremediation is reached through microbial interactions with the pollutants, resulting in compartmentalization, immobilization and concentration of the pollutants in the environment. Intensive understanding of the function and structure of the community of microbes is essential for achieving both reliable and effective cleaning of environmental pollutants. However, this is not achievable for microbiologists because majority of the environmental microbes cannot be easily cultured and so their biological characteristics are not completely known. Thus, the evolution of metagenomics has helped in revolutionizing this microbiological field of study.

Metagenomics enables direct access to the native communities of microbes present in the polluted sites irrespective of their culturability. Metagenomic techniques are anticipated to enhance the recovery of novel catabolic functions and give important knowledge regarding the cleaning up and management of contaminated sites. Although the utilities of metagenomics have just been explored by us in prospects of microbial bioremediation, from the technical viewpoint, the enormity of its applications is yet to be fully utilized. Till now, a few of the metagenomic projects have been successful in understanding the microbial activities required for cleaning contaminated sites (Sar and Islam, 2011). Scarcity of knowledge in functional selection of the metagenomic libraries, shortage of appropriate hosts for expression of genes metagenomically derived under anaerobic conditions and lengthy processes of sequence analysis are a few disadvantages. Taking into account the NGS data analysis, there may be some real scientific advancements in the bioremediation field.

However, the metagenomics field is challenged by the dynamic behavior of the communities of microbes in response to the environmental alterations in the polluted sites going through bioremediation. From the data provided by metagenomics, it is difficult to reshape the exact view of complicated communities of microbes and their contribution to various biological processes. Since bioremediation includes complex microbial interactions, so it can be proposed that 'omics' techniques like metabolomics, metatranscriptomics and metaproteomics are essential to obtain extensive information regarding the characteristics of the microbial communities involved in effective bioremediation.

REFERENCES

Abatenh, E., Gizaw, B., Tsegaye, Z., & Wassie, M. (2017). Application of microorganisms in bioremediation-review. *J Environ Microbiol, 1* (1), 2–9.

Abbas, S. H., Ismail, I. M., Mostafa, T. M., & Sulaymon, A. H. (2014). Biosorption of heavy metals: a review. *J Chem Sci and Tech, 3* (4), 74–102.

Adenipekun, C. O., & Lawal, R. (2012). Uses of mushrooms in bioremediation: a review. *Biotechnol Mol Biol Rev, 7* (3), 62–68.

Azubuike, C. C., Chikere, C. B., & Okpokwasili, G. C. (2016). Bioremediation techniques-classification based on site of application: principles, advantages, limitations and prospects. *World J Microbil Biotechnol, 32* (11), 180.

Bolan, N., Kunhikrishnan, A., Thangarajan, R., Kumpiene, J., Park, J., Makino, T., et al. (2014). Remediation of heavy metal(loid)s contaminated soils – to mobilize or to immobilize? *J Hazard Mater, 266*, 141–166.

Burgess, J. E., Parsons, S. A., & Stuetz, R. M. (2001). Developments in odour control and waste gas treatment biotechnology: a review. *Biotechnol Adv, 19* (1), 35–63.

Cardenas, E., Wu, W. M., Leigh, M. B., Carley, J., Carroll, S., Gentry, T., et al. (2010). Microbial communities in contaminated sediments, associated with bioremediation of uranium to submicromolar levels. *Appl Environ Microbiol, 74* (12), 3718–3729.

Cerqueira, V. S., Peralba, M. d., Camargo, F. A., & Bento, F. M. (2014). Comparison of bioremediation strategies for soil impacted with petrochemical oily sludge. *Int Biodeterior Biodegradation, 95.*

Chen, Y., & Murrell, J. C. (2010). When metagenomics meets stable-isotope probing: progress and perspectives. *Trends Microbiol, 18* (4), 157–163.

Cowan, D., Meyer, Q., Stafford, W., Muyanga, S., Cameron, R., & Wittwer, P. (2005). Metagenomic gene discovery: past, present and future. *Trends Biotechnol, 23* (6), 321–329.

Damodaran, D., Suresh, G., & Mohan R. (2011). Bioremediation of soil by removing heavy metals using Saccharomyces cerevisiae. *2nd International Conference on Environmental Science and Technology (ICEST)*, 26–28. Singapore: IACSIT Press, pp. 22–27

Daniel, R. (2005). The metagenomics of soil. *Nat Rev Microbiol, 3* (6), 470–478.

Das, N., & Chandran, P. (2011). Microbial degradation of petroleum hydrocarbon contaminants: An overview. *Biotechnol Res Int, 2011*, Article ID: 941810.

Dasgupta, D., Ghosh, R., & Sengupta, T. K. (2013). Biofilm-mediated enhanced crude oil degradation by newly isolated *Pseudomonas* species. *ISRN Biotechnol*, Article ID: 250749.

DeLong, E. F., Preston, C. M., Mincer, T., Rich, V., Hallam, S. J., Frigaard, N. U., et al. (2006). Community genomics among stratified microbial assemblages in the ocean's interior. *Science, 311* (5760), 496–503.

Dhankher, O. P., Pilon- Smits, E. A., Meagher, R. B., & Doty, S. (2012). Biotechnological approaches for. In A. Altman, & P. M. Hasegawa (Eds.), *Plant Biotechnology and Agriculture, Prospects for the 21st Century* (pp. 309–328). Elsevier.

Dias, R. L., Ruberto, L., Calabro, A., Balbo, A. L., Panno, M. T., & Cormac, W. P. (2014). Hydrocarbon removal and bacterial community structure in on-site biostimulated biopile systems designed for bioremediation of diesel-contaminated Antarctic soil. *Polar Biol, 38*, 677–687.

Dillewijn, P. v., Nojiri, H., Meer, J. R., & Wood, T. K. (2009). Bioremediation, a broad perspective. *Microb Biotechnol, 2* (2), 125–127.

Edgington, D. N., Gordon, S. A., Thommes, M. M., & Almodovar, L. R. (1970). The concentration of radium, thorium, and uranium by tropical algae. ANL-7615. *ANL Rep, 15* (1), 945–955.

Edwards, S. J., & Kjellerup, B. V. (2013). Applications of biofilms in bioremediation and biotransformation of persistent organic pollutants, pharmaceuticals/personal care products, and heavy metals. *Appl Microbiol Biotechnol, 97* (23), 9909–9921.

Fomina, M., & Gadd, G. M. (2014). Biosorption: current perspectives on concept, definition and application. *Bioresour Technol, 160*, 3–14.

Frascari, D., Zanaroli, G., & Danko, A. S. (2015). In situ aerobic cometabolism of chlorinated solvents: a review. *J Hazard Mater, 283*, 382–399.

Frutos, F. J., Escolano, O., Garcia, S., Babin, M., & Fernandez, M. D. (2010). Bioventing remediation and ecotoxicity evaluation of phenanthrene-contaminated soil. *J Hazard Mater, 183* (1–3), 806–813.

Gilbert, J. A., & Dupont, C. L. (2011). Microbial metagenomics: beyond the genome. *Ann Rev Mar Sci, 3*, 347–371.

Gomez, F., & Sartaj, M. (2014). Optimization of field scale biopiles for bioremediation of petroleum hydrocarbon contaminated soil at low temperature conditions by response surface methodology (RSM). *Int Biodeterior Biodegradation, 89*, 103–109.

Handelsman, J., Rondon, M. R., Brady, S. F., Clardy, J., & Goodman, R. M. (1998). Molecular biological access to the chemistry of unknown soil microbes: a new frontier for natural products. *Chem Biol, 5* (10), 245–249.

He, Z., Nostrand, J. D., Deng, Y., & Zhou, J. (2011). Development and applications of functional gene microarrays in the analysis of the functional diversity, composition, and structure of microbial communities. *Front Environ Sci Eng China, 5*, 1–20.

Hobson, A. M., Fredrickosn, J., & Dise, N. B. (2005). CH_4 and N_2O from mechanically turned windrow and vermicomposting systems following in-vessel pre-treatment. *Waste Manag, 25* (4), 345–352.

Hohener, P., & Ponsin, V. (2014). In situ vadose zone bioremediation. *Curr Opin Biotechnol, 27*, 1–7.

Huertas, M. J., Duque, E., Marqués, S., & Ramos, J. L. (1998). Survival in soil of different toluene-degrading Pseudomonas strains after solvent shock. *Appl Environ Microbiol, 64*, 38–42.

Isildak, O., Turkekul, I., Elmastas, M., & Tuzen, M. (2004). Analysis of heavy metals in some wild-grown edible mushrooms from the Middle Black Sea Region, Turkey. *Food Chem, 86* (4), 547–552.

Jonsson, C. M., Paraiba, L. C., Mendoza, M. T., Sabater, C., & Carrasco, J. M. (2001). Bioconcentration of the insecticide pyridaphenthion by the green algae Chlorella saccharophila. *Chemosphere, 43* (3), 321–325.

Kao, C. M., Chen, C. Y., Chen, S. C., Chien, H. Y., & Chen, Y. L. (2008). Application of in situ biosparging to remediate a petroleum-hydrocarbon spill site: field and microbial evaluation. *Chemosphere, 70* (8), 1492–1499. doi: 10.1016/j.chemosphere.2007.08.029.

Klaus-Joerger, T., Joerger, R., Olsson, E., & Granqvist, C. (2001). Bacteria as workers in the living factory: metal-accumulating bacteria and their potential for materials science. *Trends Biotechnol., 19* (1), 15–20. doi: 10.1016/s0167-7799(00)01514-6.

Kosaric, N. (2001). Biosurfactants and their application for soil bioremediation. *Food Technol. Biotechnol., 39* (4), 295–304. https://www.davuniversity.org/images/files/study-material/biosurfactants-Supplemental.pdf.

Kumar, A., Bisht, B. S., Joshi, V. D., & Dhewa, T. (2011). Review on bioremediation of polluted environment: a management. *Int J Environ Sci, 1* (6), 1079–1093.

Kumar, R., & Goyal, D. (2010). Waste water treatment and metal (Pb^{2+}, Zn^{2+}) removal by microalgal based stabilization pond system. *Indian J Microbiol, 50* (1), 34–40.

Lal, S., Lal, R., & Saxena, D. M. (1987). Bioconcentration and metabolism of DDT, fenitrothion and chlorpyrifos by the blue-green algae *Anabaena* sp. and *Aulosira fertilissima. Environ Pollut, 46* (3), 187–196.

Lam, K. N., Cheng, J., Engel, K., Neufeld, J. D., & Charles, T. C. (2015). Current and future resources for functional metagenomics. *Front Microbiol, 6*, 1196.

Lei, A. P., Wong, Y. S., & Tam, N. F. (2002). Removal of pyrene by different microalgal species. *Water Sci. Technol., 46* (11–12), 195–201.

Li, C. H., Wong, Y. S., & Tam, N. F. (2010). Anaerobic biodegradation of polycyclic aromatic hydrocarbons with amendment of iron(III) in mangrove sediment slurry. *Bioresour Technol, 101* (21), 8083–8092.

Liu, W., Wang, A., Cheng, S., Logan, B. E., Yu, H., Deng, Y., et al. (2010). Geochip-based functional gene analysis of anodophilic communities in microbial electrolysis cells under different operational modes. *Environ Sci Technol, 44* (19), 7729–7735.

Loutseti, S., Danielidis, D. B., Economou-Amilli, A., Katsaros, C., Santas, R., & Santas, P. (2009). The application of a micro-algal/bacterial biofilter for the detoxification of copper and cadmium metal wastes. *Bioresour Technol, 100* (7), 2099–2105.

Lovley, D. R. (2003). Cleaning up with genomics: applying molecular biology to bioremediation. *Nat Rev Microbiol, 1* (1), 35–44.

Luan, T. G., Jin, J., Chan, S. M., Wong, Y. S., & Tam, N. F. (2006). Biosorption and biodegradation of tributyltin (TBT) by alginate immobilized *Chlorella vulgaris* beads in several treatment cycles. *Process Biochem, 41* (7), 1560–1565. https://doi.org/10.1016/j.procbio.2006.02.020.

Malla, M. A., Dubey, A., Yadav, S., Kumar, A., Hashem, A., & Allah, E. F. (2018). Understanding and designing the strategies for the microbe-mediated remediation of environmental contaminants using omics approaches. *Front Microbiol, 9*, 1132.

Manikandan, N., Prasath, C. S. S., & Prakash, S. (2011). Biosorption of uranium and thorium by marine micro algae. *Indian J Mar Sci, 40* (1), 121–124.

Megharaj, M., Vekateswarlu, K., & Rao, A. S. (1987). Metabolism of monocrotophos and quinalphos by algae isolated from soil. *Bull Environ Caontam Toxicol, 39* (2), 251–256.

Mei, L., Xitao, X., Renhao, X., & Zhili, L. (2006). Effects of strontium-induced stress on marine microalgae Platymonas subcordiformis (Chlorophyta: Volvocales). *Chin J Ocean Limnol, 24*, 154–160.

Mirete, S., de Figueras, C. G., & González-Pastor, J. E. (2007). Novel nickel resistance genes from the rhizosphere metagenome of plants adapted to acid mine drainage. *Appl Environ Microbiol, 73* (19), 6001–6011.

Mustafa, Y. A., Abdul- Hammed, H. M., & Razak, Z. A. (2015). Biodegradation of 2,4-dichlorophenoxyacetic acid contaminated soil in a roller slurry bioreactor. *Soil Air Water, 43* (8), 1241–1247.

Mustapha, M. U., & Halimoon, N. (2015). Microorganisms and biosorption of heavy metals in the environment: a review paper. *J Microb Biochem Technol, 7* (5), 253–256.

Narro, M. L., Cerniglia, C. E., Baalen, C. V., & Gibson, D. T. (1992). Evidence for an NIH shift in oxidation of naphthalene by the marine cyanobacterium *Oscillatoria* sp. strain JCM. *Appl Environ Microbiol, 58* (4), 1360–1363.

Nikolopoulou, M., Pasadakis, N., Norf, H., & Kalogerakis, N. (2013). Enhanced ex situ bioremediation of crude oil contaminated beach sand by supplementation with nutrients and rhamnolipids. *Mar Pollut Bull, 77* (1–2), 37–44.

Niu, G. L., Zhang, J. J., Zhao, S., Liu, H., Boon, N., & Zhou, N. Y. (2009). Bioaugmentation of a 4- chloronitrobenzene contaminated soil with *Pseudomonas putida* ZWL73. *Environ Pollut, 157* (3), 763–771.

Nocelli, N., Bogino, P. C., Banchio, E., & Giordano, W. (2016). Roles of extracellular polysaccharides and biofilm formation in heavy metal resistance of rhizobia. *Materials (Basel), 9* (6), 418.

Pacwa-Płociniczak, M., Plaza, G. A., Piotrowska-Seget, Z., & Cameotra, S. S. (2011). Environmental applications of biosurfactants: recent advances. *Int J Mol Sci, 12* (1), 633–654.

Paudyn, K., Rutter, A., Rowe, K. R., & Poland, J. S. (2008). Remediation of hydrocarbon contaminated soils in the Canadian Arctic by landfarming. *Cold Reg Sci Technol, 53*, 102–114.

Pessoa-Filho, M., Barreto, C. C., Junior, F. B., Fragoso, R. R., Costa, F. S., Mendes, I. d., et al. (2015). Microbiological functioning, diversity, and structure of bacterial communities in ultramafic soils from a tropical savanna. *Antonie Van Leeuwenhoek, 107* (4), 935–349.

Philp, J. C., & Atlas, R. M. (2005). Bioremediation of contaminated soils and aquifers. In R. M. Atlas, & J. C. Philp (Eds.), *Bioremediation: Applied Microbial Solutions for Real-World Environmental Cleanup* (pp. 139–236). Washington: American Society for Microbiology (ASM) Press.

Pokhrel, D., & Viraraghavan, T. (2006). Arsenic removal from an aqueous solution by a modified fungal biomass. *Water Res, 40* (3), 549–552.

Popova, M., Martin, C., & Morgavi, D. P. (2010). Improved protocol for high-quality co-extraction of DNA and RNA from rumen digesta. *Folia Microbiologica, 55* (4), 368–372.

Prakash, T., & Taylor, T. D. (2012). Functional assignment of metagenomic data: challenges and applications. *Brief Bioinform, 13* (6), 711–727.

Rao, M. A., Scelza, R., Scotti, R., & Gianfreda, L. (2010). Role of enzymes in the remediation of polluted environments. *Jr Soil Sci Plant Nutr, 10* (3), 333–353.

Rayner, J. L., Snape, I., Walworth, J. L., Harvey, P. M., & Ferguson, S. H. (2007). Petroleum–hydrocarbon contamination and remediation by microbioventing at sub-Antarctic Macquarie Island. *Cold Reg Sci Technol, 48* (2), 139–153. https://doi.org/10.1016/j.coldregions.2006.11.001.

Riesenfeld, C. S., Schloss, P. D., & Handelsman, J. (2004). Metagenomics: genomic analysis of microbial communities. *Annu Rev Genet, 38*, 525–552. doi: 10.1146/annurev.genet.38.072902.091216.

Saharan, B., Sahu, R., & Sharma, D. (2011). A review on biosurfactants: fermentation, current developments and perspectives. *J Genet Eng Biotechnol, 29*, 1–39. https://astonjournals.com/manuscripts/Accepted/GEBJ-29acc7-11-11.pdf.

Sanscartier, D., Zeeb, B., Koch, I., & Reimer, K. (2009). Bioremediation of diesel-contaminated soil by heated and humidified biopile system in cold climates. *Cold Reg Sci Technol, 55* (1), 167–173.

Santos, A. L., Galdino, A. C., Mello, T. P., Ramos, L. d., Branquinha, M. H., Bolognese, A. M., et al. (2018). What are the advantages of living in a community? A microbial biofilm perspective! *Mem Inst Oswaldo Cruz, 113* (9), e180212. doi: 10.1590/0074-02760180212.

Sar, P., & Islam, E. (2011). Metagenomic approaches in microbial bioremediation of metals and radionuclides. In T. Satyanarayana et al. (Eds.), *Microorganisms in Environmental Management: Microbes and Environment*. Springer.

Shiau, B. J., Sabatini, D. A., & Harwell, J. H. (1995). Properties of food grade (edible) surfactants affecting subsurface remediation of chlorinated solvents. *Environ Sci Technol, 29* (12), 2929–2935.

Siddhapura, P. K., Vanparia, S., Purohit, M. K., & Singh, S. P. (2010). Comparative studies on the extraction of metagenomic DNA from the saline habitats of Coastal Gujarat and Sambhar Lake, Rajasthan (India) in prospect of molecular diversity and search for novel biocatalysts. *Int J Biol Macromol, 47* (3), 375–379.

Silva-Castro, G. A., Uad, I., Rodriguez-Calvo, A., Gonzalez-Lopez, J., & Calvo, C. (2014). Response of autochthonous microbiota of diesel polluted soils to land-farming treatments. *Environ Res, 137*, 49–58.

Singh, N., Megharaj, M., Gates, W. P., Churchman, G. J., Anderson, J., Kookana, R. S., et al. (2003). Bioavailability of an organophosphorus pesticide, fenamiphos, sorbed on an organo clay. *J Agric Food Chem, 51* (9), 2653–2658.

Smets, B. F., & Barkay, T. (2005). Horizontal gene transfer: perspectives at a crossroads of scientific disciplines. *Nat Rev Microbiol, 3* (9), 675–678. doi: 10.1038/nrmicro1253.

Srivastava, S., & Thakur, I. S. (2006). Isolation and process parameter optimization of Aspergillus sp. for removal of chromium from tannery effluent. *Bioresour Technol, 97* (10), 1167–1173.

Stenuit, B., Eyers, L., Schuler, L., George, I., & Agathos, S. N. (2009). Molecular tools for monitoring and validating bioremediation. In A. Singh, R. C. Kuhad, & O. P. Ward (Eds.), *Advances in Applied Bioremediation* (pp. 339–353). Berlin/Heidelberg: Springer. https://www.springer.com/gp/book/9783540896203.

Sui, H., & Li, X. (2011). Modeling for volatilization and bioremediation of toluene-contaminated soil by bioventing. *Chin J Chem Eng, 19* (2), 340–348.

Sutherland, W. J., Freckleton, R. P., Godfray, H. C., Beissinger, S. R., Benton, T., Cameron, D. D., et al. (2013). Identification of 100 fundamental ecological questions. *J Ecol, 101* (1), 58–67.

Thapa, B., Kumar, A. K., & Ghimire, A. (2012). A review on bioremediation of petroleum hydrocarbon contaminants in soil. *Kathmandu Univ J Sci Eng Technol, 8* (1), 164–170.

Thomas, T., Gilbert, J., & Meyer, F. (2012). Metagenomics – a guide from sampling to data analysis. *Microb Inform Exp, 2*, 3.

Thome, A., Cecchin, I., Reginatto, C., & Colla, L. M. (2010). Bioventing in a residual clayey soil contaminated with a blend of biodiesel and diesel oil. *J Environ Eng, 140* (11), 1–6.

Tripathi, M., Singh, D. N., Vikram, S., Singh, V., & Kumar, S. (2018). Metagenomic approach towards bioprospection of novel biomolecule(s) and environmental bioremediation. *Annu Res Rev Biol, 22* (2), 1–12.

Turnbaugh, P. J., Ley, R. E., Hamady, M., Fraser-Liggett, C. M., Knight, R., & Gordon, J. I. (2007). Feature the human microbiome project. *Nature, 449*, 804–810. doi: 10.1038/nature06244.

Tyson, G. W., Lo, I., Baker, B. J., Allen, E. E., Hugenholtz, P., & Banfield, J. F. (2005). Genome-directed isolation of the key nitrogen fixer *Leptospirillum ferrodiazotrophum* sp. nov. from an acidophilic microbial community. *Appl Environ Microbiol, 71* (10), 6319–6324. doi: 10.1128/AEM.71.10.6319-6324.2005.

van der Wal, A., Norde, W., Zehnder, A. J., & Lyklema, J. (1997). Determination of the total charge in the cell walls of Gram-positive bacteria. *Colloids Surf B, 9* (1–2), 81–100.

Velasquez, L., & Dussan, J. (2009). Biosorption and bioaccumulation of heavy metals on dead and living biomass of Bacillus sphaericus. *J Hazard Mater, 167* (1–3), 713–716. doi: 10.1016/j.jhazmat.2009.01.044.

Wackett, L. P., & Hershberger, C. D. (2001). *Biocatalysis and Biodegradation: Microbial Transformation of Organic Compounds*. Washington, DC: ASM Press.

Wang, C. L., Ozuna, S. C., Clark, D. S., & Keasling, J. D. (2002). A deep-sea hydrothermal vent isolate, *Pseudomonas aeruginosa* CW961, requires thiosulfate for Cd^{2+} tolerance and precipitation. *Biotechnol Lett, 24* (8), 637-641. doi: 10.1023/A:1015043324584.

Weiner, J. A., DeLorenzo, M. E., & Fulton, M. H. (2004). Relationship between uptake capacity and differential toxicity of the herbicide atrazine in selected microalgal species. *Aquat Toxicol, 68* (2), 121–128. doi: 10.1016/j.aquatox.2004.03.004.

Wengel, M., Kothe, E., Schmidt, C. M., Heide, K., & Gleixner, G. (2006). Degradation of organic matter from black shales and charcoal by the wood-rotting fungus *Schizophyllum commune* and release of DOC and heavy metals in the aqueous phase. *Sci Total Environ, 367* (1), 383–393. doi: 10.1016/j.scitotenv.2005.12.012.

Yan, H., & Pan, G. (2004). Increase in biodegradation of dimethyl phthalate by *Closterium lunula* using inorganic carbon. *Chemosphere, 55* (9), 1281–1285. https://doi.org/10.1016/j.chemosphere.2003.12.019.

Yee, N., & Fein, J. (2001). Cd adsorption onto bacterial surfaces: a universal adsorption edge? *Geochim Cosmochim Acta, 65* (13), 2037–2042. https://doi.org/10.1016/S0016-7037(01)00587-7.

Yooseph, S., Sutton, G., Rusch, D. B., Halpern, A. L., Williamson, S. J., Remington, K., et al. (2007). The Sorcerer II Global Ocean Sampling expedition: expanding the universe of protein families. *PLoS Biol, 5* (3), e16. doi: 10.1371/journal.pbio.0050016.

Yoshikawa, M., Zhang, M., & Toyota, K. (2016). Integrated anaerobic-aerobic biodegradation of multiple contaminants including chlorinated ethylenes, benzene, toluene, and dichloromethane. *Water Air Soil Pollut, 228* (1), 25. doi: 10.1007/s11270-016-3216–1.

Zangi-Kotler, M., Ben- Dov, E., Tiehm, A., & Kushmaro, A. (2015). Microbial community structure and dynamics in a membrane bioreactor supplemented with the flame retardant dibromoneopentyl glycol. *Environ Sci Pollut Res Int, 22* (22), 17615–17624. doi: 10.1007/s11356-015-4975–8.

11 Metagenomics for Studying Microbes in Wastewater Treatment Plants

Anand Thirunavukarasou
Sri Ramachandra Institute of Higher Education and Research
B-Aatral biosciences Private Limited

Sweety Kaur
Sri Ramachandra Institute of Higher Education and Research

Harvinder Kour Khera
Tata Institute for Genetics and Society

CONTENTS

DOI: 10.1201/9781003354147-11

11.1 INTRODUCTION

The rapid expansion of the global population with simultaneous industrialization and economical increment has led to the production of huge tons of solid garbage and wastewater annually. Over >80% of this wastewater reverts into the ecological environment without undergoing any treatment, thereby holding along with large amounts of contaminants that consist of microbes, heavy toxic metals, and organic and inorganic compounds resulting in adversely impacting physicochemical properties and microbial configuration of naturally available freshwater [1]. Different wastewater treatment plants (WWTPs) are employed in various countries to enhance the physicochemical property and microbial configuration of effluents using a chain of events like primary treatment, secondary treatment and clarification, tertiary treatment, disinfection, and processing of sludge. Metagenomic exploration of public WWTPs incorporating traditional and membrane bioreactors discovered that *Pseudomonas*, *Bacteroides*, *Aeromonas*, *Prevotella*, and *Cloacibacterium* have dominated the influent water samples; however, the disparity was noticed in bacterial species amid membrane bioreactor-treated effluents and traditional settling tank-treated effluents. The population of *Pseudomonas*, *Acinetobacter*, and *Aeromonas* load declined in major effluents using different steps of WTTPs. Altogether, the microbial uniformity of the final treated water improved with WTTP volumes, and the composition of the influent sample was impacted by different environmental parameters, especially with a change in temperature [2].

Metagenomics is a newly emerged frontline in science that deals with the genetic evaluation of the genomic material (DNA) samples directly isolated and collected from various environmental conditions. This term was devised by Handelman and his collaborators in 1998 to analyze a group of identical components, just in case of the statistical notion of meta-analysis [3]. This concept has transformed our knowledge of surrounding ecology as it provides altogether identification and evaluation of the environmental microbial consortium. Due to metagenomics, a wide range of uncultured microbes have been discovered that were hidden from conventionally existing techniques [4]. The extraction of unique chemical components with their associated biochemical roles from uncultured, cultured, known, and unknown microbes (that encompass >99% of the microbial diversity) can be cataloged by the application of metagenomics, therefore broadening the idea of the science community. Despite the monistic phylogenetic study that depends on single gene diversity, metagenomics regulates the multi-modal genomic composition of microorganisms, hence facilitating enhanced genetic information and taxonomic resolution [5]. Metagenomics helps in the construction of microbial evolutionary profile structure, associates phylogenetic functions, and identification of viruses because of its genomic diversity and ability to detect shared genetic linkage that is difficult to detect using the single gene targeting method [6]. In recent years, next-genome sequencing (NGS) techniques have replaced conventional Sanger sequencing, prior employed as an essential tool for shotgun sequencing. 454/Roche and Solexa systems are immensely employed for studying metagenomic models collected from a large range of environments [7]. Despite the development of sequencing techniques, there are still many parameters responsible for biases and errors, but metagenomics provides an accurate and unbiased source to examine the functional ability of ecological microbiota. The first extensive metagenomic study conducted shotgun sequencing for two viral strains isolated from the surface of seawater where 65% of the viral sequences were novelistic, and among them, the dominant population was encompassing 2%–3% of the constructed sequences [8]. The first microbial metagenomic project was conducted by Venter and his colleagues in the year 2004 to investigate microbial consortiums under the surface water of the Sargasso Sea where he discovered 1.2 million unknown genes using metagenome sequencing [9]. The distinctive impact of NGS has stimulated a significant increase in the number of metagenomics studies with about 300 research studies including clinical, genetically engineered, and ecological groups under progress or completed [10]. Despite the output of next-generation, sequencing cannot be randomly employed on any isolated environmental samples without the development of a suitable sequencing approach. The designed strategy applied should consider research-based interrogation and alignment of the desired community; otherwise, it can immensely hinder the downstream process and absolute accomplishment of the metagenomic study. Recent strategies permit us to interpret the complexity of microbiota, their kinetics, and their role in the ecosystem. These distinct metagenomics approaches clarify principal questions like which microbes are residing with respect to taxonomic diversity and their responsible functions with respect to functional metagenomics study [11]. This review chapter deals with the concept of metagenomics for studying various microbial communities involved in wastewater treatment, different methods involved in the collection, sequencing, and analysis of the samples followed by the major application and role of metagenomics in wastewater treatment plants.

11.2 METHODS INVOLVED IN METAGENOMIC DATA ANALYSIS

Metagenomic studies are constituted on direct genomic DNA extraction from environmental samples like in wastewater treatment aspect; it will be activated sludge (aerobic/anaerobic) and is regarded as a significant tool for interpreting the genomic and metabolic diverseness of composite ecology [12]. By the implementation of metagenomics technology, one can attain the metabolic profile of complex ecological samples and the discovery of novel microbial bioactive compounds and enzymes by constructing metagenomic libraries from the extracted genetic material (DNA/RNA) samples. In metagenomics, screening of the isolated biomolecules from ecological samples is carried out using two fundamental strategies, i.e., function-based screening and sequence-based screening. Among both approaches, genomic DNA is fragmented and the construction of metagenomic libraries using expression vectors is carried out for the gene expression. These constructed genetic libraries can further be applied for the discovery of novelistic compounds and biomolecules that can be encoded by one gene or less-size operon, while larger inserts demand extraction of large gene clusters that can encode complex metabolic pathways comprising many genes. Following the construction of a library, the generated metagenomic clones can further be employed to modify a heterologous host (Figure 11.1) [13].

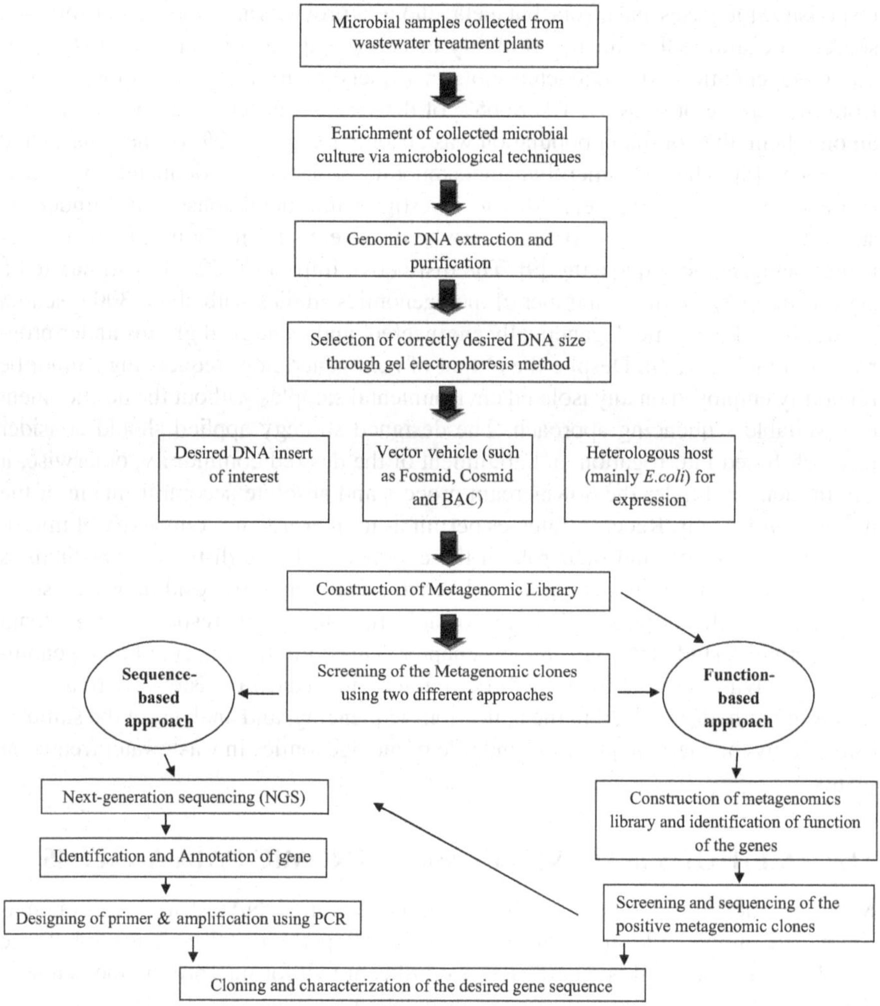

FIGURE 11.1 Workflow of metagenomic data analysis of the microbial communities in wastewater treatment plants.

11.2.1 Sampling from WWTPs

Sampling can be done by the collection of samples from both influents and effluents from a WWTP that must have a large volume capacity of around ~2,000–3,000 m^3 per day. Influent samples can be collected from crude and untreated sludge and effluent samples can be collected from the finished treated water tank that can be further employed for various agriculture and domestic purposes [14]. Followed by sampling, the next step comprises total genomic DNA isolation from microbial consortium inhabiting the collected sample.

11.2.2 Total Genomic DNA Extraction, Confirmation, and Storage

Isolation of genomic DNA is a primary and significant step for conducting metagenomics data analysis of the collected microbial consortium. As the DNA property can be largely impacted through varying sampling procedures, it is required to select an appropriate sampling method, especially when dealing with sewage treatment plants and bioreactor samples. Before DNA isolation, the collected sample can be subjected to centrifugation for around 10 minutes and obtained supernatant must be discarded. Considering pellet for further processing, isolation of the total genomic DNA can be done using various commercial kits available that account for >90% of all relevant studies. Alternatively, C-TAB-based DNA isolation is another method that can also be performed, accounting for around 6% of all related studies [15]. After extraction, quantification of the isolated DNA is done at absorbance wavelengths of 260 and 280 nm. Followed by quantification, the purified and concentrated DNA will be stored in TE buffer at −20°C till further processing.

11.2.3 Construction of the Metagenomic Library

The isolated, purified, and concentrated DNA will further be subjected to metagenomic library construction. It comprises fragmentation of a specific sequence, followed by cloning of that specific DNA sequence, incorporation within a definitive vector depending upon the size of the insert, and insertion into a heterologous host. This leads to the construction of metagenomic libraries of a few hundred base pairs [16].

11.2.4 NGS Method

NGS technologies can carry out both qualitative and quantitative analyses of microbial consortiums collected from WWTPs in a comparatively rapid and cost-effective manner. The most employed DNA sequencing technology is based upon Roche GS-FLX 454 pyrosequencing platform, accounting for an additional 90% of the analyzed studies. Followed by other NGS technologies including the Solexa (Illumina-based) sequencing platform, ABI SOLiD™ short-read DNA sequencing platform technique, and Ion PGM™ Hi-Q™ sequencing kit combined with Ion PGM™ sequencer functioned by Torrent Suite™ software. 454 pyrosequencing technology is acquiring expanding application exhibiting; however, the reference data applied for sequential profiling of collected microbial consortium is yet restricted. Hence, there is a need of conducting studies relevant to the significant analysis of the metagenome of microorganisms along with the formation of a definitive storage platform for sequencing (Table 11.1) [15].

11.2.4.1 Solexa (Illumina) Genome Sequencing

This sequencing technique was developed in the year 2006 and was broadly in use by researchers for its cost-effectiveness. Different variants that have also been developed to attain accurate sequencing are GA I and II, MiSeq, NextSeq 500, HiSeq 2500, and X Ten. The MiSeq version provides 2–300 base-pair length reads but was unable to

TABLE 11.1
Different Types of Sequencing Methods with Their Respective Features and Potential Applications

Sequencing Method	Developed by	Salient Features	Applications	References
Solexa (Illumina) genome sequencing	Shankar Balasubramanian and David Klenerman (Cambridge University) in 2006	This method works on reversible dye terminators that allow the identification of single bases when introduced into DNA strands	Classification of structure and conformation of microbial consortium in WWTPs; metagenomic profiling of antibiotic-resistant genes and mobile genetic elements; for a metagenomic insight of nitrogen metabolism in tannery WWTPs with enlarged microbial consortium; tracking motility of soil bacterial community and detection for potential microbial agents during bioremediation	[17]
Pyrosequencing	Pal Nyren (KTH Royal Institute of Technology)	This sequencing is based on the luminometric detection of pyrophosphate that is released during primer-directed DNA polymerase-catalyzed nucleotide integration and is convenient for the detailed classification of nucleic acids. Moreover, it provides precision, rigidity, analogous processing, and easy automation	Classification of a microbial consortium with their functional genes from tannery WWTPs; pathogenic bacteria residing in sludge treatment plants and microbial diversity during bioremediation of contaminated coastal sediments and oil-enriched regions	[18]
SOLiD sequencing	Life Technologies, USA	This method employs an emulsion PCR strategy with small magnetic beads to do fragment amplification for sequencing, and dependent upon the ligation-based sequencing	Application in evaluation of microbial diversity and function extracted from Anammox WWTPs.	[19]
Ion torrent semiconductor sequencing	Developed by Ion torrent Systems and then owned by Life Technologies USA	This method is dependent on the identification of H^+ ions released during DNA polymerization	Classification of microbial communities (especially bacterial) in nitrifying triggered sludge and filamentous bacteria in WWTPs	[20]
Nanopore sequencing	Developed by Oxford Nanopore Technology Limited	Very effective molecular technique for sequencing of the genome without using amplification or any chemical labeling of the sample	Classification of carbapenemase-encoding plasmids in enterobacteria extracted from WWTPs and linkage between resistome phenotype and genotype of coliform bacteria in municipal sewage	[21]

read very smaller read-sizes, so scientists developed a more advanced version, i.e., Roche 454 platform to Illumina sequencer that operates sequencing using construction method by the termination. HiSeq 2500 develops a maximum of 4 billion fragments each of 125 bases size that can be used for single read in a paired-end mode dye and its more current advanced version is HiSeq X Ten which includes a combination of ten HiSeq equipment to attain high-throughput data. A comparatively smaller-sized sequencing has also been introduced, named NextSeq 500 [22].

11.2.4.2 Pyrosequencing

Different bioactive metabolic compounds can be isolated from the collected metagenomic samples with the help of enormous therapeutic possibilities, including malacidin, minimide, and fluoroquinolone. The first high-throughput sequencing introduced was GS-20 in 2005 which carries out sequencing by synthesis method in a picotiter platform which gives an output of about 20 mega-bases in a single run and an average read size of 100 base pairs. Based on the pyrosequencing mechanism that includes nucleotide triphosphate (NTP) and additional nucleotides complementary to the desired sequencing strand can further be analyzed by the release of pyrophosphate. It is the more highly advanced version developed was GS-FLX Titanium+ which develops about 850 mega-bases within a single run and exhibits a mean read size of 700–750 base pairs, and it was most appropriate for the 16S rRNA sequencing technique because it can develop highly variable fragments of 16S rRNA. Although GS-FLX versions were stopped from commercializing by the end of 2016 due to high cost, frequent error rate and sample DNA amount needed was more in contrast to other high-throughput platforms [23].

11.2.4.3 Sequencing by Oligonucleotide Ligation and Detection (SOLiD)

The SOLiD technique was developed in the year 2006 by Life Technologies as a platform for sequencing that exploits the implementation of sequencing by ligation method. One of its versions, i.e., 5500 xl, creates around 300 gigabytes of metagenomics data along with about 3 billion base pairs in a single run providing a reading size of 75 base pairs. Despite providing such a huge amount of metagenomics data and less single-base sequencing, its main pitfall is smaller read size plus its cost for a single run [24].

11.2.4.4 Ion Torrent Semiconductor Sequencing

Ion torrent technologies were the first company to develop small-scale sequencers in the manner of personal genome machine for scientists in the year 2010 and had attained a positive response, therefore now had become the flashpoint for researchers to conduct sequencing study with relatively very less expenditure. This sequencing is mainly performed in a microtiter platform where DNA fragments are added to the beads when the DNA is complemented to the sequencing target strand, thereby releasing protons that alter the pH which can be detected by the connected detector. This sequencing creates around 10 gigabases of metagenomic data with 50 million reads in just a single run providing a read size of 200 base pairs. The recent Ion S5 version can generate around 15 gigabases of metagenomic output with 60–80 million reads within a single run of a size of around 200 base pairs [25].

11.2.4.5 Nanopore Sequencing

Oxford Nanopore Sequencing exploits a high-tech strand sequencing method that can sequence complete DNA strands by identifying the alteration in supplied electric current when passes through small nanopores constituted of proteins. MinION mk1B is a condensed sequencer that can connect with any kind of computer for rapid metagenomics data analysis. Some other of its versions are PromethION which provides 144,000 nanopore channels for sequencing, SmidgION can be easily used using a smartphone for rapid analysis and VolTRAX can be regulated by USB after loading the sample. The relatively larger read size eliminates the need of doing shotgun sequencing, hence bringing a revolution in the range of sequencing [26].

11.2.5 Analysis of Sequenced Metagenomic Data

The elucidation of the metagenomic output is prepared clearly to process the mixture of genomes and contigs of varying sizes. There are many stages for the analysis of metagenomics data including the processing of low-quality reads, elimination of less-complex reads, identification, and annotation of genes. At first, the reads of low quality are processed with utilities that are compatible with the version of sequencing to be employed for sequencing. Some utilities are FASTX-Toolkit which is used for pre-processing of FASTA data, and FastOC is a quality check utility that processes the raw high-throughput data. Elimination of less-complex reads is done by application of DUST utility and then the sequences that will be sharing >95% similarity will be eliminated using tools like MG-RAST (MetaGenomic Rapid Annotation using Subsystems Technology) that eliminates the reads which are similar to the model organisms' genomes like human, fly, cow, and mouse. Then, identification of genes is done in which "calling of genes" is done permits identification of genes present in the contigs. Utilities like Metagene, FragGeneScan, and MetaGeneMark can identify coding DNA sequence genes by applying information of codon to detect intron or exon regions of the contigs. Then, the next step includes function allocation of genes that is accompanied by using a similarity-search method where the desired sequence is compared with the sequence database that exhibits information of a list of annotated genes [27].

In gene annotation, the initial step includes a collection of crude NGS-based metagenomic outputs, followed by filtration of poor contigs and removal of duplicates. The extracted contigs will be involved in the construction of small-sized reads. Then, these small-sized contigs are allocated with their defined structural and functional properties and the entire annotation will be conducted. Acquired annotated metagenomics output can be further stored, administered, and denoted on a website for the reference of the scientific community [28].

11.3 APPLICATION OF METAGENOMICS IN DIVERSIFIED FIELDS

Analysis of different ecosystems like soil and marine with the application of metagenomics has ultimately proved that yet a large amount of undiscovered genomic diversity inhabits. Metagenomes collected from both soil and water were evaluated via different global sample collection assays, thereby revealing undiscovered

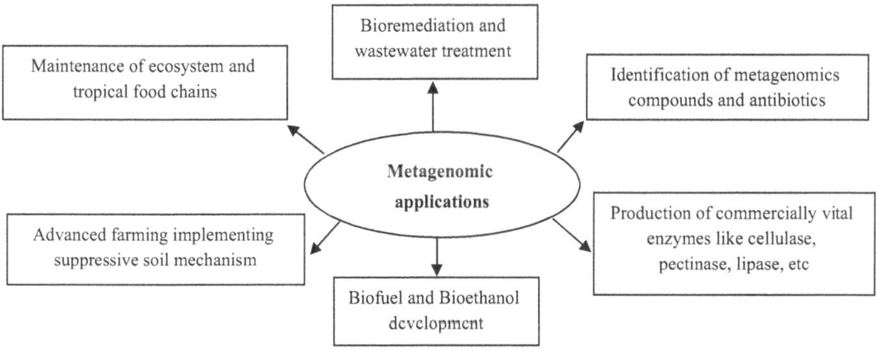

FIGURE 11.2 Application of metagenomics in diversified fields.

communities of prokaryotes, unicellular eukaryotes, and viruses that had ultimately generated a huge amount of genomic data which be studied for ecology, biotechnology, and phylogenetic-based potential applications as depicted in Figure 11.2 [29]. With the implementation of ecological metagenomics, we can get an idea of the parameters that can be responsible for contamination from different pollutants on microorganisms, therefore providing potential bioremediation solutions [30]. Using metagenomics, identification of microbial communities linked with agriculture and food can be done [31]. Mammalian anchor diversified microbial communities that inhabit the epidermis and inside the host such as skin, mucosal membrane, gastrointestinal tract, etc. Studies have identified the microbial consortium residing in the gut in a series of physiological processes with a vital role in ingestion of food, metabolic pathways, adaptiveness, immunity, and regulation of complete well-being of the host [32]. In clinical aspects, metagenomics had played a reframed and drastic impact as well. Metagenomic profiling can be depicted as a non-invasive, comparatively cheaper, and instant biomarker for diagnosis and in the case of infectious diseases, interpretation of antimicrobial resistance complements an additional layer of related information collected from clinical specimens [33]. Moreover, these metagenomics methods also affect human health with potential applications including checking antimicrobial resistance in food sources supplied by bacterial whole-genome sequencing [34]. Metagenomics permits the classification of diversified microbial communities of environments, comprising human body regions, at unique resolution points without requiring any previous knowledge and culturing [35].

11.4 APPLICATION OF METAGENOMICS IN WASTEWATER TREATMENT PLANTS

Sewage-polluted water plants comprised of many pathogenic microorganisms and viruses such as *Salmonella*, *Vibrio cholera*, *Bacillus*, and *E.coli* may result in severe water-borne gastrointestinal infections [36]. Conventionally used bacteriological indicators to evaluate the water microbial quality, i.e., coliform bacteria are usually unable to show pathogenic viruses that can inhabit for longer duration in water [37–39]. Microbiological and virological assessment of water quality is a crucial parameter in wastewater treatment plants and identification of such pathogenic microbes is vital for processing of wastewater treatment and human safety. Productive monitorization and pathogenic control is a crucial approach in the overall wastewater treatment process and identification and classification of pathogenic organisms residing in varying water sources are important for its observation and consecutive supervision. Using metagenomics, identification and characterization of microbes and viruses can be done including sequencing technology and investigation of sample genomic material without culturing the samples [40]. With the help of metagenomics, researchers have developed a huge amount of viral-genomic data, whose relevant sequences are not available in databases, thereby revealing that ecological viral consortiums are comparatively different from already discovered and classified viruses (Table 11.2) [41].

Jadeja et al. [42] conducted metagenomics analysis for the assessment of microbial consortium from effluent WWTPs, in which they assessed the microbial ecosystem of a WWTP where more than 200 industrial effluents arrive. The screening was carried out and it was observed that around 30 degradative metabolic pathways were involved in wastewater. Outcomes were stressed for the execution of the bioremediation strategy using 4-methylphenol, 2-chlorobenzoate, and 4-chlorobenzoate as molecules of interest. This study reflects the importance to describe the microbial communities that activated sludge to repress its covered potential.

Lu et al. [43] investigated microbial consortiums of polluted estuarine remnants of rivers Oujiang and Jiaojiang in East China. The chemical pollutants extracted from the remnants showed that the microbial consortium abundance is diversifying from non-contaminated regions to contaminated remnants. Polycyclic aromatic hydrocarbons and nitrobenzene are related to the leading bacterial population of Gammaproteobacteria. Using gene annotation, it was identified that different enzymatic profiles were also abundant such as 2-oxoglutarate synthase, acetolactate synthase, inorganic diphosphatase, and aconitate hydratase in estuarine sediments that ultimately enhanced in its level.

TABLE 11.2
Role of Metagenomics in Wastewater Treatment in a Progressive Manner

Steps of Analysis	Role of Metagenomics in Wastewater Treatment	Outcome
Step 1	Analysis of influent wastewater	Identification and classification of pathogenic organisms
Step 2	Evaluation of water quality	Amount and diversity of antibiotic resistance genes; identification of bio-indicators within the water sample
Step 3	Assessment of bioremediation and phage therapy	Discovery of novel microbial and viral genes responsible for bioremediation
Step 4	Additional metagenomics application	Interpretation of unrevealed genetic information and identification and production of novel bioactive compounds and enzymes

One more study reveals the diversity, complexity, and dispersion of biodegradation genes present in activated sludge collected from two WWTPs implementing metagenomics analysis [44]. The most abundant genes revealed were P450 genes involved in the process of biodegradation that was identified using the conventional polymerase chain reaction (PCR) technique. It was observed that about 87 bacterial genera were very capable of degrading chemical pollutants like Proteobacteria (60%), Bacteriodetes (17%), Mycobacterium (23%), and other Actinobacteria (9%). The existing strategy can be applied to investigate an activated sludge for other organic pollutants and to attain the efficacy of effluent WWTPs.

REFERENCES

1. Singh AK, Chandra R. Pollutants released from the pulp paper industry: Aquatic toxicity and their health hazards. *Aquat Toxicol.* 2019; 211:202–216.
2. Tong J, Tang A, Wang H, Liu X, Huang Z, Wang Z, Zhang J, Wei Y, Su Y, Zhang Y. Microbial community evolution and fate of antibiotic resistance genes along with six different full-scale municipal wastewater treatment processes. *Bioresour Technol.* 2019; 272:489–500.
3. Handelsman J, Rondon MR, Brady SF, Clardy J, Goodman RM. Molecular biological access to the chemistry of unknown soil microbes: a new frontier for natural products. *Chem Biol.* 1998; 5(10):245–249.
4. Bashir Y, Pradeep Singh S, Kumar Konwar B. Metagenomics: an application-based perspective. *Chin J Biol.* 2014.
5. Quince C, Walker AW, Simpson JT, Loman NJ, Segata N. Shotgun metagenomics, from sampling to analysis. *Nat Biotechnol.* 2017; 35(9):833–844.
6. Kristensen DM, Mushegian AR, Dolja VV, Koonin EV. New dimensions of the virus world were discovered through metagenomics. *Trends Microbiol.* 2010; 18(1):11–19.
7. Luo C, Tsementzi D, Kyrpides N, Read T, Konstantinidis KT. Direct comparisons of Illumina vs. Roche 454 sequencing technologies on the same microbial community DNA sample. *PLoS One.* 2012; 7(2):e30087.
8. Breitbart M, Salamon P, Andresen B, Mahaffy JM, Segall AM, Mead D, Azam F, Rohwer F. Genomic analysis of uncultured marine viral communities. *Proc Natl Acad Sci.* 2002; 99(22):14250–14255.
9. Venter JC, Remington K, Heidelberg JF, Halpern AL, Rusch D, Eisen JA, Wu D, Paulsen I, Nelson KE, Nelson W, Fouts DE. Environmental genome shotgun sequencing of the Sargasso Sea. *Science.* 2004; 304(5667):66–74.
10. Pagani I, Liolios K, Jansson J, Chen IM, Smirnova T, Nosrat B, Markowitz VM, Kyrpides NC. The Genomes OnLine Database (GOLD) v. 4: status of genomic and metagenomic projects and their associated metadata. *Nucleic Acids Res.* 2012; 40(D1):571–579.
11. Ghosh A, Mehta A, Khan AM. Metagenomic analysis and its applications. *Encyclopedia of Bioinformatics and Computational Biology.* 2019; Vol. 3.
12. Lam KN, Cheng J, Engel K, Neufeld JD, Charles TC. Current and future resources for functional metagenomics. *Front Microbiol.* 2015; 6:1196.
13. Madhavan A, Sindhu R, Parameswaran B, Sukumaran RK, Pandey A. Metagenome analysis: a powerful tool for enzyme bioprospecting. *App Biochem Biotechnol.* 2017; 183(2):636–651.
14. Yasir M. Analysis of microbial communities and pathogen detection in domestic sewage using metagenomic sequencing. *Divers.* 2021; 13(1):6.
15. Zhang L, Loh KC, Lim JW, Zhang J. Bioinformatics analysis of metagenomics data of biogas-producing microbial communities in anaerobic digesters: A review. *Renew SusEnerg Rev.* 2019; 100:110–126.

16. Dias R, Silva LC, Eller MR, Oliveira VM, De Paula SO, Silva CC. Metagenomics: Library construction and screening methods. *V Metagenomics: Methods, Applications, and Perspectives*. 2014; Vol. 5(3):28–34.
17. Hong YH, Ye CC, Zhou QZ, Wu XY, Yuan JP, Peng J, Deng H, Wang JH. Genome sequencing reveals the potential of Achromobacter sp. HZ01 for bioremediation. *Front Microbiol*. 2017; 8:1507.
18. Peng M, Zi X, Wang Q. Bacterial community diversity of oil-contaminated soils assessed by high throughput sequencing of 16S rRNA genes. *Int J Environ Res Public Health*. 2015; 12(10):12002–12015.
19. Rosselli R, Romoli O, Vitulo N, Vezzi A, Campanaro S, De Pascale F, Schiavon R, Tiarca M, Poletto F, Concheri G, Valle G. Direct 16S rRNA-seq from bacterial communities: a PCR-independent approach to simultaneously assess microbial diversity and functional activity potential of each taxon. *Sci Rep*. 2016; 6(1):1–2.
20. Gwin CA, Lefevre E, Alito CL, Gunsch CK. Microbial community response to silver nanoparticles and Ag+ in nitrifying activated sludge revealed by ion semiconductor sequencing. *Sci Tot Environ*. 2018; 616:1014–1021.
21. Xia Y, Li AD, Deng Y, Jiang XT, Li LG, Zhang T. MinION nanopore sequencing enables correlation between resistome phenotype and genotype of coliform bacteria in municipal sewage. *Front Microbiol*. 2017; 8:2105.
22. Levy SE, Myers RM. Advancements in next-generation sequencing. *Annu Rev Genomics Hum Gent*. 2016; 17:95–115.
23. Harrington CT, Lin EI, Olson MT, Eshleman JR. Fundamentals of pyrosequencing. *Arch Pathol Lab Med*. 2013; 137(9):1296–1303.
24. Goodwin S, McPherson JD, McCombie WR. Coming of age: ten years of next-generation sequencing technologies. *Nat Rev Genet*. 2016;17(6):333.
25. Lahens NF, Ricciotti E, Smirnova O, Toorens E, Kim EJ, Baruzzo G, Hayer KE, Ganguly T, Schug J, Grant GR. A comparison of Illumina and Ion Torrent sequencing platforms in the context of differential gene expression. *BMC genomics*. 2017; 18(1):1–3.
26. Wanunu M. Nanopores: A journey towards DNA sequencing. *Phys Life Rev*. 2012; 9(2):125–158.
27. Singh B, Roy A. Metagenomics and drug-discovery. *Metagenomics: Techniques, Applications, Challenges and Opportunities*. 2020:133–145. Springer: Singapore.
28. Taş N, de Jong AE, Li Y, Trubl G, Xue Y, Dove NC. Metagenomic tools in microbial ecology research. *Curr Opin Biotechnol*. 2021; 67:184–191.
29. Thompson LR, Sanders JG, McDonald D, Amir A, Ladau J, Locey KJ, Prill RJ, Tripathi A, Gibbons SM, Ackermann G, Navas-Molina JA. A communal catalogue reveals Earth's multiscale microbial diversity. *Nature*. 2017; 551(7681):457–463.
30. Ghosh S, Das AP. Metagenomic insights into the microbial diversity in manganese-contaminated mine tailings and their role in biogeochemical cycling of manganese. *Sci Rep*. 2018; 8(1):1–2.
31. Liu J, Cade-Menun BJ, Yang J, Hu Y, Liu CW, Tremblay J, LaForge K, Schellenberg M, Hamel C, Bainard LD. Long-term land use affects phosphorus speciation and the composition of phosphorus cycling genes in agricultural soils. *Front Microbiol*. 2018; 9:1643.
32. D'Argenio V, Salvatore F. The role of the gut microbiome in the healthy adult status. *Clinicachimicaacta*. 2015; 451:97–102.
33. Wilson MR, Naccache SN, Samayoa E, Biagtan M, Bashir H, Yu G, Salamat SM, Somasekar S, Federman S, Miller S, Sokolic R. Actionable diagnosis of neuroleptospirosis by next-generation sequencing. *N Engl J Med*. 2014; 370(25):2408–2417.
34. Chiu CY, Miller SA. Clinical metagenomics. *Nat Rev Genet*. 2019; 20(6):341–355.

35. Escobar-Zepeda A, Vera-Ponce de León A, Sanchez-Flores A. The road to metagenomics: from microbiology to DNA sequencing technologies and bioinformatics. *Front Genet.* 2015; 6:348.
36. Bofill-Mas S, Rusiñol M. Recent trends on methods for the concentration of viruses from water samples. *Curr Opin Environ Sci Health.* 2020; 16:7–13.
37. Aw TG, Howe A, Rose JB. Metagenomic approaches for direct and cell culture evaluation of the virological quality of wastewater. *J Virol Methods.* 2014; 210:15–21.
38. Shah MP. *Removal of Emerging Contaminants through Microbial Processes.* 2021. Springer.
39. Shah MP. *Advanced Oxidation Processes for Effluent Treatment Plants.* 2020. Elsevier.
40. Hayes S, Mahony J, Nauta A, Van Sinderen D. Metagenomic approaches to assess bacteriophages in various environmental niches. *Viruses.* 2017; 9(6):127.
41. Rosario K, Breitbart M. Exploring the viral world through metagenomics. *Curr Opin Virol.* 2011; 1(4):289–297.
42. Jadeja NB, Purohit HJ, Kapley A. Decoding microbial community intelligence through metagenomics for efficient wastewater treatment. *FuncIntegr Genomics.* 2019; 19(6):839–851.
43. Lu XM, Chen C, Zheng TL. Metagenomic insights into effects of chemical pollutants on microbial community composition and function in estuarine sediments receiving polluted river water. *Microbial Ecol.* 2017; 73(4):791–800.
44. Fang H, Cai L, Yu Y, Zhang T. Metagenomic analysis reveals the prevalence of biodegradation genes for organic pollutants in activated sludge. *Bioresour Technol.* 2013; 129:209–218.

12 Diversity and Interaction of Microbes in Biodegradation

Aditya Ruikar and Hitesh S. Pawar

CONTENTS

DOI: 10.1201/9781003354147-12

12.1 INTRODUCTION

Contamination can be defined as the occurrence of a substance, impurity, or undesirable material that spoils, corrupts, infects and makes that material, physical body, natural environment or workplace unfit and restricts its use. According to the Oxford Dictionary, 'Contaminant' is a polluting or poisonous substance which makes something impure.[1] Contaminants mainly enter into the environment *via* various routes such as from exhaust of industries, oil and/or chemical spillage, automobiles, roads, parking spaces, garages, natural calamities, wastewater treatment, etc. However, there are several unpredictable sources and anthropogenic activities from which contaminants can enter the environment. These pollutants are significantly involved in the natural pathways and resist the breakdown and synthesis of important chemicals in the natural pathway of that system. Contaminants can be absorbed by various organisms in the lower food chain which can result in the accumulation of hazardous compounds in the food chain. The presence of such chemicals in environmental sources and food chains such as air, water and/or soil can lead to serious ill effects on the entire living community and has resulted in the extinction of important species in the past. In 2015, it was reported that the leaching of contaminants into the air, water and soil caused pollution, which resulted in one in six premature deaths that has killed over nine million people worldwide.[2]

Chemical contamination is one of the major contaminations which mainly occurs due to the leaching of harmful chemicals in the environment. It seems to be tackled with moderate success in the developed world with treatment controls for sites of leaching and education in wastewater contamination. Majorly, contaminants present in the environment can be classified into (i) organic contaminants and (ii) inorganic contaminants. Organic contaminants are chemical contaminants which have a carbon (C)-, hydrogen (H_2)-, oxygen (O)-, nitrogen (N_2)- and sulphur (S)-based backbone. Generally, organic containments are vapour- or gaseous-phase compounds like volatile organic compounds (VOCs) which includes different solvents used for synthesis, biological and/or chemical precursors and intermediates, petrochemicals, etc. and also solid organic compounds which includes sludge in wastewater treatment plants, settled bottoms, resins, fine chemicals, natural waxes, paper, plastic, wood, food substances, etc.[3] Chlorinated dioxins/furans (PCDD/Fs), polychlorinated biphenyls (PCBs) and polycyclic aromatic hydrocarbons (PAHs) are most commonly known as persistent organic pollutants (POPs) which pose a great risk to humans and of the environment due to their bioaccumulation in food webs and cause toxicity to the environment.[4] Environment increase in the overall quantity of POPs in the due to some major factors such the massive disposal of PAH, irrational use of brominated flame retardant (BFR) materials, high amount of pollution from chemical plants, greater combustion of fossil fuels, widespread and overuse of pesticides, and overall increase in air, water and soil pollution.[5] Recently, since the past two decades, attention has been given to the byproducts or waste of pharmaceuticals, personal care products and biologically active substances which are spherically referred to as 'Emerging Organic Contaminants'.[6]

Minerals such as fluorine, heavy metals, nitrides, etc., are considered inorganic contaminants present in nature. The predominance of such minerals in the environment is increasing due to natural causes such as active volcanoes, soil erosion and weathering of rocks and so on, as well as some human activities such as irresponsible mineral utilisation by industries, discharge of wastewater effluent, mining, etc., which lead to the addition of such minerals into the natural environment.[7] It has been observed that, unlike many organic pollutants, inorganic contaminants are not susceptible to degradation. Major processes like redox reactions, complexation, adsorption and precipitation/dissolution reactions determine the transport and bioavailability of the contaminants into the natural habitat.[8] The presence of even trace amounts of these contaminants threatens and is vulnerable, causing diseases or even death of humans, aquatic organisms and all the environmental flora and fauna.

Bioremediation is the process of degradation of contaminants into harmless products by using microbes such as bacteria or fungi. A wide variety of microbes are found in the contaminated environment, which may or may not help in the bioremediation of the contaminants. Bacterial species such as *Achromobacter* sp., *Bacillus cereus*, *Burkholderia* sp., *Mycobacterium* sp., *Pseudomonas* sp., *Rhodanobacter* sp. and *Sphingomonas* sp. are commonly found in the degradation of organic contaminants,[9] whereas heavy metals are one of the most difficult inorganic contaminants for remediation, but bacterial species such as *Flavobacterium*, *Pseudomonas*, *Bacillus*, *Arthrobacter*, *Corynebacterium*, *Methosinus*, *Rhodococcus*, *Mycobacterium*, *Stereum hirsutum*, *Nocardia*, *Methanogens*, *Aspergillus niger*, *Pleurotus ostreatus*, *Rhizopus arrhizus* and *Azotobacter* help in effective bioremediation of such heavy metals which are major inorganic contaminants.[10] It is clear that microbes play a very crucial role in the removal of contaminants from the environment and contribute significantly to the reduction of pollution. Microbial diversity means the types/classes/genera or species of microbe present in a particular environment. Microbial diversity plays a key role in bioremediation of contaminants due to the fact that different microbes can be used for degradation of certain chemicals according to the conditions available.

However, the study of microbial diversity as well as their interaction with the contaminants and their degradation plays a crucial role in the integration and advancement of potential bioremediation methods. The in-depth understanding of microbial diversity and interaction provides a future pathway to explore microbial species or to engineer microbial species in the effective bioremediation process. The present book chapter mainly focused to highlights the concepts of microbial bioremediation and their diversity for the removal of organic and inorganic contaminants. The microbial diversity which can be commonly found in bioremediation techniques has been explained in detail. Moreover, the microbial diversity involved in the degradation of the contaminants and their interaction is explained in detail.

12.2 MICROBIAL DIVERSITY IN BIOREMEDIATION TECHNIQUES

Bioremediation is the use of biological substances, i.e., use of microorganisms for degradation or removal of contaminants at the contaminated site. Bioremediation involves certain methods which are used for effective use of microbes for degradation of contaminants, and hence, each method involves similar microbes which are capable of degrading specific contaminants. Hence, let us have a look at the microbial diversity which is involved in different methods of bioremediation.

12.2.1 Bioaugmentation

Bioaugmentation is the process of addition of pre-grown microbial cultures for enhancement of microbial populations at a site for cleaning of contaminants and for reduction of clean-up time as well as the cost. It is mainly used in the sludge treatment process, wastewater treatment, soil remediation and petroleum clean-up.[11] Various microbial species are involved in the bioaugmentation process. Patil et al. (2020) have reported the predominance of *Rheinheimera*, *Kocuria*, *Ralstonia*, *Pseudomonas* and *Ruminococcaceae* in textile dye wastewater and are also involved in the biodegradation of dye by benzoate, aminobenzoate, chloroalkane and chloroalkene degradation pathways.[12] In a review by Tyagi et al. (2011), several species of fungus were proved to be effective strains that can be used for the bioaugmentation process which mainly includes species such as *Phanerochaete chrysosporium*, *Cuuninghamella* sp., *Alternaria alternate (Fr.) Keissler*, *Penicillium chrysogenum* and *Aspergillus niger*, whereas bacterial species such as *Bacillus* sp., *Zoogloea* sp. and *Flavobacterium* have been seen to significantly increase the biodegradation rates of the Polyaromatic hydrocarbons and diesel contaminated soil.[13] Diesel oil which is one of the organic contaminants in the soil was found to be degraded by around 75% by microbial strains of *Arthrobacter*, *Pseudomonas*, *Rhodococcus erythropolis*, *Bacillus cereus* and *Exiguobacterium* spp. using the bioaugmentation method for bioremediation.[14] Microbial consortiums of *Mycobacterium fortuitum*, *Bacillus cereus*, *Microbacterium* spp., *Gordonia polyisoperivorans*, *Microbacteriaceae* bacterium and *Fusarium oxysporum* were isolated from a PAH-contaminated soil and it has been observed that the above microbial species can degrade the PAH contaminants successfully and can be used for natural bioaugmentation of the PAH-contaminated environment.[15] Chloroaniline which is an important intermediate and a raw material for various organic reaction is left untreated after the use and is disposed of carelessly in the wastewater from the chemical industries. Bioaugmentation of such contaminated water using chloroaniline-degrading bacteria, i.e., *Comamonas testosteroni* I2 *gfp* leads to fewer quantities of chloroaniline and better quality of wastewater that can be released safely without harming the environment.[16] Several commercial bioaugmentation bacterial cultures are available in the market. Roetech® has different *Bacillus* strains which are pre-cultured and are available in powder form for various purposes of degradation.[17] Other products such as Microcat®, BioEase® and BioPlus® are some of the other products which are available commercially for bioaugmentation.[18–20]

12.2.2 Biostimulation

Biostimulation is a process of bioremediation in which the environment is modified in such a way that it stimulates the growth of existing bacteria which are capable of bioremediation. Biostimulation is mainly carried out by adding or restricting the use of nutrients such as carbon, phosphorous, nitrogen and oxygen, which leads to the rapid growth of a particular type of microbe in the environment which can effectively grow and degrade the contaminated environment. The advantage of biostimulation is that the process is carried out by well-suited and well-distributed microbes already present in the environment.

Impairment in the catabolic activity, as well as bioremediation capacity, was observed in microbes present in hydrocarbon-rich environments when the deficiency of nutrients was induced. Hence, Sarkar et al. (2016) found that after amendment in the nitrogen content, the bacterial diversity of the petroleum sludge refinery drastically changed and the predominance of β-(unclassified MOB121, *Comamonadaceae* members *Azovibrio*) and γ-(*Pseudoxanthomonas*) *Proteobacteria; Archaebacteria* [*Methanobacteria* (*Methanobacterium*), *Methanomicrobia* (*Methanosaeta*)] and *δ Proteobacteria* (*Geobacter*) was observed and this stimulation caused greater degradation of petroleum hydrocarbon which lead to effective bioremediation.[21] Also, Fulekar et al. (2012) have shown that biostimulation of the microbial consortium by proper mixing and aeration leads to proper adaption and effective bioremediation of heavy metals by species such as *Pseudomonas*, *Bacillus* and *Escherichia coli*.[22] In estuarine water, the temperature has been always the normal room temperature, and hence, the microbial species which can tolerate such low temperatures are important for bioremediation of such water sources. Species or organisms such as *Thalassolituus*, *Cycloclasticus*, *Roseobacter* and *Oleispira antarctica* have been found to be viable at 4°C and give out maximum degradation of oil the water resources, which indicated that if these species are biostimulated by keeping the temperature low, and then they can be used as potential degraders on oil spills.[23] Microbes such as *Bacillus*, *Coprothermobacter*, *Rhodobacter*, *Pseudomonas*, *Desulfosporosinus T78*, *Methanobacterium*, *Methanosaeta* and *Achromobacter* when biostimulated using nitro-phosphate amendment gave around 50% increase in degradation of TPH hydrocarbons present in an oil sludge, which suggested that metabolism and nutrient used for stimulation plays an important role in bioremediation of such oily sludge.[24] Low-voltage currents were subjected to microbial species such as *Pseudomonas* sp., *Nesterenkonia* sp., *Bacillus* sp. and *Brevibacillus* sp., which resulted in increased denitrification of inorganic nitrogenous compounds.[25] Glycerol injection into aquifers contaminated with chlorinated ethenes led to an increase in the microbial count of *Firmicutes*, *Bacteriodetes* and *Dehalococcoides mccartyi*, thereby leading to a decrease in the concentration of chlorinated ethenes in the aquifers.[26] From the above examples, it is clear that changing/addition/modification of certain environmental factors such as temperature, feed concentration and feed modification lead to effective stimulation of microbial consortium in the environment, leading to effective bioremediation of the contaminants.

12.2.3 Biosparging

Biosparging is a bioremediation process in which air is sparged below the contamination zone at a minimal flow rate. This sparged air promotes oxygenation in the reactor due to close spacing, thereby promoting aerobic degradation of the contaminants. VOCs such as BTEX are effectively cleaned using the biosparging method of bioremediation.

Biosparging involves the provision of air to already present microbes, and hence, the microbes which are used in bioaugmentation act in a more efficient way when combined with biosparging. For example, an increase in the decay rates of propane gas by *Rhodococcus ruber* ENV425 was observed when biosparging was used along with bioaugmentation.[27] As reported by Kao et al. (2008), microbial species such as

C. magnetobacterium, *F. bacterium* and *B. bacterium* are also present at the site of biosparging when used for degradation of BTEX.[28] It is also seen that the combination of a zeolite and permeable reactive barrier containing three bacterial strains of *Variovorax* sp. *OT16*, *Pseudomonas balearica OT17* and *Ornithinibacillus* sp. *OT18* efficiently removed the BTEX mixture.[29] Pentachlorophenol (PCP), a commonly found organic contaminant, was found to be degraded by *Burkholderia cepacian*, *Flavobacterium chlorophenolicum* and *Herbaspirullum* species, whereas *Pseudomonas*, *Aquaspirullum* and *Rhodocista* spp. were also found effective in the biosparging method of bioremediation.[30]

12.2.4 Bioventing

Bioventing is a method of bioremediation in which the microbes already present in the environment are stimulated by addition of oxygen or certain nutrients to promote in-situ remediation of the contaminants. Adsorbed fuel residuals and VOCs are the main contaminants whose degradation is assisted by the bioventing process. *Pseudomonas* sp., *Actinobacteria*, *Acidobacteria* and *Alphaproteobacteria* were the predominant species observed in kerosene-contaminated soil which was enhanced by air sparging.[28] It is seen that bioventing enhances the rates of degradation of many contaminants and leads to faster degradation rates by the innate bacterial species.[31–33]

12.2.5 Bioreactor Bioremediation

In the bioreactor bioremediation process, pre-engineered bioreactors are used for degradation of contaminants. These bioreactors are provided with a suitable environment such as optimum temperature, optimum pH and all the necessary nutrients for degradation of contaminants. Microbial bioreactors in the remediation of contaminants are generally preferred due to the advantages provided by the instrument as well as the process. Advantages like a suitable and controlled environment whereby it is possible to control each and every intricate parameter in the degradation process as well as the flexibility of design according to the need are provided by the bioreactor bioremediation process, which makes it one of the most attractive processes for bioremediation of contaminants.

It was reported by Lewis (1993) that species such as *Pseudomonas fluorescens*, *Pseudomonas stutzeri* and *Alcaligenes* can be used to degrade PAH in cresoate using bioreactor bioremediation.[34] Hexachlorocyclohexane was degraded by almost 94.5%, 78.5% and 66.1% in 30 days by using white rot fungi (*Bjerkandera adusta*) in a bioreactor.[35] Mercury is a heavy metal which is toxic to human beings, and hence, its bioremediation can be done using *Cupriavidus metallidurans* MSR33 in a rotary drum bioreactor.[36] Chlorphenol bioremediation using a pilot-scale bioreactor in aerobic and anaerobic conditions was done by Echartea et al. (2019), and it was found that species like *Sedimentibacter*, *Advenella*, *Petrimonas* and *Povalibacter* were in abundance at anaerobic conditions, whereas aerobic conditions had an abundance of *Luteimonas*, *Muricauda*, *Arenibacter* and *Gelidibacter*.[37] Wastewater bioremediation can be done successfully using an immobilised consortium; hence, Memon et al. (2020) immobilised the microbial consortium containing *Enterobacter cloacae*, *Proteus vulgaris*,

Proteus penneri, *Bacillus megaterium* and *Bacillus licheniformis* on a dry coconut coir matrix, which lead to effective bioremediation of wastewater in a lab-scale column bioreactor.[38] A novel two-phase bioreactor was developed by Lytras et al. (2017) containing *Pediococcus acidilactici* which successfully removed hexavalent chromium ions from the wastewater.[39]

12.2.6 Biofiltration

In biofiltration, microbes are attached to a granular media and microbes help in the biotransformation of all the contaminants such as organic chemicals, nitrogen, phosphorus and dissolved metal constituent. Providing in-depth filtration as well as biodegradation.

It was reported by Forss (2017) that the textile wastewater when subjected to biofiltration contained azoreductase enzymes which mainly belonged to bacterial genera of *Dysgonomonas* and *Pseudomonas* and also to fungi such as *Gibberella* and *Fusarium*.[40] In the liquid sample biofilter used for reverse osmosis, a predominance of *Dehalococcoides*, *Hyphomonas*, *Erythrobacter*, *Trichodesmium* and *Nitrospira* was found, and it was suggested that the microbial communities such as *Hyphomonas*, *Erythrobacter* and *Sphingomonas* were chlorine-resistant and also were strong organic degraders under chlorinated seawater conditions.[41] Hexane is a common organic impurity present in the air causing air pollution. Pseudomonas species such as *Pseudomonas stutzeri* can be used in a biofiltration membrane for the degradation of hexane from the air.[42] *Sphingomonas* sp. *D3K1* was used by Cho et al. (2009) in composite rock-wool biofilter for degradation and removal of BTEX chemicals.[43] Dichloromethane which is another organic contaminant is degraded using a biotrickling filter containing *Pandoraea pnomenusa* cultures and has been proven to be an effective microbe in the degradation of dichloromethane.[44]

12.2.7 Land Farming

Land farming is a method of bioremediation in which soil samples having greater concentrations of contaminants are mixed with certain chemicals such as bulking agents and suitable nutrients and are provided with the necessary oxygenation by periodic tilling of the soil. Microbes in the soil and oxygen help in the degradation, transformation or immobilisation of contaminants in the soil. Controlled soil conditions are necessary for changing and optimising the degradation rates of the contaminants. The land farming bioremediation method has proven to be successful in the degradation of petrochemicals and volatile organic contaminants.

Microbial species of *Enterobacter* and *Ochrobactrum* were detected in the land-farmed soil after the treatment period of 14 months, whereas the genus *Alcaligenes* appeared in significant numbers only within the 10-month-old land-farmed soil.[45] Thirty-four specific bacteria including the TPH-degrading bacteria, i.e., *Pseudomonas* sp., *Pseudoxanthomonas* sp., *Rhodocyclaceae bacterium* and *Variovorax* sp. and diesel and fuel oil-degrading bacteria including *Acinetobacter* sp., *Pseudomonas aeruginosa*, *Pseudomonas* sp., *Alcaligenaceae bacterium* and *Burkholderia* sp. was reported by Wang et al. (2016) for land farming bioremediation of diesel- and fuel-contaminated soil. This indicates that the above eight bacterial species played important roles in the TPH biodegradation processes, causing a decrease in TPH concentrations.[46] Bacterial families of Sp*inghomonadaceae*, *Xanthobacteriaceae*, *Pseudomonadaceae*, *Ectothiorhodo*s*piraceae*, *Xanthomonodaceae* and *Comamonodaceae* were found to be the most abundant in the land-farmed soil containing oily sludge contaminants as the major contaminant in the soil.[47]

12.3 INTERACTIONS AND DEGRADATION OF ORGANIC CONTAMINANTS

Degradation of organic compounds is mainly carried out by using (i) physical degradation, (ii) chemical degradation and (iii) biological degradation as mentioned below.

12.3.1 Physical Methods

Physical methods of degradation are the methods which are basically used for preliminary degradation of contaminants by physical means. These techniques mainly include methods such as photocatalytic degradation, use of metal oxychlorides which produce reactive oxygen radicals for degradation and use of ultrasound-assisted degradation of organic contaminants.

12.3.2 Chemical Methods

Chemical methods of degradation generally include methods such as photo-electrocatalysis, use of dyes to remove the chemicals, use of UV or ozone for degradation and use of carbon-based material such as graphene oxide along with a metal catalyst to form a hybrid material for catalysis in order to degrade the contaminants. These are common methods which are used in industries.

12.3.3 Biological Methods

The abovementioned processes of degradation have their own advantages, but recently, bioremediation has been extensively used for effective remediation of organic compounds. Bioremediation includes use of microbes for treatment of contaminated site. Microbes interact with such environments and help in degradation of harmful chemicals. Microbes in bioremediation include bacteria, fungi and other microbes which can cause degradation of organic contaminants.

12.3.3.1 Microbial Diversity

A major organic contaminant known as PAHs is mainly observed in petroleum-contaminated soil. For remediation of such compounds, a variety of microbes are involved such as *Achromobacter* sp., *Bacillus cereus*, *Burkholderia* sp., *Mycobacterium* sp., *Pseudomonas* sp., *Rhodanobacter* sp., *Sphingomonas paucimobilis*, etc. PAHs mainly contain alkyl-substituted naphthalene, phenanthrenes and dibenzothiophenes. Degradation of naphthalene is mainly done by bacteria which use it as a source of carbon, and they include species such as *Alcaligenes*, *Burkholderia*,

Mycobacterium, *Polaromonas*, *Pseudomonas*, *Ralstonia*, *Rhodococcus*, *Sphingomonas* and *Streptomyces*. Degradation of phenanthrene is extensively studied and involves various bacterial strains such as *Acidovorax*, *Arthrobacter*, *Brevibacterium*, *Burkholderia*, *Comamonas*, *Mycobacterium*, *Pseudomonas* and *Sphingomonas* species. Petroleum fractions generally contain PAHs like dibenzothiophene and its higher molecular weight analogues which are neither pyrogenic nor biogenic in nature. Microbial strains such as *Pseudomonas* sp. NCIB 9816-4 are mainly involved in the degradation of dibenzothiophene.[48]

VOCs are natural or manmade hydrocarbons which have a vapour pressure greater than 102 kPa and a boiling range of 50°C–100°C. These compounds are deadly toxic to human beings and other living creatures which can have detrimental effects on the environment. Generally, heterotrophic microbial strains are used in the filters because they use VOCs as a carbon source or consume them in their catabolic pathways.[49] Among Proteobacteria, the genera *Alcaligenes*, *Acinetobacter*, *Burkholderia*, *Pseudomonas*, *Xanthobacter* and *Hyphomicrobium* are used for their degradation abilities with regards to aromatic compounds and/or halogenated compounds whereas the genera *Corynebacterium* and *Rhodococcus*, which belong to Gram positive bacterium, are also involved in the degradation of styrene. Oxygenated organic contaminants are found to be degraded by microbial species of *Rhodococcus* and *Gordonia*.[50] BTEX is one of the major volatile organic contaminants, and it has been observed that microbial species such as *Burkholderia vietnamiensis G4*, *Pseudomonas mendocina KR1*, *Pseudoxanthomonas*, *P. putida F1*, *R. jostii RHA1* and *Thauera* sp. degrade BTEX contaminants aerobically, whereas *Aromatoleum aromaticum EbN1*, *Azoarcus* sp. *T*, *Thauera aromatica K172*, *Geobacter grbiciae TACP-2T*, *Desulfobacula toluolica Tol2*, *Desulfobacterales*, *Coriobacteriaceae* and *Desulfosarcina* sp. *PP31* are involved in the anaerobic degradation of BTEX.[51]

Chlorinated derivative of methane are volatile compounds which commonly involves gases such as carbon tetrachloride (CT), chloroform (CF) and dichloromethane (DCM). Microorganisms, such as *M. trichosporium OB3b*, *Nocardioides* sp. *CF8* and *P. mendocina KR1*, degrade CF and use methane, butane and toluene, respectively, as carbon and energy sources. Methylotrophic bacteria such as *Ancylobacter*, *Bacillus*, *Chryseobacterium*, *Hyphomicrobium*, *Methylopila helvetica DM1*, *M. trichosporium OB3b*, *N. europaea* and *Methylobacterium* are known for degradation of DCM. *Acetobacterium woodii DSM1030*, *Geobacter metallireducens* and *G. sulfurreducens* are some of the microbes involved in degradation of CT.[51]

Oil sludge is a mixture of various components which include sediments, water, oil and hydrocarbons that are commonly found during processes such as crude oil refining, oil storage vessel cleaning and waste treatment. Oil sludge seems to have a complex chemical composition that varies depending on the source. The existence of alkanes, aromatics, asphaltenes, resin and a high content of aromatic hydrocarbons is most significant in the sludge.[52] 13 species were characterised in a study reported by Obi et al. (2016) which included *Stenotrophmonas*, *Pseudomonas*, *Bordetella*, *Brucella*, *Bacillus*, *Achromobacter*, *Ochrobactrum*, *Advenella*, *Mycobacterium*, *Mesorhizobium*, *Klebsiella*, *Pusillimonas* and *Raoultella* species.[53]

Traditional wastewater treatment plants are considered to be less efficient in degradation of all organic compounds found in the wastewater, especially polar organic contaminants are either partially removed or remain unchanged by these treatment plants.[54] Aerobic bacteria such as *Pseudomonas* sp., *Acinetobacter* sp., *Alcaligenes* sp., *Flavobacterium*, *Cytophaga* group, *Xanthomonas* sp., *Nocardia* sp., *Mycobacterium* sp., *Corynebacterium* sp., *Arthrobacter*, *Comamonas* sp. and *Bacillus* sp. have been reported in petrochemical wastewater, which serves in the degradation of various components in the wastewater.[55] Acidobacteria, which are associated with the removal of polychlorinated biphenyls and petrochemical compounds, as well as Actinobacteria such as *Nocardioides*, ibuprofen degraders and *Illumatobacter*, were mainly encountered in wastewater treatment and bioremediation· Bacteroidetes such as *Terrimonas* and *Flavobacterium* are involved in degradation of benzo[a]pyrene and ibuprofen, whereas, Proteobacteria such as *Sphingomonas*, *Sphingobium*, *Novosphingobium*, *Arenimonas* and *Pseudomonas* which are involved in degradation of polyaromatic compounds are also considered to major flora in the waste water.[54]

Untreated textile effluent has a toxic and carcinogenic effect, which is a major source of pollution. The most critical factor and main concern in meeting environmental requirements is the decolourisation and detoxification of synthetic dye effluents. Many bacterial species, including Gram-positive strains such as *Bacillus subtilis*, *Clostridium perfringens*, *Proteus* sp., *Pseudomonas aeruginosa* and *Pseudomonas putida* are involved in the degradation of textile dyes. Gram-negative bacterial strains such as *Klebsiella pneumonia*, *Enterococcus* sp. and *Escherichia coli*, on the other hand, showed promising decolorising efficacy on various dyes. Bacterial strains like *Xenophilus*, *Vibrio logei* and *P. nitroreducens* have enzymes that can lead to the splitting of –N=N– bonds in the dye, thereby forming amines, which is utilised by the bacterial species as a source of carbon and energy for growth and regulation.[56]

Pesticides are organic chemicals that are evidently intended to increase agricultural yield, soil productivity and product quality, minimise crop pest losses and control insect vectors in order to prevent the outbreak of human and animal epidemics.[57] Soil microbes play a vital role in litter degradation, growth of plants, nutrient cycling, and pollutant and pesticide degradation. A large number of bacterial strains, including *Arthrobacter*, *Pseudomonas*, *Ralstonia*, *Rhodococcus*, *Bacillus*, *Nocardiopsis*, *Cryptococcus*, *Acetobacter*, *Sphingomonas*, *Burkholderia cepaeia* and *Alcaligenes*, are isolated from various parts of the world. Because they recycle most natural waste materials into harmless compounds, these microorganisms are appropriately referred to as nature's biodegraders and/or scavengers.[58]

Emerging organic contaminants (EOCs) are one of the significant concerns for well-being of humans and the ecosystem. These contaminants are derived from products of daily consumption which are used in relatively small quantities but due to their use by a majority of people and on several occasions lead to release of such products into the environment in significant amounts. As a result of this, proper degradation and remediation of these pollutants are critical. Pharmaceuticals and personal care products (PPCPs), pesticides, veterinary medicines, fire retardants and other industrial products are among the various EOCs.

EOC biodegradation is a result of the cometabolism of both heterotrophic as well as autotrophic microbes. Only heterotrophic microbes can observe EOC metabolism.

Autotrophic ammonia-oxidising bacteria (AOB) and ammonia-oxidising archaea (AOA) co-metabolise a wide range of EOCs using non-specific enzymes like ammonia monooxygenase (AMO). Microbial species such as *Sphingomonas Ibu-2*, *Novosphingobium JEM-1*, *Delftia tsuruhatensis*, *P. aeruginosa*, *Nitrosomonas*, *Nitrosopira* and *Nitrosococcus* are observed to be involved in the degradation of EOC (Table 12.1).[59]

TABLE 12.1
Summary of Microbial Diversity Involved in Degradation of Organic Contaminants

Sr No.	Type of Organic Contaminant	Microbes Involved in Degradation
1.	PAH	*Achromobacter* sp., *Bacillus cereus*, *Burkholderia* sp., *Mycobacterium* sp., *Pseudomonas* sp., *Rhodanobacter* sp., *Sphingomonas paucimobilis*, *Alcaligenes*, *Polaromonas*, *Ralstonia*, *Streptomyces*, *Acidovorax*, *Brevibacterium*.
2.	Volatile organic compounds	*Alcaligenes*, *Acinetobacter*, *Burkholderia*, *Pseudomonas*, *Xanthobacter*, *Hyphomicrobium*, *Corynebacterium*, *Rhodococcus*, *Gordonia*, *Pseudoxanthomonas* sp., *Thauera* sp., *Aromatoleum aromaticum EbN1*, *Desulfobacula toluolica Tol2*.
3.	Chlorinated methanes	*M. trichosporium OB3b*, *Nocardioides* sp. *CF8*, *P. mendocina*, *Ancylobacter*, *Bacillus*, *Chryseobacterium*, *Hyphomicrobium*, *Methylopila helvetica DM1*, *N. europaea*, *Acetobacterium woodii DSM1030*, *Geobacter metallireducens* and *G. sulfurreducens*.
4.	Oil sludge	*Stenotrophomonas acidaminiphila*, *Bacillus megaterium*, *Bacillus cibi*, *Pseudomonas aeruginosa*, *Bacillus cereus*, *Bordetella*, *Brucella*, *Ochrobactrum*, *Advenella*, *Mycobacterium*, *Mesorhizobium*, *Klebsiella*, *Pusillimonas* and *Raoultella* sp.
5.	Waste water treatment	*Pseudomonas* sp., *Acinetobacter* sp., *Alcaligenes* sp., *Flavobacterium*, *Cytophaga group*, *Xanthomonas* sp., *Nocardia* sp., *Mycobacterium* sp., *Corynebacterium* sp., *Arthrobacter*, *Comamonas* sp., *Bacillus* sp. and *Terrimonas* sp.
6.	Untreated textile effluent	*Bacillus subtilis*, *Clostridium perfringens*, *Proteus* sp., *Pseudomonas aeruginosa* and *Pseudomonas putida*, *Klebsiella pneumonia*, *Enterococcus* sp., *Escherichia coli*, *Xenophilus*, *Vibrio logei* and *P. nitroreducens*.
7.	Pesticides	*Arthrobacter*, *Pseudomonas*, *Ralstonia*, *Rhodococcus*, *Bacillus*, *Nocardiopsis*, *Cryptococcus*, *Acetobacter*, *Sphingomonas*, *Burkholderia cepaeia* and *Alcaligenes*
8.	Emerging organic contaminants (EOC)	*Sphingomonas Ibu-2*, *Novosphingobium JEM-1*, *Delftia tsuruhatensis*, *P. aeruginosa*, *a Nitrosomonas*, *Nitrosopira* and *Nitrosococcus*.

12.3.3.2 Interactions of Microbes

Traditional remediation techniques are easy to use but involve great economic as well as technical challenges which can be difficult to overcome. Due to this, bioremediation is considered to be a promising approach for removal of contaminants from contaminated sites. Generally, in the bioremediation process, characterisation and detailed analysis of contaminated sites are necessary in order to check whether or not bioremediation strategies are efficient enough to replace traditional remediation techniques.

Several strategies have been developed in recent years to find whether or not the bioremediation strategies are relevant for the clean-up of contaminated sites. Various, microbial and molecular methods are majorly used for checking the relevance of the bioremediation process. In such analysis, it has been found that interactions between plants and microorganisms are critical for the biotransformation of organic chemicals, various processes affecting the bioavailability of compounds, and also for the stability of the affected ecosystem. Organic contaminants can be used as carbon and energy sources by prokaryotic and eukaryotic microorganisms, or they can be co-metabolised in the presence of suitable growth substrates. A co-metabolic process is generally driven by plant-based carbon due to the oligotrophic nature of the soil itself. Contaminants can donate electrons, thereby becoming oxidised under both aerobic and anaerobic conditions, with various electron acceptors other than oxygen enabling anaerobic respiration processes.

Mycorrhizae have been extensively studied for plant–microbe interaction, and pure cultures of ectomycorrhiza fungi have been shown to degrade various chloroaromatic, PAHs and explosives such as 2,4,6-trinitrotoluene (TNT), and an improved compound degradation has been occasionally documented in ectomycorrhiza fungi symbiosis with plants. Under aerobic conditions, just about all fungi target organic contaminants oxidatively, but fungal growth on contaminants seems to be restricted to only monoaromatic (e.g., phenol, p-cresol, toluene) or aliphatic (e.g., n-alkanes) structures.

Unlike fungi, bacteria use more complex enzymes for contaminant transformation, as well as alternate electron acceptors, allowing them to operate outside of oxygenated areas.[60] It is evident from many research studies that when microbial species interact with one another or if co-cultures are prepared, then the rate and amount of degradation are seen to increase. For example, *Stenotrophomonas maltophilia* when combined with the fungus *Penicillium janthinellum* was seen to degrade a significant amount of PAH at faster rates. Furthermore, mixed bacterial cultures of *Pseudomonas putida*, *Flavobacterium* sp. and *Pseudomonas aeruginosa* outperformed single bacterial strains in PAH degradation. It was discovered that a mixed culture of *B. cereus*, *P. putida*, *P. fluorescens* and *Achromobacter* sp. degraded 2,4-dinitrotoluene (2,4-DNT) 50 times faster than any individual strain.[61]

12.3.3.3 Degradation Pathways

Bioremediation of organic pollutants involves a variety of degradation mechanisms and enzymatic reactions. The proposed pathways of PAH degradation cited by Seo and Keum (2009) in their review describe the formation of salicylate molecule which is further degraded into acetyl Co-A and pyruvate which are energy-forming molecules for Eukaryotes. *Pseudomonas putida G7* which is one of the bacteria involved in degradation of naphthalene has been well characterised for a catabolised enzyme system which is encoded by NAH7 plasmid and mainly consists of two operons, the first of which encodes for the enzymes involved to metabolise naphthalene to salicylate and the second one encodes for lower catabolic pathway encoding the enzymes necessary for the metabolism of salicylate through the catechol meta-cleavage pathway to pyruvate and acetaldehyde. Phenanthrene degradation is initiated by 3,4-dioxygenation for the formation of cis-3,4-dihydroxy-3,4-dihydrophenanthrene, leading to final products of 4,5-dihydroxy-benzoic acid and 1,4-dihydroxy-benzoic acid, while catabolism of dibenzothiophene is catalysed by separate enzymes in two pathways, including initial sulphur oxidation (4S pathway) mainly by flavin-containing monooxygenases known as DszA and DszB and the Kodama pathway including deoxygenation on the side chain of the phenanthrene, leading to the formation of 2-thio benzoic acid.[48]

Chlorinated ethenes are one of the common VOCs found in the environment and are degraded by oxidation. Aerobic bacteria degrade these compounds using a variety of oxygenases; methanotrophs such as *Methylomonas methanica 68-1*, *Methylocystis* sp. *SB2* and *Methylosinus trichosporium OB3b* use methane monooxygenases, Aromatic compound degraders, such as *Burkholderia vietnamiensis G4* and *Pseudomonas putida F1*, use toluene monooxygenases and dioxygenases and *Nocardioides* sp. *CF8* and *Thauera butanivorans* use butane monooxygenases. *Mycobacterium aurum L1* oxidises VC with growth and uses an alkene monooxygenase to degrade cis-DCE, trans-DCE and 1,1-DCE without growth. Two microbes, *Polaromonas* sp. JS666 and *Rhodococcus jostii RHA1*, are known to oxidise cis-DCE with growth. Anaerobic bacteria degrade the chlorinated ethenes by dechlorination which is done by bacterial species of *Dehalococcoides* sp., *Desulfitobacterium strains*, *Sulfurospirillum multivorans* and *Dehalobacter restriticus.*

The degradation of BTEX compounds is a complex process and involves many enzymatic reactions. Following are the degradation pathways involved in the degradation of BTEX (Table 12.2, Figure 12.1).

Methane, toluene and butane monooxygenases oxidise CF to phosgene through trichloro methanol. Aerobic growth-linked DCM degradation mainly relies on glutathione, and DCM is dechlorinated and transformed into formaldehyde. In addition, microbes that degrade DCM as non-growth substrates have also been isolated. *M. trichosporium OB3b* and *N. europaea* degrade DCM using a methane monooxygenase and ammonia monooxygenase, respectively. CT is dechlorinated under anaerobic conditions, and this process is mediated by cofactors such as corrinoid, coenzyme F430, iron compounds, cytochromes and humic substances. Acetogens, iron reducers and methanogens degrade CT with cofactors.[51]

Biodegradation of these pesticides by various microbes is mainly done by two mechanisms, i.e., Phase I and Phase II. Phase I includes mechanisms such as oxidation, reduction, hydrolysis and conjugation. These are basic mechanisms which are applicable to all the microbes involved in the degradation. The oxidation process of degradation is carried out by cytochrome P450 enzymes present in the microbes, whereas hydrolysis is mainly carried out by non-specific esterases present in the microbes. Hydrolases, esterases (also hydrolases) and mixed-function oxidases (MFOs) are the major enzyme systems involved in pesticide degradation in the first metabolism stage, as well as the glutathione S-transferase (GST) system in the second metabolism stage.[62]

TABLE 12.2
Pathways of Degradation of BTEX

Name of Compound	Product	Enzyme	Microbe Involved
Benzene	Phenol	Toluene monooxygenases	*Burkholderia vietnamiensis G4, Pseudomonas mendocina KR1* and *Pseudoxanthomonas spadix, Pseudomonas species OX1* and *Ralstonia pickettii PKO1*
		Ammonia monooxygenase	*Nitrosomonas europaea*
		Benzene monooxygenase	*Pseudomonas aeruginosa JL104*
	Cis-1,2-dehydrobenzene diol	Toluene-2,3-dioxygenase	*Pseudomonas putida F1*
		Ring hydroxylating dioxygenase	*Rhodococcus jostii RHA1*
Ethyl benzene	Cis-1,2-dihydro-3-ethylcatechol	Toluene-2,3-dioxygenase	*Pseudomonas putida F1*
		Ring hydroxylating dioxygenase	*Rhodococcus jostii RHA1*
	Styrene	Ammonia monooxygenase	*Nitrosomonas europaea*
	1-phenyl ethanol	Ammonia monooxygenase	*Nitrosomonas europaea*
		Naphthalene dioxygenase	*Pseudoxanthomonas spadix BD-a59*
O-Xylene	2-methyl benzyl alcohol	Xylene monooxygenase	*Pseudoxanthomonas spadix*
	2,3-dimethyl phenol	Toluene/o-xylene monooxygenase	*Pseudomonas species OX1*
		Benzene monooxygenase	*Pseudomonas aeruginosa JL104*
	3,4-dimethyl phenol	Benzene monooxygenase	*Pseudomonas aeruginosa JL104*
		Toluene monooxygenases	*Ralstonia pickettii PKO1*
M-Xylene	3-methylbenzyl alcohol	Xylene monooxygenase	*Pseudoxanthomonas spadix, Pseudomonas putida F1*
	2,4-dimethyl phenol	Benzene monooxygenase	*Pseudomonas aeruginosa JL104*
		Toluene monooxygenases	*Ralstonia pickettii PKO1, Pseudomonas species OX1*

(*Continued*)

TABLE 12.2 (*Continued*)
Pathways of Degradation of BTEX

Name of Compound	Product	Enzyme	Microbe Involved
P-Xylene	4 methyl benzyl alcohol	Ammonia monooxygenase	*Nitrosomonas europaea*
		Xylene monooxygenase	*Pseudomonas putida F1* and *Pseudoxanthomonas spadix*
	2,5-dimethyl phenol	Toluene monooxygenases	*Pseudoxanthomonas spadix* and *Ralstonia pickettii PKO1*
Toluene	o-cresol	Toluene monooxygenase	*Burkholderia vietnamiensis G4 Pseudomonas species OX1*
		Benzene monooxygenase	*Pseudomonas aeruginosa JL104*
	p-cresol	Toluene monooxygenase	*Pseudoxanthomonas spadix* and *Pseudomonas mendocina KR*, Pseudomonas species OX1 and *Ralstonia pickettii PKO1*
		Benzene monooxygenase	*Pseudomonas aeruginosa JL104*
	m-cresol	Benzene monooxygenase	*Pseudomonas aeruginosa JL104*
		Toluene monooxygenase	*Pseudomonas species OX1*
	Benzyl alcohol	Ammonia monooxygenase	*Nitrosomonas europaea*
		Xylene monooxygenase	*Pseudomonas putida mt-2*
	Toluene-cis-dehydration	Toluene-2,3-dioxygenase	*Pseudomonas putida F1*, *Thauera species DNT-1*
		Ring hydroxylating dioxygenase	*Rhodococcus jostii RHA1*

12.3.3.4 Genomics Involved in Microbial Degradation

Studies carried out recently on the microbial biodegradation mechanisms of organic pollutants which is been made possible by the development of several new methodologies. Characterisation of natural as well as engineered contaminant degrading microbes is commonly done by culture-independent methodologies, which has now become a trend in approaches for analysing and evaluating biodegradation processes. Furthermore, genomic data of a variety of microbes which have relevance to biodegradation have been documented successfully, and it is expected that the organism which is yet not reported would be documented soon, which will allow us to view these species globally for their ability of degradation of various contaminants.

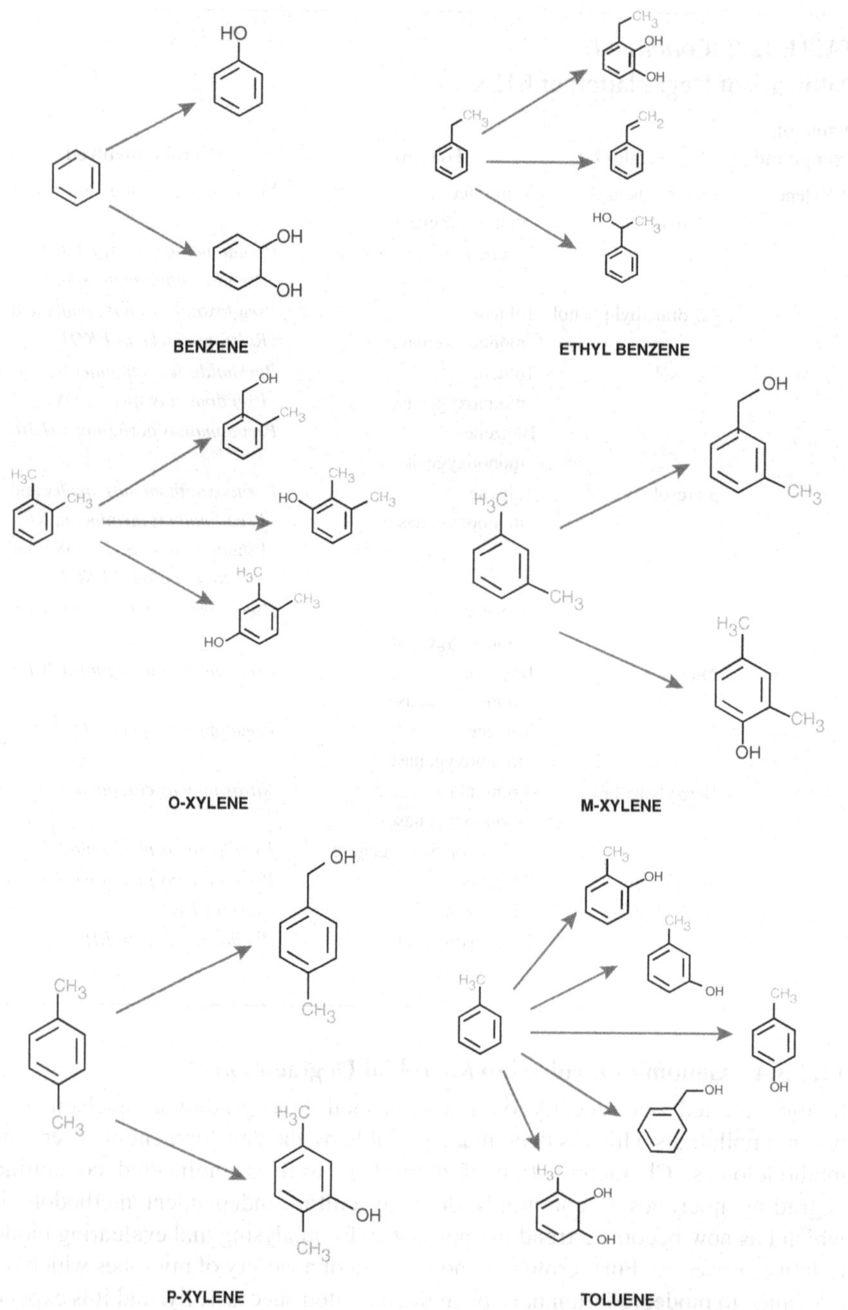

FIGURE 12.1 Degradation products of BTEX.

The genome sequence of *Pseudomonas putida* strain KT2440 is well known for its metabolic versatility. It has probable determinants for high-affinity nutrient acquisition systems (e.g., oxygenases, oxidoreductases, ferredoxins and cytochromes, dehydrogenases, sulphur metabolism proteins), efflux pumps and GSTs (typically associated with protection against toxic substrates and metabolites) and an extensive array of components, which are used by the microbe for thriving in the contaminated environment.

The genome of *Rhodopseudomonas palustris* encodes four distinct oxygenase-dependent aromatic-ring cleavage pathways, as well as 19 mono- and dioxygenase genes, 4 cytochrome P450 determinants and additional genes for nitrile hydratases, amidases, phosphonate utilisation, carboxylesterase and 16 glutathione-S-transferase genes for degradation of PAH. Several strains of *Burkholderia* have been found to have exceptional pollutant biodegradation capacities. The genome of one such organism, *Burkholderia fungorum* LB400, is made up of three massive replicons (chromosomes) and a megaplasmid. It has an unrivalled ability to kill environmentally significant polychlorinated biphenyl congeners.[63]

12.4 INTERACTION AND DEGRADATION OF INORGANIC CONTAMINANTS

Inorganic contaminants do have hazardous effects on all living organisms, due to which remediation and degradation of such contaminants are of utmost importance. Several methods of removal of inorganic minerals have been adopted which has led to a reduction in the toxicity of these contaminants.

12.4.1 Physical Methods

There are several processes involved in degradation of inorganic contaminants which mainly include photocatalysis, use of nanophotocatalysts, photodegradation by using ZnO-clay and use of graphene-based nano-spiral graphite for adsorption.[64–67]

12.4.2 Chemical Methods

Chemical methods use various reagents for the removal of inorganic contaminants. It mainly includes the use of nano-scale metallic iron for removal, electrokinetic remediation, use of clays for removal, co-precipitation and use of graphene-based nano-absorbents.[68–72]

12.4.3 Biological Methods

Although a variety of physical and chemical methods are available for removal as well as degradation of inorganic contaminants, these methods can be disadvantageous as they are not environment-friendly as well as the cost required for these methods is quite high which leads us to a more sustainable as well as an economic method, i.e., a biological method in which microbes, natural or engineered, perform the task of degradation and removal of inorganic contaminants. A variety of microbes are involved in this process, and the diversity of microbes is in many ways responsible for their practical usage in degradation of inorganic pollutants.

12.4.3.1 Microbial Diversity

Ammonia, nitrate and nitrogen are the three main elements of the biological nitrogen cycle. When the organic nitrogenous compounds are degraded in nature, ammonia is formed. Also, as we are aware that ammonia is essential for agricultural soils, farmers around the globe try to preserve the ammoniacal contents in the soil, whereas ammonia in freshwater sources above a specific limit leads to eutrophication, which can reduce the biochemical oxygen demand (BOD) of the freshwater, leading to toxicity in marine as well as terrestrial organisms. Nitrifying bacteria quickly oxidise ammonia to nitrate in any sufficiently aerated soil or water, a procedure which is helpful in wastewater treatment plants. While chemolithotrophic bacteria are assumed to be responsible for the majority of nitrification, reduced nitrogen compounds have been observed to be oxidised to nitrate by heterotrophs and even animals. Nitrification is a two-step method that demands collaboration between two bacteria groups that oxidise ammonia to nitrite or nitrite to nitrate. The ammonia-oxidising bacteria *Nitrosomonas europeae* is the most well-studied, and it can be found in soils, seawater and freshwater, whereas *Nitrosovibrio* is the least common. *Nitrosospira* has been found in both acidic soils and fresh water, although *Nitrosococcus* seems to be more essential in marine nitrification. Nitrobacter has also dominated research on nitrite oxidation to nitrate. *Nitrospira*, *Nitrococcus* and *Nitrospina* are microbes that thrive in high-salt environments, though one *Nitrospira* strain has been isolated from the soil as well. The aforementioned bacteria are autotrophic in nature. It is also observed that some of the heterotrophic bacteria such as *Thiosphaera pantotropha*, *Paracoccus denitrificans*, *Arthrobacter* and a *Corynebacterium* sp. are also involved in the degradation of ammonia as well as inorganic nitrogen. *Pseudomonads*, *Paracoccus denitrificans*, *Hyphomicrobium* and *Alcaligenes* are some of the most common bacteria which are involved in the denitrification process.[73]

The ions with partly or fully filled d-orbitals are known as heavy metals. They are elements with atomic weights ranging from 63.5 to 200.6 and a specific gravity greater than 5. Living organisms require small amounts of heavy metals such as cobalt, copper, iron, manganese, molybdenum and vanadium. Essential metals at high concentrations, on the other hand, can be toxic to living organisms. Environmental pollution caused by heavy metals is gaining popularity around the world. Heavy metals' obstinate and tenacious nature is a major threat to the environment and the lives of both plants and animals, including serious diseases in humans. Cadmium, chromium, mercury, lead, arsenic and antimony are other heavy metals that should be avoided in surface water systems. Runoff water and contaminated water sources have been observed to carry heavy metals with them. Currently, a wide range of microbes like bacteria, fungi, yeasts and algae are being investigated for their potential application in bioremediation processes, and some of them have already been used as heavy metal biosorbents. *Bacillus*, *Pseudomonas* and *Streptomyces* are strong metal biosorbents in the bacteria class. Bacterial species such as *Pseudomonas fluorescens*, *Pseudomonas aeruginosa*, *Lysinibacillus sphaericus*, *Streptomyces flavomacrosporus* and *Microbacterium profundi* are commonly involved in the biodegradation of many heavy metals. Also, fungal species such as *Aspergillus tereus*, *Penicillium chrysogenum*, *Candida utilis*, *Hansenula anomala*, *Rhodotorula mucilaginosa*, *Rhodotorula rubra*, *Saccharomyces cerevisiae* are observed to degrade the heavy metal contaminants.[74]

Chromium which is one of the inorganic metal ions is commonly found in the wastewater of leather industries and is found to be challenging to remove by simple means. Bacterial species such as *Pseudomonas fluorescence*, *Bacillus cereus* and *Bacillus decolorationis* have been found to degrade chromium ions easily.[75] Other microbial species such as *Escherichia coli*, *Desulfovibrio* sp., *Bacillus* sp., *Shewanella* sp., *Arthrobacter* sp., *Streptomyces* sp. *MC1*, *Microbacterium* sp. *CR-07* and *Pseudomonas putida* have also been known to reduce chromium ions. Enzymatic reduction of chromium is done by *Halomonas* sp. strain TA-04 under halophilic conditions.[76]

Copper (Cu) is a micronutrient that is required by all living organisms, but it can be poisonous in high concentrations. Humanly activities such as fungicide spraying and mining have led to an increase in the amount of copper in the environment, leading to copper contamination. Bioremediation of copper is mainly done in two ways, i.e., by immobilisation of copper in the environment which is commonly done by bacterial species such as *Bacillus*, *Kocuria flava* and *Sporosparcina* koreensis and the second method involves mobilisation of copper from the environment, called as bioleaching. Bioleaching of copper is performed by chemo lithotrophic bacteria mainly belonging to the genus of *Acidithiobacillus*, such as *Acidithiobacillus ferrooxidans*, *Acidithiobacillus thiooxidans*, *Acidithiobacillus caldus*, *Leptospirillum ferrooxidans* and *Sulfobacillus thermotolerans*.[77]

Zinc toxicity in humans is a growing concern because it has the potential to affect neutrophil and lymphocyte functions. Bioremediation of zinc is an important factor from an environmental as well as a medical perspective. It was reported by Carpio et al. (2016) that species such as *Pseudomonas* sp., *Escherichia* sp., *Bacillus* sp., *Acinetobacter* sp. and *Enterobacter* sp. were effective in the removal of metals such as zinc from soil and water.[78] *Bacillus megaterium EMCC 1013*, *Rhizobium rhizogenes EMCC1743*, *Rhizobium leguminosarum EMCC1130*, *Azotobacter vinelandii* and *Nocardiopsis Dassenvillei* were the five different species reported by El-barbary et al. (2018) which were able to tolerate and degrade zinc ions effectively.[79]

Increased iron ore production in recent decades to meet growing steel demand for a wide range of industrial applications has presented a major environmental challenge. That being said, since iron is such a valuable metal economically, there seems to be increasing interest among industry to reduce the environmental effect of its mining. Hence, bioremediation methods using various microbial species are explored for the biodegradation of iron. It was reported by Martins et al. (2019) that the use of *Penicillium janthinellum* and *Syncephalastrum racemosum* can effectively remove iron oxides from the soil leading to its bioremediation.[80] It was found by Gupta et al. (2020) that microbial consortiums containing *Clostridiales* and *Bacillales* members were able to degrade iron as well as sulphate by reduction.[81]

Lead (Pb) is a non-bioessential, long-lasting and hazardous heavy metal contaminant that is a major cause of environmental pollution. Environmental lead levels have increased more than 1,000-fold during the last three centuries as a consequence of widespread human activities. Lead is a chronic toxin because it is a persistent environmental pollutant that accumulates up over time, causing biomagnification at multiple tropic levels in food chains. Many microbial species such as *Enterococcus hirae*, *Helicobacter pylori*, *Escherichia coli*, *Ralstonia metallidurans*, *Cupriavidus metallidurans*, *Synechococcus PCC 7942*, *Anabaena PCC 7120*, *Oscillatoria brevis*, *Pseudomonas aeruginosa*, *Pseudomonas putida*, *Marinobacter* sp., *Pseudomonas marginalis*, *Paenibacillus jamilae*, *Vibrio Harvey*, *Providentia alcalifaciens strain 2EA*, *Klebsiella* sp. and *Acidiphilium symbioticum* H8 were reported to resist and degrade lead ions by various mechanisms.[82]

Bioremediation of other heavy metals has also been reported in the literature; for example, *Alcaligenes* sp., *Pseudomonas* sp., *Moraxella* sp. are mainly involved in the degradation of cadmium, nickel is degraded by *Bacillus subtilis* and *P. licheniformis*, silver by *Streptomyces noursei*, gold by *Aspergillus niger* and *Chlorella pyrenoidosa* and mercury and manganese by *Penicillium chrysogenum* and *Bacillus licheniformis. Saccharomyces cerevisiae* is mainly involved in the degradation of cobalt, uranium and thorium.[83]

Fluorine is found in 0.09% of the earth's surface as fluoride, and it enters drinking water supplies primarily due to industrial wastewater discharges (like semiconductor, glass and ceramic processing, coal-fired power plants, aluminium electroplating smelters, beryllium extraction plants, brick and iron works). While fluoride concentration in drinking water exceeds 0.5 mg/L, it has been associated with skeletal and dental fluorosis in people of all ages. Hence, bioremediation of fluorine is an important aspect of health and environmental safety. Microbes such as *Azolla filiculoides*, *Providencia vermicola*, *cyanobacteria* and *Aspergillus niger* are mainly reported for fluorine removal and biodegradation. Fluoroacetate-degrading bacteria such as *Acinetobacter*, *Arthrobacter*, *Aureobacterium*, *Bacillus*, *Pseudomonas*, *Streptomyces*, *Weeksella*, *Ancylobacter*, *Burkholderia*, *Cupriavidus*, *Paenibacillus*, *Ralstonia*, *Staphylococcus* and *Stenotrophomonas* are characterised and reported for removal of fluorine in the form of fluoroacetate compounds. Microbial species such as *Aspergillus* sp., *Rhizopus* sp and *Acinetobacter* sp. *RH5* are also involved in the bioremediation of fluorine and mainly degrades sodium fluoride (Table 12.3).[84]

12.4.3.2 Interactions of Microbes

Microorganisms naturally or unnaturally interact with metals as well as minerals, causing physical and chemical changes; whereas it has been observed that minerals and metals present in the environment are directly or indirectly involved in the growth and development of these microorganisms. Metals like sodium, zinc, potassium, calcium, copper, cobalt, magnesium, manganese and iron, which are vital for life, can cause toxicity in cells if their concentrations exceed the threshold. Redox reactions occurring in microbial cells require 18 micronutrients such as copper, cobalt, chromium, nickel, zinc, magnesium, iron, sodium, potassium and manganese. These micronutrients act as components of various enzymes, stabilise molecules through electrostatic interactions, and form concentration gradients and charges across cytoplasmic membranes. Metal toxicity is influenced by the physicochemical properties of a given environment as well as the chemical behaviour of metal species. Some metals, despite their toxicity, allow microorganisms to flourish in metal-polluted conditions, with various mechanisms being used to build metal resistance. As a result, heavy-metal-resistant bacteria have emerged, which have been isolated from a range of environmental sources all over the world.[85]

TABLE 12.3
Summary of Microbial Diversity Involved in Degradation of Inorganic Contaminants

Sr No.	Inorganic Contaminant	Microbes Involved in Degradation
1	Nitrides and Inorganic nitrogen	*Nitrosomonas europeae*, *Nitrosovibrio*, *Nitrosospira*, *Nitrosococcus*, *Thiosphaera pantotropha*, *Paracoccus denitrificans*, *Arthrobacter*, *Corynebacterium* sp., *Pseudomonads*, *Paracoccus denitrificans*, *Hyphomicrobium* and *Alcaligenes*.
2	Zinc	*Pseudomonas* sp., *Escherichia* sp., *Bacillus* sp., *Acinetobacter* sp., *Enterobacter* sp., *Rhizobium rhizogenes EMCC1743*, *Rhizobium leguminosarum EMCC1130*, *Azotobacter vinelandii* and *Nocardiopsis Dassenvillei*
3	Chromium	*Pseudomonas fluorescence*, *Bacillus cereus*, *Bacillus decolorationis*, *Desulfovibrio* sp., *Bacillus* sp., *Shewanella* sp., *Arthrobacter* sp., *Streptomyces* sp. *MC1*, *Microbacterium* sp. *CR-07*, *Pseudomonas putida*, *Escherichia coli* and *Halomonas* sp. *strain TA-04*.
4	Copper	*Bacillus* sp., *Kocuria flava* and *Sporosparcina koreensis*, *Acidithiobacillus* sp., *Pseudomonas* sp., *Escherichia* sp., *Acinetobacter* sp. and *Enterobacter* sp., *Rhizobium rhizogenes EMCC1743*, *Rhizobium leguminosarum EMCC1130*, *Azotobacter vinelandii* and *Nocardiopsis Dassenvillei*
5	Iron	*Penicillium janthinellum*, *Syncephalastrum racemosum*, *Clostridiales* and *Bacillales*.
6	Lead	*Enterococcus hirae*, *Helicobacter pylori*, *Escherichia coli*, *Ralstonia metallidurans*, *Cupriavidus metallidurans*, *Synechococcus PCC 7942*, *Anabaena PCC 7120*, *Oscillatoria brevis*, *Pseudomonas* sp., *Marinobacter* sp., *Paenibacillus jamilae*, *Vibrio Harvey*, *Providentia alcalifaciens strain 2EA*, *Klebsiella* sp., *Acidiphilium symbioticum H8* and *P. ficuserectae PKRS11*.
7	Cadmium	*Alcaligenes* sp., *Pseudomonas* sp., *Moraxella* sp.
8	Nickel	*Bacillus subtilis* and *P. licheniformis*
9	Manganese and mercury	*Penicillium chrysogenum* and *Bacillus licheniformis*.
10	Fluorine	*Acinetobacter*, *Arthrobacter*, *Aureobacterium*, *Bacillus*, *Pseudomonas*, *Streptomyces*, *Weeksella*, *Ancylobacter*, *Burkholderia*, *Cupriavidus*, *Paenibacillus*, *Ralstonia*, *Staphylococcus* sp., *Stenotrophomonas*, *Azolla filiculoides*, *Providencia vermicola*, *Cyanobacteria* and *Aspergillus niger*

12.4.3.3 Degradation Pathways

Microbes play a crucial role in the conversion of toxic heavy metals into non-toxic types. Microbes help with the mineralisation of organic contaminants into end products like carbon dioxide and water, as well as metabolic intermediates that are used as primary substrates for cell growth during the bioremediation process. They primarily have two functions: (i) generating degradative enzymes for the target contaminants, and (ii) opposing relevant heavy metals. Microorganisms play a role in environmental restoration in a variety of ways, including heavy metal binding, immobilisation, oxidation, transformation and volatilisation. By understanding the mechanism that governs the activity and growth of microbes in polluted sites, their metabolic potential, and their response to environmental changes, the designer microbial approach can make bioremediation more effective in specific areas.[86]

Microbial metabolism produces products like hydrogen, oxygen and hydrogen peroxide, which are used in oxidising metal as well as its reduction. Metal solubilisation or precipitation is usually accompanied by metal reduction or oxidation. Microbial metabolites may also play a role in metal solubilisation or precipitation. Microbial processing of organic acids in fermentation or inorganic acids (nitric and sulphuric acids) in aerobic oxidation helps the formation of dissolved metal chelates. The precipitation of non-dissolved phosphates, carbonates and sulphides of heavy metals such as arsenic, cadmium, chromium, copper, lead, mercury and nickel can be aided by microbial production of phosphate, hydrogen sulphide and carbon dioxide. Production of hydrogen sulphide by sulphate-reducing bacteria is generally effective for removing heavy metals and radionuclides from sulphate-containing mining drainage waters and nuclear waste. As a result, heavy metal adsorption onto the microbial surface may be substantial depending on pH. To collect uranium and other radionuclides from waste streams, biosorption is used, for example, by fungal fermentation residues. Metals can be oxidised and metals can be solubilised in metal-containing minerals like sulphides. Before landfilling or biotransformation, this method is used to bioleach heavy metals from sewage sludge. Many metals, such as arsenic and mercury, can be volatilised by methylation attributed to the prevalence of anaerobic microorganisms. Methane-producing bacteria and arsenic can be methylated by archaea and fungi to create the volatile toxic dimethyl arsine and trimethylarsine, or it can be converted by algae to the less toxic non-volatile methanearsonic and dimethylarsinic acids.[87]

12.4.3.4 Genomics Involved in Microbial Degradation

Genomic data gives us information about which genes or genetic loci are involved in the degradation of various inorganic compounds. There are several microbes which are involved in the degradation of various inorganic compounds, and due to the advancements in genetic analysis and metagenomic approaches, it is now possible to get an exact locus or exact gene sequence for the metabolic pathway proteins which are involved in the degradation of inorganic contaminants.

Ammonia-oxidising bacteria such as *Nitrosomonas europeae* have their whole genome sequenced using modern approaches and it has led us to the understanding of genes of *Nitrosomonas europeae* which are involved in the degradation of inorganic nitrogen as well as ammonia from the environment. Nitrogen reductase marker gene *nifH* has been characterised for nitrogen fixation as well as nitrogen and carbon cycling in the environment. Ammonia monooxygenase marker gene amoA has been observed for catalysing the first step of nitrification. Other genetic loci such as *napA*, *narG*, *nirK*, *nirS* and *nosZ* are predicted to be involved in the denitrification pathway.[88]

Pseudomonas species are widely involved in the degradation of both organic as well as inorganic contaminants. It has been proved that cadmium, which is one of the heavy metals, is detoxified by *Pseudomonas* species by RND-driven systems like Czc and the presence of the czcA gene has been confirmed for resistance against cadmium in *Pseudomonas* species. Two genes namely *cadA* and *cadR* have also been identified in *P. putida* 06909 which render the species resistant to cadmium.[89] Chromium is involved in oxidative damage to the cells and hence it is considered as hazardous mineral when it is beyond a certain concentration. Chromium is effluxed out from the cells by chromium efflux proteins which are mainly encoded by the chrA gene thereby providing resistance to the species against chromium toxicity. In a bacterial species of *Ochrobactrum tritici*, a chromium-resistant operon is involved in providing resistance against the chromium. The operon mainly involved the presence of one transposable element TnOtChr and four genes in chrBACF, which help in the degradation of chromium in the species. Arthrobacter sp. strain FB24 contained three genes, chrJ, chrK and chrL, which were involved in the removal and degradation of chromium.

Lead is one of the widely found inorganic contaminants in the environment. *Cupriavidus (Ralstonia) metallidurans CH34* bacterial species contain a megaplasmid named pMOL30, which encodes an operon, namely, the pbr operon, which contains several structural genes and one regulatory gene (pbrR), of which pbrT encodes a Pb(II) uptake protein, pbrA encodes a P-type Pb(II) efflux ATPase, pbrB encodes a predicted integral membrane protein of unknown function, and pbrC encodes a predicted prolipoprotein signal peptidase, which is involved in the degradation of lead. Metalothein protein encoded by BmtA in *P. aeruginosa strain WI-1* and smtAB gene in *Salmonella choleraesuis* and *Proteus penneri*, respectively, has also been observed in the bioaccumulation of lead and reducing its toxicity.

Mercury is a heavy metal which induces toxicity to living beings even if it is present in trace amounts. There are two operons which are mainly involved in degradation and resistance to mercury; they are the narrow-spectrum mer operon and broad-spectrum mer operon. Narrow-spectrum mer operon mainly contains merR, merT, merC, merF, merP and merD genes which are induced by inorganic mercuric ions and lead to resistance of the species from its toxic effects. The broad-spectrum mer operon contains genes that are identical to the operon with a narrow range in addition to the genes already present, additional genes such as merE, merG and merB are present.

Nickel has the largest no. of isotopes amongst heavy metals and is abundantly found in the environment. Nickel biodegradation involves the presence of a cnrCBA efflux pump encoded by the cnrYHXCBAT gene system present in *Cupriavidus (Ralstonia) metallidurans CH34. Achromobacter xylosoxidans 31A* contains another nickel-resistant operon, namely, the nre operon, which is responsible for nickel resistance in the species.[90]

Hence, from the above examples, it is clear that genomic studies give us a broader outlook on the actual molecular mechanisms which are involved in the bioremediation of inorganic contaminants.

12.5 FUTURE PERSPECTIVE

After a review of all bioremediation processes and the microbial diversity involved in them, it is clear that microbial degradation is by far an excellent approach for the degradation of organic and inorganic contaminants. However, the biological approach has some disadvantages such as the interaction and degradation of contaminants are based only on the microbial species involved, which can be worked on. It is still difficult to predict whether or not a bacterium or a fungus which survives in the contaminated environment will degrade the contaminants or not. For this, a detailed study of pathways involved in the degradation of contaminants and microbial interactions with the environment needs to be done. Genetic analysis and genetic manipulation for the designing of contaminant-specific microbes also need to be done. Control of spreading and degradation is also one of the parameters that should be looked after as an uncontrolled growth of any microbe can hamper the biodiversity of nature. It is evident from the research studies that the practical use of microbial species in the degradation of contaminants has 'time' as its major bottleneck phenomenon. To overcome this time barrier, the thermodynamic process as well as enzyme kinetics involved in degradation should be studied deeply in order to get better metabolic rates. Physical and chemical methods used in the degradation process have been found to manage a large amount of contamination load, whereas biological methods are only small-scale operations. As a result, metabolic engineering of microbial pathways could be a solution to the problem, increasing contamination handling by microbes and making it a large-scale process.

12.6 CONCLUSION

It can be seen that microbial degradation is the future of bioremediation. It has several advantages as follows:

(a) Microbial biodegradation is one of the best methods that can be utilised to its fullest potential for the remediation of organic as well as inorganic contaminants.
(b) Microbial remediation is a sustainable and environmentally friendly approach for the remediation of various contaminants.
(c) A variety of microbes are involved in each and every method of bioremediation leading to the site and contaminant-specific choice of microbes.
(d) Various pathways are involved in the degradation of contaminants which leads to effective and safe remediation of these contaminants.
(e) Microbes have been observed to interact with biotic and abiotic factors leading to symbiont degradation of various contaminants.
(f) Variety of genes are involved in the bioremediation process and can be used to engineer future microbes for better degradation of contaminants.

Due to the massive benefits mentioned above, it can be said that bioremediation is an effective and environmentally friendly method for removing pollutants, resulting in reduced air, water and soil contamination, leading to a better tomorrow.

REFERENCES

1. Contamination – Wikipedia. https://en.wikipedia.org/wiki/Contamination.
2. Landrigan, P. J. et al. Pollution and global health – An agenda for prevention. *Environmental Health Perspectives* vol. 126 084501 (2018).
3. Wilcox, J. B. Electrical and thermal solutions. in *Environmental Solutions* 203–211 (Elsevier Inc., 2005). doi: 10.1016/B978-012088441-4/50010-1.
4. Clarke, B. O. & Smith, S. R. Review of "emerging" organic contaminants in biosolids and assessment of international research priorities for the agricultural use of biosolids. *Environment International* vol. 37 226–247 (2011).
5. Jacob, J. A review of the accumulation and distribution of persistent organic pollutants in the environment. *International Journal of Bioscience, Biochemistry and Bioinformatics* 657–661 (2013). doi: 10.7763/ijbbb.2013.v3.297.
6. Pal, A., Gin, K. Y. H., Lin, A. Y. C. & Reinhard, M. Impacts of emerging organic contaminants on freshwater resources: Review of recent occurrences, sources, fate and effects. *Science of the Total Environment* vol. 408 6062–6069 (2010).
7. Srivastav, A. L. & Ranjan, M. Inorganic water pollutants. in *Inorganic Pollutants in Water* (Elsevier, 2020). doi: 10.1016/B978-0-12-818965-8.00001-9.
8. Schwarzenbach, R. P., Egli, T., Hofstetter, T. B., von Gunten, U. & Wehrli, B. Global water pollution and human health. *Annual Review of Environment and Resources* vol. 35 109–136 (2010).
9. Seo, J.-S., Keum, Y.-S. & Li, Q. Bacterial degradation of aromatic compounds. *International Journal of Environmental Research and Public Health* vol. 6 278–309 (2009).
10. Verma, S. & Kuila, A. Bioremediation of heavy metals by microbial process. *Environmental Technology and Innovation* vol. 14 100369 (2019).
11. Bioaugmentation – Wikipedia. https://en.m.wikipedia.org/wiki/Bioaugmentation.
12. Patil, S. M. et al. Regeneration of textile wastewater deteriorated microbial diversity of soil microcosm through bioaugmentation. *Chemical Engineering Journal* vol. 380 122533 (2020).
13. Tyagi, M., da Fonseca, M. M. R. & de Carvalho, C. C. C. R. Bioaugmentation and biostimulation strategies to improve the effectiveness of bioremediation processes. *Biodegradation* vol. 22 231–241 (2011).
14. Alisi, C. et al. Bioremediation of diesel oil in a co-contaminated soil by bioaugmentation with a microbial formula tailored with native strains selected for heavy metals resistance. *Science of the Total Environment* vol. 407 3024–3032 (2009).
15. Jacques, R. J. S. et al. Microbial consortium bioaugmentation of a polycyclic aromatic hydrocarbons contaminated soil. *Bioresource Technology* vol. 99 2637–2643 (2008).
16. Boon, N., Top, E. M., Verstraete, W. & Siciliano, S. D. Bioaugmentation as a tool to protect the structure and function of an activated-sludge microbial community against a 3-chloroaniline shock load. *Applied and Environmental Microbiology* vol. 69 1511–1520 (2003).
17. Treatments, W. *Bacterial Cultures for Wastewater Treatment We Make Water Clean.* www.roebictechnologyinc.com.
18. Water and waste management | Novozymes. https://biosolutions.novozymes.com/en/wastewater.
19. Bioaugmentation products | Monera Technologies. https://www.moneratec.com/products/.
20. Bioaugmentation wastewater treatment | SUEZ. https://www.suezwatertechnologies.com/products/wastewater-treatments/bioaugmentation-chemicals.
21. Sarkar, J. et al. Biostimulation of indigenous microbial community for bioremediation of petroleum refinery sludge. *Frontiers in Microbiology* vol. 7, 1407 (2016).

22. Fulekar, M. H., Sharma, J. & Tendulkar, A. Bioremediation of heavy metals using biostimulation in laboratory bioreactor. *Environmental Monitoring and Assessment* vol. 184 7299–7307 (2012).
23. Coulon, F., McKew, B. A., Osborn, A. M., McGenity, T. J. & Timmis, K. N. Effects of temperature and biostimulation on oil-degrading microbial communities in temperate estuarine waters. *Environmental Microbiology* vol. 9 177–186 (2007).
24. Roy, A. et al. Biostimulation and bioaugmentation of native microbial community accelerated bioremediation of oil refinery sludge. *Bioresource Technology* vol. 253 22–32 (2018).
25. Dehghani, S., Rezaee, A. & Hosseinkhani, S. Biostimulation of heterotrophic-autotrophic denitrification in a microbial electrochemical system using alternating electrical current. *Journal of Cleaner Production* vol. 200 1100–1110 (2018).
26. Atashgahi, S. et al. Geochemical and microbial community determinants of reductive dechlorination at a site biostimulated with glycerol. *Environmental Microbiology* vol. 19 968–981 (2017).
27. Lippincott, D. et al. Bioaugmentation and propane biosparging for in situ biodegradation of 1,4-dioxane. *Groundwater Monitoring & Remediation* vol. 35 81–92 (2015).
28. Kao, C. M., Chen, C. Y., Chen, S. C., Chien, H. Y. & Chen, Y. L. Application of in situ biosparging to remediate a petroleum-hydrocarbon spill site: Field and microbial evaluation. *Chemosphere* vol. 70 1492–1499 (2008).
29. Ahmadnezhad, Z., Vaezihir, A., Schüth, C. & Zarrini, G. Combination of zeolite barrier and bio sparging techniques to enhance efficiency of organic hydrocarbon remediation in a model of shallow groundwater. *Chemosphere* vol. 128555 (2020) doi: 10.1016/j.chemosphere.2020.128555.
30. Stokes, C. E. Effects of in-situ biosparging on pentachlorophenol (PCP) degradation and bacterial communities in PCP contaminated groundwater. https://search.proquest.com/openview/73d26d4bee7226f587a90c8497fafec8/1?pq-origsite=gscholar&cbl=18750&diss=y (2011).
31. Kabelitz, N. et al. Enhancement of the microbial community biomass and diversity during air sparging bioremediation of a soil highly contaminated with kerosene and BTEX. *Applied Microbiology and Biotechnology* vol. 82 565–577 (2009).
32. Sui, H., Li, X.-G. & Jiang, B. Benzene, toluene and p-xylene interactions and the role of microbial communities in remediation using bioventing. *The Canadian Journal of Chemical Engineering* vol. 83 310–315 (2008).
33. Pfiffner, S. M., Palumbo, A. v., Sayles, G. D. & Gannon, D. Microbial population and degradation activity changes monitored during a chlorinated solvent biovent demonstration. *Ground Water Monitoring & Remediation* vol. 24 102–110 (2010).
34. Lewis, R. F. SITE demonstration of slurry-phase biodegradation of pah contaminated soil. *Air and Waste* vol. 43 503–508 (1993).
35. Quintero, J. C., Lú-Chau, T. A., Moreira, M. T., Feijoo, G. & Lema, J. M. Bioremediation of HCH present in soil by the white-rot fungus Bjerkandera adusta in a slurry batch bioreactor. *International Biodeterioration & Biodegradation* vol. 60 319–326 (2007).
36. Bravo, G., Vega-Celedón, P., Gentina, J. C. & Seeger, M. Bioremediation by Cupriavidus metallidurans strain MSR33 of mercury-polluted agricultural soil in a rotary drum bioreactor and its effects on nitrogen cycle microorganisms. *Microorganisms* vol. 8, 1952 (2020).
37. Lopez-Echartea, E. et al. Bioremediation of chlorophenol-contaminated sawmill soil using pilot-scale bioreactors under consecutive anaerobic-aerobic conditions. *Chemosphere* vol. 227 670–680 (2019).
38. Memon, H., Lanjewar, K., Dafale, N. & Kapley, A. Immobilization of microbial consortia on natural matrix for bioremediation of wastewaters. *International Journal of Environmental Research* vol. 14 403–413 (2020).

39. Lytras, G. et al. A novel two-phase bioreactor for microbial hexavalent chromium removal from wastewater. *Journal of Hazardous Materials* vol. 336 41–51 (2017).
40. Forss, J., Lindh, M. v., Pinhassi, J. & Welander, U. Microbial biotreatment of actual textile wastewater in a continuous sequential rice husk biofilter and the microbial community involved. *PLoS One* vol. 12 e0170562 (2017).
41. Jeong, S. et al. Effect of engineered environment on microbial community structure in biofilter and biofilm on reverse osmosis membrane. *Water Research* vol. 124 227–237 (2017).
42. Fazaelipoor, M. H., Shojaosadati, S. A. & Farahani, E. V. Two liquid phase biofiltration for removal of n-hexane from polluted air. *Environmental Engineering Science* vol. 23 954–959 (2006).
43. Cho, E., Galera, M. M., Lorenzana, A. & Chung, W. J. Ethylbenzene, o-xylene, and BTEX removal by Sphingomonas sp. D3K1 in rock wool-compost biofilters. *Environmental Engineering Science* vol. 26 45–52 (2009).
44. Elías, A. et al. Application of biofiltration to the degradation of hydrogen sulfide in gas effluents. *Biodegradation* vol. 11 423–427 (2000).
45. Katsivela, E. et al. Bacterial community dynamics during in-situ bioremediation of petroleum waste sludge in landfarming sites. *Biodegradation* vol. 16 169–180 (2005).
46. Wang, S. Y., Kuo, Y. C., Hong, A., Chang, Y. M. & Kao, C. M. Bioremediation of diesel and lubricant oil-contaminated soils using enhanced landfarming system. *Chemosphere* vol. 164 558–567 (2016).
47. Dörr de Quadros, P. et al. Oily sludge stimulates microbial activity and changes microbial structure in a landfarming soil. *International Biodeterioration and Biodegradation* vol. 115 90–101 (2016).
48. Seo, J.-S., Keum, Y.-S. & Li, Q. Bacterial degradation of aromatic compounds. *International Journal of Environmental Research and Public Health* vol. 6 278–309 (2009).
49. Berenjian, A., Chan, N. & Malmiri, H. J. Volatile organic compounds removal methods: A review. *American Journal of Biochemistry and Biotechnology* vol. 8 220–229 (2012).
50. Malhautier, L., Khammar, N., Bayle, S. & Fanlo, J. L. Biofiltration of volatile organic compounds. *Applied Microbiology and Biotechnology* vol. 68 16–22 (2005).
51. Yoshikawa, M., Zhang, M. & Toyota, K. Biodegradation of volatile organic compounds and their effects on biodegradability under co-existing conditions. *Microbes and Environments* vol. 32 188–200 (2017).
52. Ubani, O., Atagana, I. H. & Thantsha, S. M. Biological degradation of oil sludge: A review of the current state of development. *African Journal of Biotechnology* vol. 12 6544–6567 (2013).
53. Obi, L. U., Atagana, H. I. & Adeleke, R. A. Isolation and characterisation of crude oil sludge degrading bacteria. *SpringerPlus* vol. 5 1946 (2016).
54. Posselt, M. et al. Bacterial diversity controls transformation of wastewater-derived organic contaminants in river-simulating flumes. *Environmental Science and Technology* vol. 54 5467–5479 (2020).
55. Shokrollahzadeh, S., Azizmohseni, F., Golmohammad, F., Shokouhi, H. & Khademhaghighat, F. Biodegradation potential and bacterial diversity of a petrochemical wastewater treatment plant in Iran. *Bioresource Technology* vol. 99 6127–6133 (2008).
56. Bhatia, D., Sharma, N. R., Singh, J. & Kanwar, R. S. Biological methods for textile dye removal from wastewater: A review. *Critical Reviews in Environmental Science and Technology* vol. 47 1836–1876 (2017).
57. Satish, G. P., Ashokrao, D. M. & Arun, S. K. Microbial degradation of pesticide: A review. *African Journal of Microbiology Research* vol. 11 992–1012 (2017).
58. Verma, J. P., Jaiswal, D. K. & Sagar, R. Pesticide relevance and their microbial degradation: A-state-of-art. *Reviews in Environmental Science and Biotechnology* vol. 13 429–466 (2014).

59. Tran, N. H., Urase, T., Ngo, H. H., Hu, J. & Ong, S. L. Insight into metabolic and cometabolic activities of autotrophic and heterotrophic microorganisms in the biodegradation of emerging trace organic contaminants. *Bioresource Technology* vol. 146 721–731 (2013).
60. Fester, T., Giebler, J., Wick, L. Y., Schlosser, D. & Kästner, M. Plant-microbe interactions as drivers of ecosystem functions relevant for the biodegradation of organic contaminants. *Current Opinion in Biotechnology* vol. 27 168–175 (2014).
61. Mikesková, H., Novotný, C. & Svobodová, K. Interspecific interactions in mixed microbial cultures in a biodegradation perspective. *Applied Microbiology and Biotechnology* vol. 95 861–870 (2012).
62. Laura, Ma., Snchez-Salinas, E., Dantn Gonzlez, E. & Luisa, M. Pesticide biodegradation: Mechanisms, genetics and strategies to enhance the process. in *Biodegradation – Life of Science* (InTech, 2013). doi: 10.5772/56098.
63. Pieper, D. H., Martins Dos Santos, V. A. P. & Golyshin, P. N. Genomic and mechanistic insights into the biodegradation of organic pollutants. *Current Opinion in Biotechnology* vol. 15 215–224 (2004).
64. Augugliaro, V., Litter, M., Palmisano, L. & Soria, J. The combination of heterogeneous photocatalysis with chemical and physical operations: A tool for improving the photoprocess performance. *Journal of Photochemistry and Photobiology C: Photochemistry Reviews* vol. 7 127–144 (2006).
65. Tahir, M. B. et al. Role of nanophotocatalysts for the treatment of hazardous organic and inorganic pollutants in wastewater. *International Journal of Environmental Analytical Chemistry* (2020) doi: 10.1080/03067319.2020.1723570.
66. Sasikala, S. P., Nibila, T. A., Babitha, K. B., Peer Mohamed, A. A. & Solaiappan, A. Competitive photo-degradation performance of ZnO modified bentonite clay in water containing both organic and inorganic contaminants. *Sustainable Environment Research* vol. 1 1 (2019).
67. Park, C. M. et al. Potential utility of graphene-based nano spinel ferrites as adsorbent and photocatalyst for removing organic/inorganic contaminants from aqueous solutions: A mini review. *Chemosphere* vol. 221 392–402 (2019).
68. Scott, T. B., Popescu, I. C., Crane, R. A. & Noubactep, C. Nano-scale metallic iron for the treatment of solutions containing multiple inorganic contaminants. *Journal of Hazardous Materials* vol. 186 280–287 (2011).
69. Wen, D., Fu, R. & Li, Q. Removal of inorganic contaminants in soil by electrokinetic remediation technologies: A review. *Journal of Hazardous Materials* vol. 401 (2021).
70. Srinivasan, R. Advances in application of natural clay and its composites in removal of biological, organic, and inorganic contaminants from drinking water. *Advances in Materials Science and Engineering* vol. 2011 (2011).
71. Denmark, I. S. et al. Removal of inorganic mercury by selective extraction and coprecipitation for determination of methylmercury in mercury-contaminated soils by chemical vapor generation inductively coupled plasma mass spectrometry (CVG-ICP-MS). *Analytica Chimica Acta* vol. 1041 68–77 (2018).
72. Kim, S. et al. Aqueous removal of inorganic and organic contaminants by graphene-based nanoadsorbents: A review. *Chemosphere* vol. 212 1104–1124 (2018).
73. Cole, J. A. Biodegradation of inorganic nitrogen compounds. in *Biochemistry of Microbial Degradation* 487–512 (Springer: Netherlands, 1994). doi: 10.1007/978-94-011-1687-9_15.
74. Gupta, A. et al. Microbes as potential tool for remediation of heavy metals: A review. *Journal of Microbial & Biochemical Technology* vol. 8 364–372 (2016).
75. Narendra Kumar Ahirwar, G. G. Biodegradation of chromium contaminated soil by some bacterial species. *International Journal of Science and Research (IJSR)* vol. 4 1024–1029 (2015).

76. Focardi, S., Pepi, M. & Focardi, S. E. Microbial reduction of hexavalent chromium as a mechanism of detoxification and possible bioremediation applications. in *Biodegradation – Life of Science* (InTech, 2013). doi: 10.5772/56365.
77. Cornu, J. Y., Huguenot, D., Jézéquel, K., Lollier, M. & Lebeau, T. Bioremediation of copper-contaminated soils by bacteria. *World Journal of Microbiology and Biotechnology* vol. 33 1–9 (2017).
78. Mejias Carpio, I. E. et al. Biostimulation of metal-resistant microbial consortium to remove zinc from contaminated environments. *Science of the Total Environment* vol. 550 670–675 (2016).
79. El-barbary, T. A. A. & El-Badry, M. A. Bioremediation potential of Zn (II) by different bacterial species. *Saudi Journal of Biomedical Research (SJBR)* vol. 3 144–150 (2018).
80. de Martins, B. A., Lima, M. T. N. S., Barreto, D. L. C. & Takahashi, J. A. Green bioremediation of iron ions by using fungal biomass. *Chemical Engineering Transactions* vol. 74 325–330 (2019).
81. Gupta, A. & Sar, P. Characterization and application of an anaerobic, iron and sulfate reducing bacterial culture in enhanced bioremediation of acid mine drainage impacted soil. *Journal of Environmental Science and Health – Part A Toxic/Hazardous Substances and Environmental Engineering* vol. 55 464–482 (2020).
82. Naik, M. M. & Dubey, S. K. Lead resistant bacteria: Lead resistance mechanisms, their applications in lead bioremediation and biomonitoring. *Ecotoxicology and Environmental Safety* vol. 98 1–7 (2013).
83. Meenambigai, P., Vijayaraghavan, R., Gowri, R. S., Rajarajeswari, P. & Prabhavathi, P. Biodegradation of heavy metals – A review. *International Journal of Current Microbiology and Applied Sciences* vol. 5 375–383 (2016).
84. Singh, A. & Gothalwal, R. A reappraisal on biodegradation of fluoride compounds: Role of microbes. *Water and Environment Journal* vol. 32 481–487 (2018).
85. Prabhakaran, P., Ashraf, M. A. & Aqma, W. S. Microbial stress response to heavy metals in the environment. *RSC Advances* vol. 6 109862–109877 (2016).
86. Verma, S. & Kuila, A. Bioremediation of heavy metals by microbial process. *Environmental Technology and Innovation* vol. 14 100369 (2019).
87. Bharagava, R. N., Purchase, D., Saxena, G. & Mulla, S. I. Applications of metagenomics in microbial bioremediation of pollutants. in *Microbial Diversity in the Genomic Era* 459–477 (Elsevier, 2019). doi: 10.1016/B978-0-12-814849-5.00026-5.
88. Levy-Booth, D. J., Prescott, C. E. & Grayston, S. J. Microbial functional genes involved in nitrogen fixation, nitrification and denitrification in forest ecosystems. *Soil Biology and Biochemistry* vol. 75 11–25 (2014).
89. Chellaiah, E. R. Cadmium (heavy metals) bioremediation by Pseudomonas aeruginosa: A minireview. *Applied Water Science* vol. 8 154 (2018).
90. Das, S., Dash, H. R. & Chakraborty, J. Genetic basis and importance of metal resistant genes in bacteria for bioremediation of contaminated environments with toxic metal pollutants. *Applied Microbiology and Biotechnology* vol. 100 2967–2984 (2016).

13 Metagenomics
A Pathway for Searching in Microbial Contexts

Aditi Nag
Dr. B. Lal Institute of Biotechnology

Bhavuk Gupta
NIT, Rourkela

Sudipti Arora
Dr. B. Lal Institute of Biotechnology

CONTENTS

DOI: 10.1201/9781003354147-13

13.1 INTRODUCTION

Wastewater processing of different types of wastewater (e.g., domestic, sewage, industrial) usually employs many processes. These may include treatments which are mechanical as well as biological. During this process, solid particles and liquid are separated and treated by applying different approaches. The biological approach is considered to be a key for the degradation of pollutants of chemical nature such as those which cause toxicity or those which fall under the category of xenobiotics. Activated sludge (AS), which is produced by the collection of the solid components of wastewater, is composed of both aerobic and anaerobic microorganisms which have the inherent ability to degrade carbon compounds, like petroleum-based products, and organic compounds like benzo-compounds, toluene, etc. (Shchegolkova et al., 2016).

The AS process which is perhaps a century old is still popular in wastewater treatment plants (WWTPs) for removing compounds which are biodegradable and organic. This process thus prevents these compounds from being discharged in plant effluents and causes the oxygen depletion of receiving waters (Modin et al., 2016). The AS process is majorly used across the globe for the treatment of industrially and domestically generated wastewater since this process caters to the need and necessity of high effluent quality within a limited space availability (Khan et al., 2007). AS can be explained as a process of sewage treatment in which air or oxygen is forcefully fed into sewage liquor to develop a biological floc thus reducing the organic content of the sewage (Dharaskar, 2015). It is a complex mix of microbiological and biochemical interventions employing different sorts of organisms; mostly bacteria. The particles which are components of wastewater are called 'activated' when they get attached to different organisms including lower microbes like bacteria or higher-order microbes like fungi, and protozoa. and are different from those microbes present in primary sludge which usually feed upon the components present in the wastewater (Bhargava, 2016). In the AS process, bacteria involved secrete some adhesive substances that coagulate the minute particles carried in sewage (Ahansazan et al., 2014). AS is produced in either raw or settled wastewater by promoting conditions for the growth of organisms such as bacteria in aeration tanks and dissolved oxygen is present. AS thus consists of sludge particles, harbouring living organisms. The AS process is perhaps one of the most complex microbial systems harbouring a variety of bacteria, viruses, protozoa, fungi, algae, and protozoa. In this complex ecosystem, about 95% of the total number of microbes is by bacterial load, therefore bacteria play a crucial part in wastewater treatment (Xu et al., 2018). Thus, it can be argued that identifying the functional microbes, especially bacteria, would help greatly to improve the performance of wastewater treatment systems (Zhang et al., 2018).

13.2 OPERATION OF ACTIVATED SLUDGE

Reliability of working by a WWTP can be defined as the probability of good performance, over a specified period of time and under applied operating conditions.

Taking into consideration plant performance, reliability is determined by the percentage of the plant's operating time at which the quality of effluent meets specified permit requirements (Andraka, 2020). The AS process is a continuous-flow system and aerobic, which has a mass of different types of activated microorganisms that have the ability to degrade the organic matter. To briefly summarise the whole process, in an aeration tank, post-primary treatment water is pumped so that an active mass of microbes can be mixed in it. These microbes may vary from bacteria to protozoans essentially having the capacity to degrade organic waste materials into CO_2, water or any other by-products. The bacteria involved in this method usually belong to Gram-negative species. These vary from those which can oxidise carbon to nitrogen, or floc and non-floc formers, or are aerobes to facultative anaerobes. The members from protozoa may include cilia bearing flagella bearing or amoebas. Pumping or diffused oxygen supply helps to maintain aerobic conditions as well as homogeneity of the tank contents. Once a specific duration has passed, the mixed liquor is pumped into the next clarifier tank called the secondary tank so that the sludge can settle down and free the effluent to be discharged (Figure 13.1). The process also includes recycling a portion of the settled sludge back to the aeration tank for maintaining the required AS concentration to the optimum. The process also eliminates a portion of the settled sludge to maintain the required solids retention time for effective organic removal (Bhargava, 2016). The AS process, if carried out under proper conditions, is very efficient. It removes almost 85%–95% of the solids and reduces the biochemical oxygen demand (BOD). The efficiency of this system mostly depends on factors such as wastewater physical characteristics. Toxic wastes that enter the treatment system can disturb biological activity. Wastes that contain soaps or detergents can cause frothing and thereby can create aesthetic or nuisance problems. In areas where industrial and sanitary wastes are being mixed and treated it decreases the treatment yield. Industrial wastewater must often be pretreated to remove the toxic chemical components before it goes into the AS treatment process (Nielsen et al., 2004).

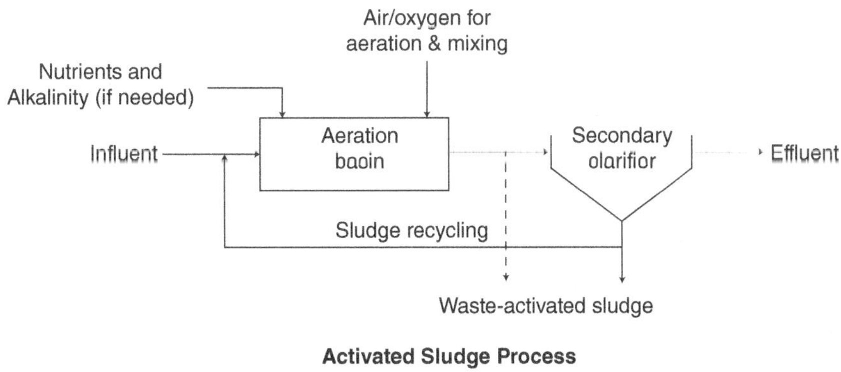

FIGURE 13.1 Basic principle of AS process.

13.3 EVOLUTION IN AS PROCEDURES

13.3.1 Conventional Complete Mix AS Process

The AS process has undergone many changes as well as various adaptations to be optimised for different conditions. These adaptations vary from being modifications to the very principles of the process to those which have evolved as empirical solutions to specific challenges in the execution. Some of the types are briefly mentioned as follows: (i) **The conventional AS process** in which the effluent from primary sedimentation tanks is aerated in the aeration tank. A reasonably steady rate of aeration is maintained in the tank leading to oxygen-limited reactions near inlets. Hence, the process efficiency is reduced. (Scholz, 2006). However, this method gives effective buffering against toxic conditions and therefore is particularly suitable for the treatment of effluents from industries. (ii) In addition to the conventional systems there are many other methods existing such as (a)**Series or Plug Flow System**, in which the AS is reused and the waste which is to be treated is introduced at one end of a long tank with many of aerators and a mixture of raw and pre-settled wastewater is withdrawn at the other end. This method is highly common for treating domestic sewage. One of the advantages of using this method is that a progressively reducing substrate level produces better settling characteristics of the sludge. (b) **Tapered Aeration**, which is just a moderation of the plug flow system where the rate of oxygen supply throughout the length of the channel is adjusted to create a gradient of oxygen supply from the inlet to another end. Hence, the proportion of total air is at a higher proportion when pumped towards the inlet end and the rate of supply is reduced toward the channel outlet. This system is reliable for the treatment of strong, readily biodegradable wastes containing no toxic substances. (c) **Step Feed AS Process,** another one of the modified plug flow processes in which a balance between the oxygen requirements and supply is maintained by adding the waste to the aeration tank at different points along the length of the channel. (d) **High-Rate AS Process** or the modified aeration process where the treatment plant operates at much lower microorganism-to-waste ratios than the conventional process such that the oxygen demands are somewhat less than in conventional processes, but the rate of oxygen requirement becomes higher per unit of Mixed liquor volatile suspended solids (MLVSS). However, in contrast to high-rate ASP, (e) **Extended Aeration** is a process characterised by a very high microbe-to-food ratio leading to low net sludge yield, higher hydraulic retention and higher mixed liquor suspended solids (MLSS). (f) **Contact Stabilisation** exploits the rapid reduction in BOD of the waste which is left untreated by biosorption and bio-flocculation with the reused AS. The main advantage of this method above others is that the re-aeration of sludge takes place in a smaller tank. Thus the volume of tanks and thus the capital costs are highly reduced. (g) **Oxidation Ditches** are a low-costing variation of the extended aeration process which can be operated in a continuous manner or with intermittent frequency. Sewage either flows into one part of the ditch continuously and gets mixed with the AS and circulated around the ditch by the action of the aerators or in intermittent-flow plants, where the ditch hosts both biological oxidation and settling process. (h) **Deep Shaft Process,** this variation has been developed by research in the process of enhancing the transfer of oxygen in the AS process. This can be achieved by increasing the oxygen partial pressure by increasing the total pressure applied to the system. Therefore, this system is able to maintain aerobic biological activity at a higher rate of the AS process. It is a highly efficient process.

The AS process can be a highly complicated system which enlists the help of microflora for treatment purposes. In order to perform functions, such as removal of organic carbon, nitrification and denitrification, neutralisation of organic waste, enhanced biological phosphorus removal, etc. several groups of heterotrophs and autotrophs of microflora are being sustained together. Efforts have made it possible to predict their metabolic activities through computational modelling and be enhanced by genetic engineering or the process of bioaugmentation. In order to ensure optimum process performance, it is now possible to individually control these microbes by elaborate parameter manipulation of the system. The goal has now become to meet new waste management objectives. So the continual progress of AS relies on a never-ending quest for more science to face and tackle new problems (Orhon, 2014).

13.4 MICROBIAL COMPOSITION OF ASP

In the AS process, microorganisms are blended with wastewater. The industrial and domestic wastewater treatment AS process has been extensively used because of its high microbial diversity and activity which results in the removal of most organic pollutants and elements. It is reported that the composition and diversity of the microbial community had the greatest impact on the stability and performance of the wastewater treatment systems (Xu et al., 2018). Table 13.1 shows the different bacteria reported in AS of different WWTPs. When coming in contact with the organic waste in the wastewater, microorganisms consume them as food. AS is composed of aerobic and anaerobic microorganisms such as bacteria, archaea, fungi, and protists. It is capable of degrading organic compounds, including petroleum products, toluene, and benzopyrene (Shchegolkova et al., 2016). In addition, the bacteria and other microflora also develop a sticky layer of slime around their cell wall that enables them to clump together forming bio-solids or sludge that separates them from the liquid phase. The successful removal of waste is accomplished by the microorganisms present in the wastewater. The microorganisms of AS form well-organised structural and functional communities, AS floccules, in which microbial cells and organic and inorganic particles are bound by extracellular polymeric compounds (Kallistova et al., 2014). Flocculation (clumping) is a characteristic of the microorganisms found in inactivated sludge enabling them to form a solid mass large enough to settle to the bottom of the settling tank. As the flocculation characteristics of the sludge improve, hence settling and wastewater treatment also improve.

AS consists of numerous constituents such as bacteria, extracellular polymeric substances (EPSs), and organic and inorganic particles. Together with other physical factors, e.g., conditions such as shear, these constituents are responsible for floc structure and floc properties. These properties, on the other hand, largely determine the sludge properties such as flocculation, settling and dewaterability (Nielsen et al., 2004). The bacterial cells make up only a minor part of the organic material in the flocs. Different methods based either on respiration rates or cell count 71 and average

TABLE 13.1
Showing Types of Bacteria Reported in Different Types of Sludge

Reported Organism Type	Type of Sludge	Reported by
Proteobacteria Acidobacteria Comamonadaceae Flavobacterium Nitrospira	Domestic	Xu et al. (2018)
Tetrasphaera		
Caldilinea	Municipal wastewater	Shchegolkova et al. (2016)
Opitutus Thiothrix, Planctomtyces, Prosthecobacter, Pasteuria		
Caldilinea, Prosthecobacter Byssovorax, Nitrospira, Flavobacterium	Municipal wastewater and petroleum	Shchegolkova et al. (2016)
Luteolibacter, Flavobacterium, cloacibacterium, Acinetobacter	Municipal wastewater and slaughterhouse	Shchegolkova et al. (2016)
Proteobacteria, Bacteroidetes, Saccharibacteria, Planctomycetes, Armatimonadetes, Chlamydia, Chlorobi, Cyanobacteria, Fibrobacteres, Verrucomicrobia	Chemical industries wastewater	Yang et al. (2020)
Proteobacteria, Bacteroidetes, Saccharibacteria, Planctomycetes, Elusimicrobia, Ignavibacteriae, Latescibacteria, Parcubacteria, Spirochaetes	Chemical industries and municipal wastewater	Yang et al. (2020)
Proteobacteria, Bacteroidetes, Saccharibacteria, Planctomycetes, Chlamydia, Chlorobi, Chloroflexi, Elusimicrobia	Municipal wastewater	Yang et al. (2020)
Ignavibacteriae, Latescibacteria		
Nitrifying bacteria, planctomycetes, acinetobacter, Gordonia, Sphaerotilus	Moscow wastewater	Kallistova et al. (2014)

content of carbon or nitrogen have been used to estimate this fraction, and values of 5%–20% have been recorded (Frølund et al., 1996). Industrial wastewater reduces the efficiency of the AS process due to the presence of toxins or chemicals. Hence, this process is most efficient for domestic wastewater treatment due to the presence of a high organic load (Yang et al., 2020).

13.5 METAGENOMICS

The past quarter of a century has seen radical shifts in the views and experimentations in the field of microbiology. It is now a commonly accepted fact that most microbes are not able to grow readily in pure culture. This has forced the hand of microbiologists to look for the need for an additional lens to their microscopes so to speak in order to look at the microbial world more broadly. Scientists have now acknowledged the extent of our ignorance about the range of microbial diversity present in this realm. This change in view has led to a revolution in the microbiological thought process and prompted a search for an additional more powerful tool which can act as their lens (a timeline of key events is given in Figure 13.2). This change in thinking that most of the microbes are non-culturable was prompted by another, equally important realisation: microorganisms which underpin most of the geochemical cycles, waste treatment plants and many human health conditions have escaped detection completely and these processes have been credited to be driven by inorganic processes and stress, respectively (Handelsman, 2004). Researchers developed a new technique called metagenomics. Metagenomics is the study of the genomes of the members of a microbial community. It includes cloning and analysing the genomes without culturing the organisms *in vitro* in the community. This technique has the opportunity to describe the planet's diverse microbial inhabitants, many of which cannot yet be cultured *in vitro*. It uses the technique of sequencing at a massive scale also called the next-generation sequencing (NGS), as the conventional methods have presented some challenges during transferring the information about all the microorganisms present in a natural context to the *in vitro* (Garrido-Cardenas & Manzano-Agugliaro, 2017). In another word, when genomics starts dealing with all the microbes present in a given environment instead of dealing with just one, it becomes metagenomics. Thus using metagenomics one can characterise the entire microbial community present in an environmental sample; it can be called the community genome (Vieites et al., 2010). Microbes have important roles to play in various ecosystems; however, many remain to be characterised in detail. Metagenomics provides different tools to study uncultured species. This innovative field offers a new approach to studying microbial communities as entire units, without cultivating them as individual members. Nowadays metagenomics has applications in the field of bioinformatics since with the introduction of technology and computational biology, now we are able to create increasingly more competent models to tackle the problems of DNA sequencing and genome classification (Tonkovic et al., 2020). The metagenomics approach deals with the extraction of DNA from a community so that all of the genomes of organisms in the community are pooled at the same time. These genomes are usually fragmented and cloned into an organism that can be cultured to create 'metagenomic libraries', and these libraries are then subjected to analysis based on DNA sequence or on functions conferred on the surrogate host by the metagenomic DNA.

13.6 TIMELINE

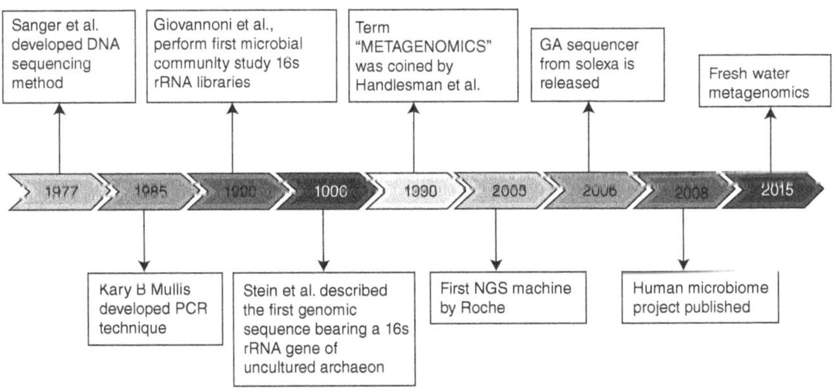

FIGURE 13.2 Some major discoveries in the field of metagenomics.

13.7 TECHNIQUES USED IN METAGENOMICS

In order to gain an understanding of the functioning of the microbial world metagenomics mainly focuses on investigations on three levels which are interconnected including sample processing, DNA sequencing of the microbes, and functional analysis on the basis of information obtained (Vieites et al., 2010). Since the metagenomic samples are collected from the target environment itself one can assume that in theory, these samples represent an entire community genome. This genome can thus be processed in a similar way as that of a whole genome of any pure culture of bacteria. This approach provides ecologists to gain a comprehensive understanding of the whole microbial communities' context (Chakraborty et al., 2014). It also provides access to the functional gene composition of microbial communities and thus gives a much broader description than phylogenetic surveys, which are often based only on the diversity of one gene, for instance, the 16S rRNA gene which is most widely used for prokaryotes. Metagenomics provides genetic information on potentially novel enzymes, genomic linkages and phylogeny for uncultured organisms, and evolutionary profiles of community function and structure. It can also be complemented with metagenomic, meta-transcriptomic or meta-proteomic analysis to characterise organisms. Metagenomics is also a powerful tool for generating novel hypotheses of microbial function (Thomas et al., 2012). For removal of most of the organic material and nutrients from the sewage wastewater; microorganisms present in AS or biofilms are thought to be the key component and therefore they are central to each biological WWTP. Even though the microbial populations have been targets of extensive studies for many years, it was not possible until recently to estimate the true range of wastewater communities as molecular and metagenomics approaches are just developed.

13.7.1 Sequencing Technology

The technique used to find out in which specific order the four nucleic acid bases (adenine, thymine, guanine, and cytosine) tandemly present, is known as DNA sequencing which encodes different translating and non-translating transcripts of a living cell. Genomes are known to contain almost all of the information regarding the transcripts and proteins required to be expressed for the normal life of an individual, as well as, the complete set of noncoding sequences in a cell which might have some role in the regulation of expression of such transcripts, or evolution, etc. Further, the studies also reveal that genome sequences thus obtained may be highly conserved in their patterns of similarity to each other. These similarities are identifiable by scanning for local and global similarity across the sequences and thus there is a need for the sequence of the entire genome of any individual or organism (Shetty et al., 2019). Therefore, sequencing technology does precisely that and deciphers the DNA sequences in order to determine the particular order of the base pairs of the nucleotides present in a DNA molecule or genome. The order of nucleotide bases in polynucleotide chains finally contains the information for the hereditary and biochemical properties of life. Therefore, the concern to measure or identify such sequences is imperative to biological research (Heather & Chain, 2016). Nowadays, DNA sequencing technologies have become sophisticated enough to predict the primary structure of genes present in genomes, identify genetic variations present across the homologs, in predicting gene function as well as detect mutations which might lead to disease. Traditional sequencing techniques merely started with a setup consisting of chemical reactions or applications of enzymes but during the last 4 decades, many technological sophistications have been introduced to make DNA sequence read outputs more reliable and much faster. Previously sequencing was costly and took more time but now it's cheaper and also takes less time. Now, sequencing is done by highly automated machines. Approximately for 30 years now, the most used methods have been based on the dideoxy-nucleotide chain termination principle, commonly called 'Sanger sequencing' after the name of the inventor of this method and thus, currently has gained the status of the most applied technique in the sequencing. The added advantages of its high efficiency and low radioactivity have prompted the rapid development of commercialised versions of the technique (Kchouk et al., 2017; Knetsch et al., 2019). Further advancements were done after the launch of the human genome project. The completion of the first human genome drafts was accomplished using advanced DNA sequencing technology which prompted even more interest to develop modern and faster strategies for DNA sequencing, known as 'high-throughput sequencing' or 'next-generation sequencing' (Pareek et al., 2011). The latest technology advancements include whole genome sequencing (WGS) and fourth-generation sequencing; these techniques are highly advanced techniques as they use computational biology methods. Now it is the genomic era where we are trying to connect everything with our genome and so it is desirable to have better sequencing technologies.

13.7.1.1 First-Generation Sequencing

Invention of polymerase chain reaction (PCR) during the period of 1970–1980 provided a great amount of boost to the development of molecular biology, especially in the field of genetics and genomics. Application of PCR to the approaches of sequencing enabled the development of the first DNA sequencing technology (Shetty et al., 2019). The first method developed was simple and enzyme-based. Sanger sequencing and Maxam–Gilbert sequencing are considered as first-generation sequencing methods. They dominated genomics for nearly 40 years. They advanced genomic research by folds and paved a path for subsequent sequencing technologies (Shetty et al., 2019). For first-generation sequencing, fragment ladders of the sequence are generated by two methods, (i) by enzymatically extending a primer hybridised to a pool of template molecules and introducing specific T, C, G and A terminations along the template and (ii) by base-specific cleavages, were introduced. Fragment ladders are then separated by gel electrophoresis. Products are revealed by radioactive labelling and exposure to a film or by the introduction of dyes fluorescent in nature into the termination reaction and imaging. In a capillary sequencer, this technique can also be achieved during sequencing in real time (Gut, 2013).

13.7.1.1.1 Maxam–Gilbert Sequence Method

Maxam–Gilbert is one of the first-generation sequencing techniques which is also known as the method of sequencing by chemical degradation and represents the group of the oldest sequencing platforms (Kchouk et al., 2017). Since the core step of sequencing involves the cleavage of DNA fragments at certain specific bases this sequencing method is also popularly referred to as the chemical cleavage method. In 1977, Allan Maxam along with Walter Gilbert developed this technique based on the partial chemical modifications which were specific for different nucleobases in the DNA which ultimately leads to the cleavage of the fragment near modified bases. Post its discovery this method rapidly rose in popularity as it was possible to directly use the purified DNA but this popularity was not long-lasting due to its technical complexity and risks involved (Shetty et al., 2019). Procedures include chemical treatment that generates cleavage of DNA sugar–phosphate backbone such that these fragments are generated only in one or maximum of two of the four nucleotide bases cleavage four reactions mixture (C, T+C, G, A+G). Since the fragments are tagged at an end this leads to a series of tagged fragments which are then separated as per their length along the electrophoresis gel. The sequencing is done on the given DNA without any amplifications or cloning. This technique is considered dangerous because of the use of toxic and radioactive chemicals (Kchouk et al., 2017).

13.7.1.1.2 Sanger Sequencing

This was the first DNA sequencing strategy called the Sanger chain termination method. WGS DNA sequencing is a capillary-based, semi-automated version of the original Sanger strategy (Cao et al., 2017). The Sanger chain termination method is based on the semiconservative way of DNA amplification where both the strands act as a template for forming new sequences however these templates are the fragments which need to be sequenced. Chemical modifications are done to the nucleotides used for this technique such that they become dideoxy-nucleotides (ddNTPs). These ddNTPs are generated for each of the four bases viz. ddG, ddA, ddT, and ddC; and mixed with dNTPs for the amplification reactions. The dNTPs in the mixture will allow for strand elongation of the newly forming strands however, once a ddNTP base gets incorporated into the elongating strand, any further elongation is prevented and thus a fragment of unique length is generated. Then, these unique fragments with different sizes are separated according to their size on gel slab electrophoresis and visualised by X-ray tagging or UV exposure. This method of DNA sequencing was widely used for 3 decades and even today for single or low-throughput DNA sequencing. A major drawback is that it does not allow the sequencing of complex genomes such as the plant species genomes, and the sequencing is extremely expensive and time-consuming (Kchouk et al., 2017).

13.7.1.2 Second-Generation Sequencing (SGS)

NGS also called second-generation sequencing has gained the capacity to sequence more amount of DNA like in gigabases in a high-throughput and highly efficient manner that was not possible with first-generation sequencing methods (Frese et al., 2013). In contrast to the traditional first-generation techniques where the individual analysis of DNA fragments (one fragment per capillary) was done; sequencers of second-generation are capable of analysing millions of DNA templates in parallel. This is achieved usually by putting single template DNA samples into many individually designed miniaturised reaction vessels for sequencing by amplification (Bayés et al., 2012). The read lengths of previous NGS were relatively short, due to the small sequencing colonies and progressive signal deterioration of about 35–500 bp. Technical and chemical refinements were used to increase the read lengths. NGS technology has successfully contributed to elucidating the sequence and function of the human genome (Frese et al., 2013). Following the human genome project, 454 sequencers were launched in 2005 and Solexa released its Genome Analyser the next year, followed by sequencing by oligo ligation detection (SOLiD) provided by Agencourt, which are the three most typical massively parallel sequencing systems in the NGS that shared good performance on throughput, accuracy, and cost compared with Sanger sequencing. The founder companies were then purchased by other different companies, like in 2006 Agencourt was purchased by Applied Biosystems, and in 2007, 454 was purchased by Roche and Solexa was purchased by Illumina. After years of evolution, these three systems exhibit better performance and their own advantages in terms of read length, accuracy, and applications (Liu et al., 2012). With NGS it is now possible to investigate microbial populations by looking at their composition and functions in diverse environments like high temperature, etc., with unprecedented resolution and throughput (Boers et al., 2019). A few variations are as follows.

13.7.1.2.1 Roche/454 Sequencing

This was the first NGS technology released in 2005 with the name of a sequencer called "454 Genome Sequencer (Life Sciences, today Roche)". This was a great initial success which was due to the association with emulsion PCR, a new amplification strategy emerged, and pyrosequencing technology. Emulsion PCR is an innovative methodology which uses small water droplets scattered in a lipid solution, where individual DNA fragments are amplified (Del Vecchio et al., 2017). In this technique, the DNA templates are amplified by emulsion PCR and the sequences are obtained by iterative pyrosequencing (Bayés et al., 2012). The technique involves fragmentation of DNA, then ligated to adapters and mixed with micro-beads containing complementary adapters. DNA fragments-beads complexes are emulsified in droplets containing PCR reactants so that each droplet contains a single copy of the DNA fragment which is to be amplified. Afterwards, a standard PCR reaction amplifies DNA. In the end, every bead carries hundreds of thousands of amplified fragments on its surface. Samples are then loaded onto the wells of a picotiter plate to perform many thousands of pyrosequencing reactions in parallel by sequentially adding, one at a time, the four deoxynucleotides. A CCD camera detects light signals. This technology is error-prone to homopolymeric stretches, with consequent mistakes in the estimated length and introduction of "indel" errors. The two most recent platforms, GS Junior and GS FLX have greatly improved the output, with significant read lengths of 400 and 1,000 nucleotides, respectively (Del Vecchio et al., 2017). The parallelisation can increase the yield of sequencing efforts by orders of magnitudes, allowing researchers to completely sequence a genome (Heather & Chain, 2016).

13.7.1.2.2 Ion Torrent Sequencing

Ion Torrent technology is based on original chemical–physical principles, different from those characterising the other next-generation platforms. This technique was commercialised in 2010 by Life Technologies (Del Vecchio et al., 2017). Ion Torrent introduced an instrument that uses similar underlying methods as the Roche system for clonal amplification and parallel sequencing but has different detection methods. The specification that is being provided moves this instrument into the realm of the Roche FLX junior (Bayés et al., 2012). It is a semiconductor-based technology where minimal pH changes are produced by the release of hydrogen ions as a by-product of nucleotide incorporation which is detected by a detector. This is done by using an "Ion chip", structured into two parts, to deliver reactants and communicate directly with a proton detector for nucleotide identification during the reaction of incorporation. Ion Torrent recognises added nucleotides without the use of fluorescence. In fact, the instrument recognises one nucleotide at a time and incorporation's specificity is guaranteed by the detection and release of H+ ions. In this technique, the most frequent errors are due to phasing. Nonetheless, the error rate is very low. Continuous improvements in this technology have increased read length from the initial 100–200 nucleotides. Now Ion Torrent throughput made an even greater jump making its read length from 10 Mb to the current maximum of 15 Gb (Del Vecchio et al., 2017).

13.7.1.2.3 Illumina/Solexa Platform

As second-generation technologies were rapidly coming to dominate modern DNA and RNA sequencing efforts. Illumina sequencing (also known as Solexa) had played an increasingly prominent role (Cox et al., 2010, p. 485). The Solexa sequencing platform was commercialised in 2006, with Illumina acquired by Solexa in early 2007. The principle is based on sequencing-by-synthesis chemistry, with novel reversible terminator nucleotides for each of the four bases labelled with a different fluorescent dye, and DNA polymerase enzyme able to incorporate them. DNA fragments are ligated at both ends to adapters and after denaturation fragments are immobilised at one end on solid support creating a 'bridge' structure through hybridising with its free end to the complementary. In the mixture which contains the PCR amplification reagents, the adapters present on the surface act as primers for the following PCR amplification. PCR reaction is used to amplify the DNA fragments (termed DNA 'polonies', resembling cell colonies) that are created on the surface. Then four different kinds of nucleotides (ddATP, ddGTP, ddCTP, ddTTP) contain different fluorescent dyes which are cleavable and a removable blocking group which complements the template one base at a time. The signal is mostly captured by a charge-coupled device (CCD) (Liu et al., 2012). The sequence read length which is achieved in the repetitive reactions is about 35 nucleotides. The sequence of at least 40 million polonies can be simultaneously determined in parallel. This results in a very high sequence throughput, in the order of gigabases per support (Ansorge, 2009).

13.7.1.2.4 By Ligation: ABI/SOLiD Technique

Introduced by Applied Biosystems through the name SOLiD system in late 2007. The key distinguishing feature of this platform is the difference in its sequencing chemistry. It uses a DNA ligase instead of a DNA polymerase enzyme. Principles follow that set of four fluorescently labelled octamers whose fourth and fifth bases are encoded by the attached fluorescent group and are hybridised to the template. The probes that match the templates are then ligated to a primer oligonucleotide in a template-directed reaction. The fluorescent readout identifies the fixed bases.

A cleavage reaction removes the last four bases with the fluorescent tag and hybridisation and ligation cycles are repeated to determine in successive cycles bases 9 and 10, 14 and 15 and so on. After completion of a defined number of cycles, a new primer is annealed that is offset from the first primer by one base (Bayés et al., 2012). Procedures follow, DNA fragments are ligated to the adapters then they are bound to beads. A water droplet in oil emulsion contains the amplification reagents and only one fragment bound per bead, DNA fragments on the beads are amplified by the emulsion PCR. After DNA denaturation, the beads are deposited onto a glass support surface. Sequences can be determined in parallel for more than 50 million bead clusters, resulting in a very high throughput of the order of gigabases per run. Applied Biosystems also produced an updated version in 2008, the SOLiD 2.0 platform, which increased the output of the instrument from 3 to 10 Gb per run. This change reduced the overall run time of a fragment library on the new system to 4.5 from 8.5 days on the existing machine (Ansorge, 2009).

13.7.1.3 Third-Generation Sequencing (TGS)

The scenario was changing again since new innovative approaches and technical advances in technology had allowed another step forward for sequencing techniques and thus, the TGS technologies came. In comparison to the previous generation of technology, TGS provide two crucial and tightly bound advantages, (i) the ability to sequence single molecules, thus avoiding one of the main sources of bias which gets introduced during library preparation & the PCR amplification step, and (ii) the increased read length of their output (Lavezzo et al., 2016). Different types of sequencers were Pacific Biosystems, and Oxford Nanopore. Sequencers were distributed by Pacific Biosystems (PacBio) using single-molecule real time (SMRT) sequencing technique, whereas Oxford Nanopore Techniques (ONT) developed a device which performed nanopore sequencing (Bleidorn, 2015). This new generation of sequencing technology had the potential to exploit more fully the high catalytic rates and high processivity of DNA polymerase also, avoid any biology or chemistry altogether which led to radically increase read length (from tens of bases to tens thousands of bases per read) and decreased time of result (from days to hours or minutes). The promises then of this new, offered many advantages over previously sequencing technologies which were (i) higher throughput, (ii) faster turnaround time, (iii) longer read lengths to enhance de novo assembly, (iv) higher consensus accuracy to enable rare variant detection, (v) small amounts template or sample, and (vi) low cost, even for human genome now less than 100 dollars (Schadt et al., 2010).

13.7.1.3.1 Single-Molecule Real-Time Sequencing (SMRT)

It was released in early 2011 by PacBio named PacBio RS sequencer, which worked on SMRT technology. Initially, average read lengths were relatively short (1.5 kb) and had high error rates (13%). Now the technology has strongly improved over recent years. Average read lengths now have increased more than tenfold and the throughput per run has also increased by 100-fold owing to the development of improved sequencing chemistry and the release of a new sequencer, the Sequel. This machine generates about tenfold more sequence data and is twofold less expensive. However, the 'single-pass' error rate has remained roughly the same since the beginning

(13%), but molecules of up to 1–2 kb can now be sequenced many times owing to the circular templates and increased polymerase processivity, strongly improving overall accuracy. Moreover, increased throughput has led to a sharp reduction in the cost of sequencing per base. For genomic DNA library preparation, PacBio is being commercialised a 'SMRTbell template prep kit' and an 'express' variant for rapid library preparation with an approximately 3-hour workflow (van Dijk et al., 2018). The molecular approach of the SMRT sequencing technique is based on monitoring polymerase activity while incorporating differently labelled nucleotides (dNTPs) into the DNA strand. Each nucleotide carries a base-specific fluorescent label with it on its phosphate group, which is released after being incorporated by the polymerase. Incorporated nucleotides are detected by real-time imaging at the time of strand synthesis. The whole process takes place in a small well which is surrounded by aluminium walls called zero-mode waveguide (ZMW). Single DNA polymerase molecules are attached to the surface of these wells, from where their activity can be monitored. With a diameter of 70 nm and a depth of 100 nm these wells are extremely small, therefore so-called SMRT-cells for sequencing. For the sequencing process, fluorescently labelled nucleotides are flooded into these small cavities and their presence while floating in and out is measured as background noise (Bleidorn, 2015).

13.7.1.3.2 Nanopore Sequencing

The idea of using nanopores originated at the end of the 1980s. However, due to technical error, the first successful sequencing results were reported only in 2012 and after that 2 years later, ONT released their first nanopore sequencer, the pocket-sized version named 'MinION', in a large-scale collaborative MinION Access Programme (MAP) (van Dijk et al., 2018). It was released on the market in 2014. It is different from other platforms as it utilises nanopores for sequencing. It does not follow a sequencing-by-synthesis approach instead, an ionic current is being passed across the flow cell during sequencing through which the different nucleotide bases are being distinguished by the changes in current frequency as they pass through the nanopores. This technique requires minimal capital cost compared to other technologies like Illumina, Thermo Fisher, and Pacific Biosciences platforms. This can be utilised both in the laboratory and out in the field, is also able to fit in the palm of our hand and can be connected to either ONTs "MinIT" computer module or any computer having a USB connection. MinION permits direct, electronic analysis of single molecules in real time (Petersen et al., 2019). Through the rapid evolution of chemistry, a significant increase in throughput has been achieved. While the early chemistries produced about 184–450 million bases of sequence data per 48 hours of run, today's flow cells in combination with the latest library preparation kit versions produce up to 20 Gb of sequence data. The translocation speed has also been increased, from 30 bases per second (bps) to 450 bps using today's chemistry (van Dijk et al., 2018). Distinct advantages of this system include low instrument fabrication and operation costs due to the lack of labelled nucleotide and optical detection systems (laser and CCD camera). In addition, the Oxford Nanopore platform is compatible with direct RNA sequencing and the detection of modified bases by virtue of each individual base's characteristic ability to disturb electrical current, which should enable epigenetics applications. The disadvantage, however, is that because the template molecule is digested during sequencing, redundant sequencing and high accuracy are not possible. However, this drawback can be eliminated by simply replacing the exonuclease coupled to the nanopore with a DNA polymerase (Munroe & Harris, 2010).

13.7.2 RAPD

Random amplified polymorphic deoxyribonucleic acid (RAPD) is a PCR-based technique in which random primers mostly decamer primers are used to randomly amplify segments of DNA. This technique leads to the amplification of DNA sequences and generates a set of fingerprinting patterns (Adzitey et al., 2012). This technique leads to the generation of numbers of DNA fragments which are of varied length and their subsequent separation on agarose or polyacrylamide gels helps to exploit the complexities of the genome of a microbial community present in a typical habitat. RAPD primers target the DNA sequences with unique and repetitive sequences with the resulting amplicons to be used as a genetic fingerprint (Dash & Das, 2018). The advantages of RAPD are that it is relatively cheap, rapid, readily available, and easy to perform (Adzitey et al., 2012).

13.7.3 Ribosomal RNA Intergenic Spacer Analysis (RISA)

It is known widely that all organisms can be classified using the small subunit ribosomal RNA. Ribosomal RISA is a molecular technique that employs the polymerase chain amplification of a portion of the intergenic spacer region (ISR) which is in between the 16S (prokaryotes) and the 23S (eukaryotes) subunits of the ribosomal subunits (Edet et al., 2017). The ISR contains significant heterogeneity in both length and nucleotide sequence. By using primers annealing to conserved regions in the 16S and 23S rDNA genes, RISA profiles can be generated from most of the dominant bacteria existing in an environmental sample. RISA provides a community-specific profile, with each band corresponding at least to one organism of the original community (Adzitey et al., 2012). Specifically, it uses primers annealing to the highly conserved regions in the ribosomal subunit genes. It is capable of generating highly reproducible bacteria community profiles and the automated version allows for the analysis of more than one sample at a time. However, like microarray and other PCR-based techniques, it is subject to PCR biases and requires large quantities of DNA (Edet et al., 2017). The RISA method can be used to generate more complex fingerprints than the 16S-RFLP technique and is able to show differences between samples.

13.7.4 Fluorescence *In Situ* Hybridisation (FISH)

FISH is a versatile molecular technique that allows at source (*in situ*) phylogenetic identification and enumeration of individual microbial cells using whole cell hybridisation to oligonucleotides (18–30 nucleotides long) probes. The probes usually contain a 5′ end fluorescently labelled to a dye which allows for its detection following hybridisation by fluorescent microscope. The strength of the signal is correlated to the cellular rRNA contents and growth rates. These parameters can be used to estimate the metabolism dynamics of the community under study. Some of the disadvantages of FISH include low signal intensity, background noises, and poor accessibility of targets. However, variants such as catalysed reporter deposition (CARD) FISH, flow cytometry coupled FISH, and FISH coupled with ion mass spectrophotometry give better resolutions (Edet et al., 2017). FISH is a cytogenetic technique that uses complementary fluorescence-tagged probes to depict the presence or absence of a target sequence on a chromosome. Fluorescence microscopy is utilised for discovering the location of fluorescent probe binding to the chromosome. FISH conveniently locates the genes or parts of a gene on the chromosome in a single cell. This enables the study of structural, numerical and gene-level mutations in the chromosomes. Not quite the same as most different methods utilised for chromosome study, FISH need not be performed on cells that are effectively isolating, which makes it a flexible strategy. FISH is also used for gene mapping and identification of novel oncogenes. Also, FISH applies to the detection of infectious microbes. The latest modifications in FISH technology involve techniques for improving labelling probes' efficiency and also using high-resolution imaging systems for spot visualisation of intranuclear chromosomal arrangement and the report of RNA transcription in each cell. These convenient, however viable, methods have revolutionised cytogenetics and have shown potential as a simple method of cancer research (Singh et al., 2020).

13.7.5 Terminal Restriction Fragment Length Polymorphism (T-RFLP)

This molecular technique is similar to amplified ribosomal DNA restriction analysis but differs in the use of 5′ fluorescently labelled primer during the PCR. As a method of studying microbial diversity, it relies on DNA polymorphisms, single nucleotides change in sequences. The resulting 16S rDNA fragments from PCR are digested with restriction enzymes and electrophoresis in agarose or acrylamide gels. RFLP has been used to estimate the prokaryotic diversity in hypersaline ponds (Edet et al., 2017). The procedure involves the isolation of DNA, followed by digestion of DNA with restriction endonucleases, final step is size fractionation of the resulting DNA fragments by electrophoresis. After that DNA is transferred from the electrophoresis gel to the membrane and membrane-bound fragments are hybridised with radiolabeled and chemiluminescent probes. RFLP fingerprinting technique is regarded as the most sensitive method for strain identification and several bacterial strains have been widely studied using this technique to establish a rapid PCR-RFLP-based identification scheme for four closely related *Carnobacterium* species (Olubukola, 2003).

13.7.6 Quantitative PCR Q-PCR

Quantitative PCR (Q-PCR), or real-time PCR, has been used in microbial investigations to measure the abundance and expression of taxonomic and functional genes (Rastogi & Sani, 2011). Unlike traditional PCR, Real-time q-PCR utilises the PCR reaction to detect only targets of interest genes. Gene-specific primers and probes are used and the target is detected either by the incorporation of a double-stranded DNA (dsDNA)-specific dye or by the release of a TaqMan FRET (fluorescence resonance energy transfer) probe through polymerase 5′–3′ exonuclease activity

(Goodwin et al., 2016). The software records the fluorescence or dye which indicates an increase in amplicon concentration during the early exponential phase of amplification. This facilitates the quantification of genes (or transcripts) when they are proportional to the starting template concentration. In the case of a RNA-based target (e.g., transcripts of a gene or retroviruses) Q-PCR is coupled with a preceding reverse transcription (RT) reaction. Then, it can be used to quantify gene expression (RT-Q-PCR). Q-PCR is highly sensitive to starting template concentration and measures template abundance in a large dynamic range of around six orders of magnitude. Q-PCR has also been successfully used in environmental samples for quantitative detection of important physiological groups of bacteria such as ammonia oxidisers, methane oxidisers, and sulfate reducers (Rastogi & Sani, 2011).

13.7.7 Pulsed-Field Gel Electrophoresis (PFGE)

PFGE is an agarose gel electrophoresis technique used for separating larger pieces of DNA by applying electrical current that periodically changes direction (three directions) in a gel matrix, unlike the conventional gel electrophoresis where the current flows only in one direction. In PFGE, intact chromosomes are digested using restriction enzymes or restriction endonucleases to generate a series of DNA fragments of different sizes (also known as restriction fragments length polymorphisms, RFLPs) and patterns specific for a particular species or strain. Using this method provides benefits as this technique increases reproducibility as well as the power of discrimination and type-ability. However, genetic instability has proven to affect PFGE; moreover, this technique has limited availability and may take at least 3 days to complete a single test. Another concern about using this method may be its cost as compared to other methods like RAPD, plasmid analysis, etc. (Adzitey et al., 2012).

13.8 METATRANSCRIPTOMICS

In recent years questions have been asked to find out the 'active members' of the microflora involved in bioremediation and wastewater treatments. In order to understand and search for the active microbes and the molecular mechanisms, the focus has now shifted to studying the whole transcription profile instead of reading whole genomes. Metatranscriptomics thus can be defined as the large-scale sequencing of mRNAs allowing observation of gene expression patterns. Metatranscriptomics is relatively a new field, first mentioned around 2008 and since then it has grown vigorously (Westreich et al., 2018). By focusing on what genes are being expressed by the entire microbial community, metatranscriptomics helps us in the identification of the active functional gene profile of a microbial community. The metatranscriptome analysis shed light on the gene expression in a given sample at a given instant and under specific conditions by analysing the total mRNA (Aguiar-Pulido et al., 2016). Now high-throughput techniques have developed in conjunction with computer-automated analysis approaches of metatranscriptomics offering a novel and complete method for looking for the organisms present and also the transcriptional activity occurring within the population at any chosen specific point in time (Westreich et al., 2018). Functional metatranscriptomics even allows researchers to characterise genes expressed by different eukaryotic microorganisms directly from the environment. The steps involve extraction and analysis of mRNA to provide information on the regulation and expression profiles of complex communities. The transcriptomics-oriented technique is thus based on the extraction of environmental RNA instead of DNA (Mukherjee & Reddy, 2020). The sequencing and in-silico analysis of mRNA is now being routinely applied to study complex microbial communities in diverse ecosystems. The typical goal of metatranscriptomics is to taxonomically classify transcripts, predict gene functions, quantify their abundances, and relate these to environmental data in order to study the impact of environmental conditions on microbial communities in different habitats (Toseland et al., 2014).

13.8.1 Denaturing Gradient Gel Electrophoresis (DGGE)

DGGE has been applied in the field of microbiological studies for several applications including the analysis of microbial diversity and its dynamics in the microbial population present in complex habitats. The analysis of the relative intensity of PCR products' bands obtained in this method allows rapid analysis of change in the microbial community and information about its composition. In bacteria, the hypervariable region (V3) of rRNA is the most studied. The disadvantages of this technique are (i) less than 1% of the target DNA in the total population is hardly detected by DGGE, (ii) separation of small fragments (500–700 bp), and (iii) biases are always associated with PCR amplifications which get incorporated into the analysis (Bailón-Salas et al., 2017).

13.8.2 Microarray

The basic principle of DNA microarray technology is to immobilise known DNA sequences which are referred to as probes in micrometre-sized spots on a solid matrix (microarray) and specifically hybridise a complementary sequence of the analyte or target DNA. It is now possible with this technique to carry out a genome-wide analysis of the entire microbial genome as it can be represented in a single array. Panicker with his associates identified two major types of DNA microarrays commonly used oligonucleotides-based and PCR product-based arrays (Edet et al., 2017). DNA microarrays are thus a way to obtain comprehensive data about microbial communities present in environmental samples in a high-throughput manner. Hybridisation of the unknown amplicons obtained from total environmental DNA to known molecular probes that are attached to the microarrays provides the information of sequence. The positive signals are detected by the use of confocal laser scanning microscopy. This technique allows samples to be rapidly evaluated with replication, which is a significant advantage in microbial community analysis (Rastogi & Sani, 2011). DNA microarrays have also been used for genetic research since the early 1980s. Microarrays still remain a widely used technique in genomic research. They are used to characterise single nucleotide polymorphism (SNPs) at costs far below NGS routines (Goodwin et al., 2016). The latest development includes the application of DNA microarrays to detect and identify bacterial species or to assess microbial diversity. This rapidly characterises the composition and functions of microbial communities because a single array contains thousands of template sequences with the possibility of very broad hybridisation with wide identification capacity. The microarrays can contain specific target genes such as nitrate reductase, nitrogenase or naphthalene dioxygenase to provide functional diversity information representing different species found in the environmental sample (Agrawal et al., 2015).

13.8.3 Temperature Gradient Gel Electrophoresis (TGGE)

TGGE is based on the same principle as DGGE but the difference is only that a temperature gradient is applied rather than a chemical denaturalisation. The sequence of different amplicons determines the melting behaviour so that sequences achieve migration to different positions of the gel. During the amplification step, a staple or "clamp" composed of guanines and cytosines (about 30–50 nucleotides) is added. This is in order that the DNA strands do not separate completely during electrophoresis. To determine the phylogenetic identities, the gel bands must be excised, reamplified, and sequenced or transferred to nylon membranes and hybridised with specific molecular probes for different taxonomic groups (Bailón-Salas et al., 2017).

13.8.4 Length Heterogeneity PCR (LH-PCR)

Length heterogeneity PCR (LH-PCR) analysis is similar to that of the T-RFLP method except that, T-RFLP detects amplicon length variations that are being produced after restriction digestion, whereas in LH-PCR microorganisms are differentiated based on natural length polymorphisms which occur due to mutation within their genes (Mills et al., 2007). Amplicon LH-PCR analyses the hypervariable regions which are present in 16S rRNA genes, therefore producing a characteristic profile. This technique utilises a fluorescent forward primer which is dye-labelled, and a fluorescent standard of internal size is run with each sample to measure the amplicon lengths in base pairs. The intensity (height) or area under the peak in the electropherogram is proportional to the relative abundance of that particular amplicon. Advantage of using LH-PCR over the T-RFLP is that the LH-PCR does not require any restriction digestion so PCR products can be directly analysed with a fluorescent detector. The limitations of the LH-PCR technique include the inability to resolve complex amplicon peaks and underestimation of diversity as phylogenetically distinct taxa may produce same-length amplicons (Rastogi & Sani, 2011).

13.9 TECHNIQUES USED FOR ANALYSIS OF MICROBIOME FOUND IN WASTE WATER

Metagenome analysis is a fruitful approach for wholesale characterisations of microbial populations which persists in waste wastewater. Previously, it was very difficult to identify each of them as most of them were unable to be cultivated *in vitro*. Now different metagenomic techniques as stated above have solved the problem of identification and characterisation of these bacteria. Table 13.2 shows the different types of bacteria reported in WWTPs of different types. It is important to study the sludge microbiome as the performance of the AS process depends on microbial communities that persist in it (Yang et al., 2011). The WWTPs treat different types of wastewater like municipal, hospital and industrial wastewater. The stage prior to biological treatment is called raw sludge, whereas the remains of sludge after biological digestion and during the air-drying stage are designated as dried sludge (DS) (Sidhu et al., 2017). Enhancement of the microbial population in treatment plants results in the efficient removal of pathogens, decomposing of organic matter, neutralisation of toxic substances, and nutrient enhancement. Changes in the complexity or diversity can result in alterations in metabolism leading to a decline in the treatment efficiency (Silva-Bedoya et al., 2016). Poorly treated wastewater can lead to increase environmental and public health risks. Treated water hygiene is essential to maintain both the biotic and abiotic environment that could support human life on Earth (Yadav et al., 2020).

TABLE 13.2
Representing Different Techniques Used to Identify Bacteria Found in Different WWTPs

Type of Waste Water	Techniques Used	Organisms Identified	Place of Experiment	Reference
Activated sludge	T-RFLP DGGE	*Betaproteobacteria* *Chloroflexi* *Verrucomicrobia* *Alphaproteobacteria* *Firmicutes* *Rhodocyclus*	China	Yang et al. (2011)
Activated sludge	Sequencing by Illumina	*Bacteroides* *Pseudomonas* *Flavobacteria* *Myxococcales* *Rhodobacteria*	India	Yadav and Kapley (2019)
Domestic & industrial	RISA TTGE	*Firmicutes* *Mucormycotina* *Sordariomycetes* *Protobacteria* *Klebsiella* *Bacillus* *Enterococcus*	Colombia	Silva-Bedoya et al. (2016)
Pretreated sludge from the primary reactor		*Epsilonproteobacteria* *Bacterodia* *Gamma proteobacteria* *Flavobacteria*	*India*	*Sidhu et al. (2017)*
Post-treated sludge from drying beds	Sequencing & PCR	*Clostridia* *Methanomicrobia* *Betaproteobacteria* *Deltaproteobacteria*		

(*Continued*)

TABLE 13.2 (*Continued*)
Representing Different Techniques Used to Identify Bacteria Found in Different WWTPs

Type of Waste Water	Techniques Used	Organisms Identified	Place of Experiment	Reference
Municipal wastewater	Sequencing by Illumina	*Firmicutes* *Betaproteobacteria* *Gammaproteobacteria* *Bacteroidetes*	Germany	Schneider et al. (2020)
Hospital wastewater		*Betaproteobacteria* *Gammaproteobacteria* *Bacteroidetes*		
AS		*Betaproteobacteria* *Alphaproteobacteria* *Actinobacteria* *chloroflexi* *Bacteroidetes*		
Municipal wastewater	Pyrosequencing & PCR of 16s rRNA	*Firmicutes* *Actinobacteria* *Bacteroidetes* *Fusobacteria* *Synergistetes*	China	Cai et al. (2013)
Influent	Sequencing by Illumina	*Alphaproteobacteria* *Clostridia* *Bacteroides* *Methanomicrobia* *Bifidobacterium* *Betaproteobacteria*	China	Giwa et al. (2019)
AS		*Alphaproteobacteria* *Clostridia* *Gammaproteobacteria* *Bacteroides*		
Return sludge		*Bifidobacterium* *Methanomicrobia* *Bacilli* *Betaproteobacteria* *Deltaproteobacteria* *Clostridia*		
Effluent		*Bacteroides* *Bifidobacterium* *Methanomicrobia* *Deltaproteobacteria* *Flavobacteria*		

REFERENCES

Adzitey, F., Huda, N., & Ali, G. R. R. (2012). Molecular techniques for detecting and typing of bacteria, advantages and application to foodborne pathogens isolated from ducks. *3 Biotech*, *3*(2), 97–107. https://doi.org/10.1007/s13205-012-0074-4.

Agrawal, P. K., Agrawal, S., & Shrivastava, R. (2015). Modern molecular approaches for analyzing microbial diversity from mushroom compost ecosystem. *3 Biotech*, *5*(6), 853–866. https://doi.org/10.1007/s13205-015-0289-2.

Aguiar-Pulido, V., Huang, W., Suarez-Ulloa, V., Cickovski, T., Mathee, K., & Narasimhan, G. (2016). Metagenomics, metatranscriptomics, and metabolomics approaches for microbiome analysis. *Evolutionary Bioinformatics*, *12s1*, EBO.S36436. https://doi.org/10.4137/ebo.s36436.

Ahansazan, B., Ahansazan, N., Ahansazan, Z., & Afrashteh, H. (2014). Activated sludge process overview. *International Journal Of Environmental Science And Development*, *5*(455), 81–85. https://doi.org/10.7763/IJESD.2014.V5.455.

Andraka, D. (2020). Reliability analysis of activated sludge process by means of biokinetic modelling and simulation results. *Water*, *12*(1), 291. https://doi.org/10.3390/w12010291.

Ansorge, W. J. (2009). Next-generation DNA sequencing techniques. *New Biotechnology*, *25*(4), 195–203. https://doi.org/10.1016/j.nbt.2008.12.009.

Bailón-Salas, A. M., Medrano-Roldán, H., Valle-Cervantes, S., Ordaz-Díaz, L. A., Urtiz-Estrada, N., & Rojas-Contreras, J. A. (2017). Review of molecular techniques for the identification of bacterial communities in biological effluent treatment facilities at pulp and paper mills. *BioResources*, *12*(2). https://doi.org/10.15376/biores.12.2.bailon_salas.

Bayés, M., Heath, S., & Gut, I. G. (2012). Applications of second generation sequencing technologies in complex disorders. *Current topics in behavioral neurosciences*, *12*, 321–343. https://doi.org/10.1007/7854_2011_196.

Bhargava, D. A. (2016). Activated sludge treatment process – concept and system design. *International Journal of Engineering Development and Research (IJEDR)*, *4*(2), 890–896. http://www.ijedr.org/papers/IJEDR1602156.pdf.

Bleidorn, C. (2015). Third generation sequencing: technology and its potential impact on evolutionary biodiversity research. *Systematics and Biodiversity*, *14*(1), 1–8. https://doi.org/10.1080/14772000.2015.1099575.

Boers, S. A., Jansen, R., & Hays, J. P. (2019). Understanding and overcoming the pitfalls and biases of next-generation sequencing (NGS) methods for use in the routine clinical microbiological diagnostic laboratory. *European Journal of Clinical Microbiology & Infectious Diseases*, *38*(6), 1059–1070. https://doi.org/10.1007/s10096-019-03520-3.

Cai, L., Ju, F., & Zhang, T. (2013). Tracking human sewage microbiome in a municipal wastewater treatment plant. *Applied Microbiology and Biotechnology*, *98*(7), 3317–3326. https://doi.org/10.1007/s00253-013-5402-z.

Cao, Y., Fanning, S., Proos, S., Jordan, K., & Srikumar, S. (2017). A review on the applications of next generation sequencing technologies as applied to food-related microbiome studies. *Frontiers in Microbiology*. https://doi.org/10.3389/fmicb.2017.01829

Chakraborty, A., DasGupta, C. K., & Bhadury, P. (2014). Application of molecular techniques for the assessment of microbial communities in contaminated sites. *Microbial Biodegradation and Bioremediation*, 85–113. https://doi.org/10.1016/b978-0-12-800021-2.00004-2.

Cox, M. P., Peterson, D. A., & Biggs, P. J. (2010). SolexaQA: at-a-glance quality assessment of Illumina second-generation sequencing data. *BMC Bioinformatics*, *11*(1), 485. https://doi.org/10.1186/1471-2105-11-485.

Dash, H. R., & Das, S. (2018). Microbiology of atypical environments (Volume 45) (Methods in microbiology, Volume 45). In V. Gurtler & J. T. Trevors (Eds.), *Chapter 4: Molecular Methods for Studying Microorganisms From Atypical Environments* (1st ed., Vol. 45, pp. 89–122). Academic Press. https://doi.org/10.1016/bs.mim.2018.07.005.

Del Vecchio, F., Mastroiaco, V., Di Marco, A., Compagnoni, C., Capece, D., Zazzeroni, F., Capalbo, C., Alesse, E., & Tessitore, A. (2017). Next-generation sequencing: recent applications to the analysis of colorectal cancer. *Journal of Translational Medicine*, *15*(1), 246. https://doi.org/10.1186/s12967-017-1353-y.

Dharaskar, S. A. (2015). Treatment of biological waste water using activated sludge process. *International Journal of Environmental Engineering*, *7*(2), 101. https://doi.org/10.1504/ijee.2015.069812.

Edet, U., Antai, S., Brooks, A., Asitok, A., Enya, O., & Japhet, F. (2017). An overview of cultural, molecular and metagenomic techniques in description of microbial diversity. *Journal of Advances in Microbiology*, *7*(2), 1–19. https://doi.org/10.9734/jamb/2017/37951.

Frese, K., Katus, H., & Meder, B. (2013). Next-generation sequencing: from understanding biology to personalized medicine. *Biology*, *2*(1), 378–398. https://doi.org/10.3390/biology2010378.

Frølund, B., Palmgren, R., Keiding, K., & Nielsen, P. (1996). Extraction of extracellular polymers from activated sludge using a cation exchange resin. *Water Research*, *30*(8), 1749–1758. https://doi.org/10.1016/0043-1354(95)00323-1.

Garrido-Cardenas, J., & Manzano-Agugliaro, F. (2017). The metagenomics worldwide research. *Current Genetics*, *63*(5), 819–829. https://doi.org/10.1007/s00294-017-0693-8.

Giwa, A. S., Ali, N., Athar, M. A., & Wang, K. (2019). Dissecting microbial community structure in sewage treatment plant for pathogens' detection using metagenomic sequencing technology. *Archives of Microbiology*, *202*(4), 825–833. https://doi.org/10.1007/s00203-019-01793-y.

Goodwin, S., McPherson, J. D., & McCombie, W. R. (2016). Coming of age: ten years of next-generation sequencing technologies. *Nature Reviews Genetics*, *17*(6), 333–351. https://doi.org/10.1038/nrg.2016.49.

Gut, I. G. (2013). New sequencing technologies. *Clinical & Translational Oncology*, *15*(11), 879–881. https://doi.org/10.1007/s12094-013-1073-6.

Handelsman, J. (2004). Metagenomics: application of genomics to uncultured microorganisms. *Microbiology and Molecular Biology Reviews*, *68*(4), 669–685. https://doi.org/10.1128/mmbr.68.4.669-685.2004.

Heather, J. M., & Chain, B. (2016). The sequence of sequencers: the history of sequencing DNA. *Genomics*, *107*(1), 1–8. https://doi.org/10.1016/j.ygeno.2015.11.003.

Kallistova, A., Pimenov, N., Kozlov, M., Nikolaev, Y., Dorofeev, A., & Aseeva, V. et al. (2014). Microbial composition of the activated sludge of Moscow wastewater treatment plants. *Microbiology*, *83*(5), 699–708. https://doi.org/10.1134/s0026261714050154.

Kchouk, M., Gibrat, J. F., & Elloumi, M. (2017). Generations of sequencing technologies: from first to next generation. *Biology and Medicine*, *09*(03). https://doi.org/10.4172/0974-8369.1000395.

Khan, M. B., Nisar, H., Ng, C. A., Lo, P. K., & Yap, V. V. (2017). Generalized classification modeling of activated sludge process based on microscopic image analysis. *Environmental Technology*, *39*(1), 24–34. https://doi.org/10.1080/09593330.2017.1293166.

Kretsch, C. W., van der Veer, E. M., Henkel, C., & Taschner, P. (2019). DNA sequencing. In E. van Pelt-Verkuil, W. van Leeuwen, & R. te Witt (Eds.), *Molecular Diagnostics*. Springer, Singapore. https://doi.org/10.1007/978-981-13-1604-3_8.

Lavezzo, E., Barzon, L., Toppo, S., & Palù, G. (2016). Third generation sequencing technologies applied to diagnostic microbiology: benefits and challenges in applications and data analysis. *Expert Review of Molecular Diagnostics*, *16*(9), 1011–1023. https://doi.org/10.1080/14737159.2016.1217158.

Liu, L., Li, Y., Li, S., Hu, N., He, Y., Pong, R., Lin, D., Lu, L., & Law, M. (2012). Comparison of next-generation sequencing systems. *Journal of Biomedicine and Biotechnology*, *2012*, 1–11. https://doi.org/10.1155/2012/251364.

Modin, O., Persson, F., Wilén, B.-M., & Hermansson, M. (2016). Nonoxidative removal of organics in the activated sludge process. *Critical Reviews in Environmental Science and Technology*, 1–38. https://doi.org/10.1080/10643389.2016.1149903.

Mukherjee, A., & Reddy, M. S. (2020). Metatranscriptomics: an approach for retrieving novel eukaryotic genes from polluted and related environments. *3 Biotech*, *10*(2), 71. https://doi.org/10.1007/s13205-020-2057-1.

Munroe, D. J., & Harris, T. J. R. (2010). Third-generation sequencing fireworks at Marco Island. *Nature Biotechnology*, *28*(5), 426–428. https://doi.org/10.1038/nbt0510-426.

Olubukola, O. B. (2003). Molecular techniques: an overview of methods for the detection of bacteria. *African Journal of Biotechnology*, *2*(12), 710–713. https://doi.org/10.5897/ajb2003.000-1127.

Orhon, D. (2014). Evolution of the activated sludge process: the first 50 years. *Journal of Chemical Technology & Biotechnology*, *90*(4), 608–640. https://doi.org/10.1002/jctb.4565.

Pareek, C. S., Smoczynski, R., & Tretyn, A. (2011). Sequencing technologies and genome sequencing. *Journal of Applied Genetics*, *52*(4), 413–435. https://doi.org/10.1007/s13353-011-0057-x.

Petersen, L. M., Martin, I. W., Moschetti, W. E., Kershaw, C. M., & Tsongalis, G. J. (2019). Third-generation sequencing in the clinical laboratory: exploring the advantages and challenges of nanopore sequencing. *Journal of clinical microbiology*, *58*(1), e01315-19. https://doi.org/10.1128/JCM.01315-19.

Rastogi, G., & Sani R. K. (2011). Molecular techniques to assess microbial community structure, function, and dynamics in the environment. In I. Ahmad, F. Ahmad, & J. Pichtel (Eds.) *Microbes and Microbial Technology*. Springer, New York, NY. https://doi.org/10.1007/978-1-4419-7931-5_2.

Schadt, E. E., Turner, S., & Kasarskis, A. (2010). A window into third-generation sequencing. *Human Molecular Genetics*, *19*(R2), R227–R240. https://doi.org/10.1093/hmg/ddq416.

Schneider, D., Aßmann, N., Wicke, D., Poehlein, A., & Daniel, R. (2020). Metagenomes of wastewater at different treatment stages in Central Germany. *Microbiology Resource Announcements*, *9*(15), e00201-20. https://doi.org/10.1128/mra.00201-20.

Scholz, M. (2006). Wetland systems to control urban runoff. In *Chapter 18: Activated Sludge Processes* (1st ed., pp. 115–129). Elsevier. https://doi.org/10.1016/B978-044452734-9/50021-9.

Shchegolkova, N., Krasnov, G., Belova, A., Dmitriev, A., Kharitonov, S., & Klimina, K. et al. (2016). Microbial community structure of activated sludge in treatment plants with different wastewater compositions. *Frontiers in Microbiology*, *7*, 90. https://doi.org/10.3389/fmicb.2016.00090.

Shetty, P. J., Amirtharaj, F., & Shaik, N. A. (2019). Introduction to nucleic acid sequencing. In N. Shaik, K. Hakeem, B. Banaganapalli, & R. Elango (Eds.) *Essentials of Bioinformatics* (Vol. I). Springer, Cham. https://doi.org/10.1007/978-3-030-02634-9_6.

Sidhu, C., Vikram, S., & Pinnaka, A. K. (2017). Unraveling the microbial interactions and metabolic potentials in pre- and post-treated sludge from a wastewater treatment plant using metagenomic studies. *Frontiers in Microbiology*, *8*, 1382. https://doi.org/10.3389/fmicb.2017.01382.

Silva-Bedoya, L. M., Sánchez-Pinzón, M. S., Cadavid-Restrepo, G. E., & Moreno-Herrera, C. X. (2016). Bacterial community analysis of an industrial wastewater treatment plant in Colombia with screening for lipid-degrading microorganisms. *Microbiological Research*, *192*, 313–325. https://doi.org/10.1016/j.micres.2016.08.006.

Singh, S., Singh, H., Rout, B., Tripathi, R. B. M., Chopra, C., & Chopra, R. S. (2020). Metagenomics: techniques, applications, challenges and opportunities. In C. Chopra, N. R. Sharma, & R. S. Chopra (Eds.) *The New Science of Metagenomics: Revealing the Secrets of Microbial Physiology* (1st ed., pp. 3–23). Springer. https://doi.org/10.1007/978-981-15-6529-8.

Thomas, T., Gilbert, J., & Meyer, F. (2012). Metagenomics – a guide from sampling to data analysis. *Microbial informatics and experimentation*, *2*(1), 3. https://doi.org/10.1186/2042-5783-2-3.

Tonkovic, P., Kalajdziski, S., Zdravevski, E., Lameski, P., Corizzo, R., & Pires, I. et al. (2020). Literature on applied machine learning in metagenomic classification: a scoping review. *Biology*, *9*(12), 453. https://doi.org/10.3390/biology9120453.

Toseland, A., Moxon, S., Mock, T., & Moulton, V. (2014). Metatranscriptomes from diverse microbial communities: assessment of data reduction techniques for rigorous annotation. *BMC Genomics*, *15*(1), 901. https://doi.org/10.1186/1471-2164-15-901.

van Dijk, E. L., Jaszczyszyn, Y., Naquin, D., & Thermes, C. (2018). The third revolution in sequencing technology. *Trends in Genetics*, *34*(9), 666–681. https://doi.org/10.1016/j.tig.2018.05.008.

Vieites, J. M., Guazzaroni, M. E., Beloqui, A., Golyshin, P. N., & Ferrer, M. (2010). Molecular methods to study complex microbial communities. In W. Streit, & R. Daniel (Eds.) *Metagenomics. Methods in Molecular Biology (Methods and Protocols)* (Vol. 668). Humana Press, Totowa, NJ. https://doi.org/10.1007/978-1-60761-823-2_1.

Westreich, S. T., Treiber, M. L., Mills, D. A., Korf, I., & Lemay, D. G. (2018). SAMSA2: a standalone metatranscriptome analysis pipeline. *BMC Bioinformatics*, *19*(1), 175. https://doi.org/10.1186/s12859-018-2189-z.

Xu, S., Yao, J., Ainiwaer, M., Hong, Y., & Zhang, Y. (2018). Analysis of bacterial community structure of activated sludge from wastewater treatment plants in winter. *BioMed Research International*, *2018*, 1–8. https://doi.org/10.1155/2018/8278970.

Yadav, B., Pandey, A. K., Kumar, L. R., Kaur, R., Yellapu, S. K., Sellamuthu, B., Tyagi, R. D., & Drogui, P. (2020). Introduction to wastewater microbiology: special emphasis on hospital wastewater. *Current Developments in Biotechnology and Bioengineering*, 1–41. https://doi.org/10.1016/b978-0-12-819722-6.00001-8.

Yadav, S., & Kapley, A. (2019). Exploration of activated sludge resistome using metagenomics. *Science of The Total Environment*, *692*, 1155–1164. https://doi.org/10.1016/j.scitotenv.2019.07.267.

Yang, C., Zhang, W., Liu, R., Li, Q., Li, B., Wang, S., Song, C., Qiao, C., & Mulchandani, A. (2011). Phylogenetic diversity and metabolic potential of activated sludge microbial communities in full-scale wastewater treatment plants. *Environmental Science & Technology*, *45*(17), 7408–7415. https://doi.org/10.1021/es2010545.

Yang, Y., Wang, L., Xiang, F., Zhao, L., & Qiao, Z. (2020). Activated sludge microbial community and treatment performance of wastewater treatment plants in industrial and municipal zones. *International Journal of Environmental Research and Public Health*, *17*(2), 436. MDPI AG. http://dx.doi.org/10.3390/ijerph17020436.

Zhang, B., Xu, X., & Zhu, L. (2018). Activated sludge bacterial communities of typical wastewater treatment plants: distinct genera identification and metabolic potential differential analysis. *AMB Express*, *8*(1). https://doi.org/10.1186/s13568-018-0714-0.

Index